D1067594

Identifying British Insects and Arachnids

Until now, individuals wishing to identify British insects have found it difficult to track down the specialist keys published in obscure literature, whereas the popular guides are often misleadingly simplistic, covering only a fraction of the species. This book bridges the gap, providing expert guidance through the taxonomic maze. It contains an introduction to each group of organisms, and over 2000 references selected as being the most useful and up-to-date for accurate identification, together with notes on their relevance and coverage. A further chapter covers the understanding and retrieval of scientific references, with advice on using libraries and other information services. This will be an essential reference book for anyone involved in insect and arachnid identification, from interested amateurs to professionals dealing with unfamiliar groups.

PETER C. BARNARD is Head of the UK Biodiversity Programme in the Entomology Department of The Natural History Museum, London, where he researches caddisfly and lacewing taxonomy. He has been assistant editor of the *Entomologist's Gazette*, and joint editor for the *Handbooks for the Identification of British Insects*.

Identifying British Insects and Arachnids:

an annotated bibliography of key works

EDITED BY

Peter C. Barnard

Department of Entomology,
The Natural History Museum

THE
NATURAL
HISTORY
MUSEUM

CAMBRIDGE
UNIVERSITY PRESS

FLORIDA STATE
UNIVERSITY LIBRARIES

JUN 1 1999

TALLAHASSEE, FLORIDA

Z
5859
.G7
I4
1999

PUBLISHED BY THE PRESS SYNDICATE OF THE UNIVERSITY OF CAMBRIDGE
The Pitt Building, Trumpington Street, Cambridge, United Kingdom

CAMBRIDGE UNIVERSITY PRESS
The Edinburgh Building, Cambridge CB2 2RU, UK   http://www.cup.cam.ac.uk
40 West 20th Street, New York, NY 10011-4211, USA   http://www.cup.org
10 Stamford Road, Oakleigh, Melbourne 3166, Australia

© The Natural History Museum, London 1999

This book is in copyright. Subject to statutory exception
and to the provisions of relevant collective licensing agreements,
no reproduction of any part may take place without
the written permission of Cambridge University Press.

First published 1999

Printed in the United Kingdom at the University Press, Cambridge

Typset in Utopia 9/13pt, in QuarkXPress™   [SE]

A catalogue record for this book is available from the British Library

Library of Congress Cataloguing in Publication data
Identifying British insects and arachnids: an annotated bibliography
of key works / edited by Peter C. Barnard.
     p.   cm.
Includes bibliographical references and index.
ISBN 0 521 63241 2 (hb)
1. Insects – Great Britain – Identification – Bibliography
2. Arachnida – Great Britain – Identification – Bibliography
I. Barnard, Peter C., (Peter Charles), 1949–   .
Z5859.G714   1999
016.5957′0941–dc21   98-36132   CIP

ISBN 0 521 63241 2 hardback

Contents

Introduction   [viii]
Acknowledgements   [xiii]

Sources of information   [1]
JULIE M.V. HARVEY

Collembola: the springtails   [21]
PETER C. BARNARD

Protura: the proturans   [23]
PETER C. BARNARD

Diplura: the two-tailed bristletails   [24]
PETER C. BARNARD

Thysanura: the silverfish and firebrats   [25]
PETER C. BARNARD

Archaeognatha: the bristletails   [26]
PETER C. BARNARD

Ephemeroptera: the mayflies or up-winged flies   [27]
STEPHEN J. BROOKS

Odonata: the dragonflies and damselflies   [30]
STEPHEN J. BROOKS

Plecoptera: the stoneflies   [33]
STEPHEN J. BROOKS

Phasmida: the stick-insects   [36]
JUDITH A. MARSHALL

Orthoptera: the grasshoppers, crickets and bush-crickets   [38]
JUDITH A. MARSHALL

Dermaptera: the earwigs   [40]
JUDITH A. MARSHALL

Blattodea: the cockroaches   [42]
JUDITH A. MARSHALL

Psocoptera: the booklice and barklice   [44]
JON H. MARTIN

Phthiraptera: the lice   [47]
CHRISTOPHER H.C. LYAL

Thysanoptera: the thrips   [51]
JON H. MARTIN

Hemiptera: the true bugs   [54]
JON H. MARTIN & MICK D. WEBB

Neuroptera: the lacewings   [76]
PETER C. BARNARD

Megaloptera: the alderflies   [78]
PETER C. BARNARD

Raphidioptera: the snakeflies   [79]
PETER C. BARNARD

Coleoptera: the beetles   [80]
PETER M. HAMMOND & STUART J. HINE

Strepsiptera: the stylops   [139]
PETER M. HAMMOND & STUART J. HINE

Mecoptera: the scorpionflies   [140]
PETER C. BARNARD

Trichoptera: the caddisflies   [141]
PETER C. BARNARD

Lepidoptera: the moths and butterflies   [145]
MARK S. PARSONS, GADEN S. ROBINSON, MARTIN R. HONEY &
DAVID J. CARTER

Diptera: the flies   [171]
NIGEL P. WYATT & JOHN E. CHAINEY

Siphonaptera: the fleas   [194]
THERESA M. HOWARD

Hymenoptera: the bees, wasps and ants   [196]
JOHN S. NOYES, MIKE G. FITTON, DONALD L.J. QUICKE,
DAVID G. NOTTON, GEORGE R. ELSE, NIGEL D.M. FERGUSSON,
BARRY BOLTON, SUZANNE LEWIS & LARAINE C. TAREL

Class Arachnida: the spiders, harvestmen, pseudoscorpions
and mites   [320]

Pseudoscorpiones: the pseudoscorpions   [322]
PAUL D. HILLYARD

Opiliones: the harvestmen   [324]
PAUL D. HILLYARD

Araneae: the spiders   [326]
PAUL D. HILLYARD

Acari: the mites and ticks   [330]
ANNE S. BAKER

Index   [345]

Introduction

The first move in identifying a British insect is often to look it up in a general colour guide, of which Michael Chinery's (1993) *Insects of Britain and northern Europe* is one of the best known and most comprehensive. But even that excellent book cannot cover more than a small fraction of the British insect species: in fact it illustrates less than 800 of the *ca.* 23 500 species currently known from the British Isles. Indeed, most of them could not be recognised from such illustrations, because the characters separating many closely related species are subtle, and often require the use of a good hand-lens at the very least. So where does one turn next for more information?

Those who are already knowledgeable about a group of insects will know all the relevant literature on that group, and for a colourful and well-known group one or two books may be all that is needed. But it is more likely that the necessary literature will consist of a handful of standard textbooks, coupled with a more-or-less extensive pile of separate papers, written by many different authors over a long period of time, and gleaned from a wide variety of scientific journals. The expert often forgets how long it has taken to accumulate all this literature, and for the beginner just starting out, or even for an experienced entomologist who is searching for keys to an unfamiliar group, the process can be slow and painful.

This book is therefore designed to point the way into the huge mass of literature which is necessary for the identification of British insects and arachnids, the latter being included because they are often studied by entomologists. It has been prepared by the staff of the Entomology Department of The Natural History Museum, which constitute the largest collection of insect systematists in the world. Previous guides to the literature have covered wider areas than the British Isles (and have therefore been less detailed) and have not given much guidance or recommendation on the value of the references. In this book almost every reference has a comment which outlines its main reason for being included and, armed with all this literature, anyone can build up their own library to equal anything that the expert uses.

But it is one thing to give a list of references to the beginner, and quite a different matter to know exactly how to gather all those books and papers together. Again, the expert can easily forget how one's favourite books have been out of print for some time, that easy access to a specialist library containing all the relevant journals is taken for granted, and that colleagues have been sending free reprints of their scientific papers over the years. Hence the reason for including a vital chapter in this book on 'Sources of information', written by the Entomology Librarian at The Natural History Museum. This shows the layman how to understand references, how to find them and obtain copies, which libraries and other information sources to consult, how to find out-of-print books, and much more.

Following that first chapter, each order of insects, together with the main groups of arachnids, has a chapter to itself. The size of each chapter is of course related to the number of references needed, which in turn is loosely linked to the number of species in each group (which ranges from two in the Thysanura to over 7000 in the Hymenoptera). In each case the number of British species is given (or a good approximation, where there is no recent checklist), then a brief account of the general biology of the group, a note on the important general references, and a summary of the higher classification down to family level (except for the Acari). For the five larger orders with over 1000 species (Hemiptera, Coleoptera, Diptera, Lepidoptera and Hymenoptera) the chapter is subdivided in accordance with the taxonomic subgroups, each of which may be further divided, with its own biological summary and lists of references. A glance at these larger orders will show the uneven coverage of the existing literature, and highlights how some groups are, for one reason or another, more popular and better known. For example, one might need only one or two books to identify the butterflies, or even many families of larger moths, but the study of many of the parasitic Hymenoptera entails the gathering of large numbers of isolated scientific papers, many not written in English. A book such as this can only represent the situation as it stands, but it is hoped that the emphasis on gaps in our knowledge may prompt some workers to try to fill them! A comparison of the chapters will also show that different groups have demanded different treatments: Lepidoptera taxonomy, except for the smaller families, is so advanced that much work now concentrates on biology, distribution and so on, and references on these topics are therefore covered in greater detail than in other chapters.

Apart from including the arachnids, some readers may be wondering why the Collembola and Protura are here as well, as many recent texts on entomology have placed them firmly outside the true insects. In fact they have been moved in and out of the Insects with amusing regularity over recent years, and their systematic relationship with the former members of the artificial group 'Apterygota', such as Diplura and Thysanura, is still far from settled. Traditionally their literature has been linked to that of entomology and, as discussed in the Collembola chapter, the most recent textbook places this group back among the insects in the broad sense, albeit in a very isolated position.

In a book of this size, which contains well over 2000 references, there are certain to be some errors and omissions, and the editor or contributors will be grateful to receive any comments or additions which could be included in a future edition. References up to the end of 1997 are included, with a few early ones from 1998 also appearing.

Entomological journals

Entomological papers appear in a wide range of scientific journals, but the main British journals devoted entirely to entomology are listed here.

Entomologist's Gazette
Entomologist's Monthly Magazine
Entomologist's Record & Journal of Variation

Entomological societies

There are a great many local natural history societies in Britain, details of which can usually be obtained from Public Libraries, but the following are the important national ones, together with a list of their main serial publications.

Amateur Entomologists' Society
Publishes *Bulletin of the Amateur Entomologists' Society.*

British Entomological and Natural History Society
The society for the field entomologist whose journal, *British Journal of Entomology and Natural History*, includes reports on field meetings and the annual exhibition as well as many articles on identification. The latter frequently includes photographs of new species and unusual aberrations. Also publishes a number of book titles.

Royal Entomological Society

Predominantly professional society; publishes *Antenna, Handbooks for the Identification of British Insects, Systematic Entomology, Physiological Entomology, Ecological Entomology, Medical & Veterinary Entomology,* and *Insect Molecular Biology* (the last five in collaboration with Blackwell Science, Oxford). Also publishes a range of book titles.

General entomological references

As this book is largely intended for the reader who is moving on from basic texts to more specialised ones, the following list of references is not comprehensive. However, some of the more important ones are given, with a note on their significance where this is not obvious from the title. Many other useful series, such as the *Naturalists' Handbooks, Handbooks for the Identification of British Insects* (published by the Royal Entomological Society) and the *AIDGAP* keys (published by the Field Studies Council), are listed in their appropriate chapters throughout this book.

Alford, D.V. 1991. *A colour atlas of pests of ornamental trees, shrubs, and flowers.* Wolfe Publishing, London.

Bevan, D. 1987. *Forest insects: a guide to insects feeding on trees in Britain.* Forestry Commission Handbook no. 1. HMSO, London.

Chinery, M. 1993. *Insects of Britain and northern Europe.* (3rd edn). Collins Field Guide, HarperCollins, London.
The best field guide of its kind.

Fry, R. & Lonsdale, D. 1991. *Habitat conservation for insects – a neglected green issue.* Amateur Entomologists' Society, Middlesex.

Gaedike, R. 1997. Bibliographie der Bestimmungstabellen europäische Insekten (1991–1995). *Nova Supplementa Entomologica* 9: 3–142.
A useful listing of European literature. See also earlier publications in the same series by Gaedike in *Beiträge zur Entomologie*: (1976) 26(1): 49–166; (1981) 31(2): 235–304; (1986) 36(1): 261–319; (1988) 38(1): 239–76; (1992) 42(1): 55–195.

Gullan, P.J. & Cranston, P.S. 1994. *The insects: an outline of entomology.* Chapman & Hall, London.
Now widely regarded as the best general textbook on entomology but does not include systematics.

Hollis, D. (Ed.) 1980. *Animal identification: a reference guide; Vol. 3, Insects.* British Museum (Natural History) and John Wiley, London and Chichester.
This work provides a list of the key references, with few comments, which enable a biologist to identify insects from around the world.

Kirby, P. 1992. *Habitat management for invertebrates: a practical handbook.* Joint Nature Conservation Committee/Royal Society for the Protection of Birds/National Power, Sandy.

Richards, O.W & Davies, R.G. 1977. *Imms' general textbook of entomology*. (10th edn) Chapman & Hall, London.

Although well known, this is now looking very outdated, and Gullan & Cranston (1994) is recommended as a general textbook.

Shirt, D. (Ed.) 1987. *British red data books: 2. Insects*. Nature Conservancy Council, Peterborough.

Summary of the status of threatened insects in Britain.

Sims, R.W., Freeman, P. & Hawksworth, D.L. 1988. *Key works to the fauna and flora of the British Isles and northwestern Europe*. 5th edn. Clarendon Press, Oxford. Systematics Association Special Volume 33.

Provides references to books and periodical papers, which help to identify organisms (including insects) from the British Isles and north-western Europe. Becoming outdated, and has only a limited range of comments and other information.

Smith, K.G.V. & Smith V. 1983. *A bibliography of the entomology of the smaller British offshore islands*. Classey, Faringdon.

Acknowledgements

The editor and contributors are grateful to following for advising on various chapters of the manuscript, by suggesting additional references and adding useful extra information. Dr David Agassiz (Lepidoptera); Dr Roger Booth, IIE (Coleoptera); Dr Mike Cox, IIE (Coleoptera); David Hollis (Psocoptera, Hemiptera); Dr John LaSalle, IIE (Hymenoptera); Dr Chris Malumphy, CSL, York (Hemiptera); Dr Andy Polaszek, IIE (Hymenoptera); Kevin Tuck (Lepidoptera); Dr Mike Wilson, National Museums of Wales (Hemiptera). We also thank Janet Camp for assistance with typing. Julie Harvey kindly read the entire typescript, and suggested numerous additions and improvements.

The illustrations are taken from Lydekker, R. (Ed.) 1896, *The Royal Natural History*.

Sources of information

JULIE M.V. HARVEY

We live in a world increasingly dominated by information that is delivered in an electronic format. Many people have access to sophisticated personal computers that can store and manipulate large amounts of data. High-speed communication networks can transmit this information around the world in a matter of seconds. Although these developments have revolutionised the process of publishing, printed books and journals still remain the predominant method of communicating scientific information. There is a vast amount of entomological research and biological data that has been published since the seventeenth century. Many of these publications are still key reference works. For the foreseeable future, researchers will need access to information in traditional printed form, combined with a growing dependence on electronic systems to locate, deliver and store information.

Finding relevant and useful references is fundamental to successful research. Whether you are looking for information inspired by an amateur interest, studying for an academic examination or conducting research, finding references can be a frustrating process. This is not helped by the fact that most people receive little or no formal guidance in how to locate sources of information.

This book provides the reader with a range of key entomological references to the identification of the British insect fauna. Each reference, annotated with useful comments, has been carefully selected by an expert in the field. However, this still leaves unanswered two basic questions, namely 'where can I find more references on a given entomological topic?' and 'how can I obtain copies of books and journal publications?' This chapter addresses both these issues and aims to guide readers to a wide range of information sources.

Finding sources of references

There is no single method for successfully finding references on a particular entomological subject. The best approach will depend on the topic, the depth of interest and the time available to pursue the research.

Bibliographies

If you have been fortunate enough to have already found a relevant book or journal article, you should have a gateway to further references on this topic. This will be in the form of a bibliography (sometimes known as 'list of references' or 'further reading'). In books, bibliographies may occur at the end of each chapter, or at the back of the book, subdivided by chapter or topic. More commonly, bibliographies appear as an alphabetical list, arranged by author. Bibliographies in journals are always printed at the end of the paper or article. It is surprising that many entomologists do not fully exploit these rich sources of information. Although bibliographies are frequently underrated, authors have included these references for a number of good reasons. It enables the author to substantiate particular points referred to in the text and gives the reader access to more detailed sources of information that could not be included in the publication. A comprehensive and accurate bibliography in a publication indicates that the author has thoroughly researched the subject. Scanning a bibliography can help to assess the quality of a publication.

There are a number of inevitable problems with any bibliography. The most obvious is that only references predating the bibliography can be included. So a useful work you have discovered, published in 1970, will only list publications up to this date. You will need other sources to find more recent publications. Secondly, most bibliographies will provide only a standard reference, without any comment or annotation about its usefulness or readability. Only when you have found the reference and read it yourself will this become apparent.

Printed abstracting and indexing services

Anyone conducting a serious and detailed review of a particular entomological subject is confronted by the sheer enormity of the size of the literature. For example, the Entomology Library at The Natural History Museum, London, houses over 90 000 bound volumes and currently subscribes to over 900 entomological journals.

Most specialist research is published as papers or articles in scientific journals. Entomological papers appear in specialist entomological journals and in a wide range of ecological, biological, medical and agricultural titles. It would be impossible for an individual to scan all the current literature, never mind the vast amount of historical material that has been published. Fortunately, a number of publications exist that carefully index scientific papers from thousands of journals. These services are known as secondary journals. They enable researchers to locate publications by author, title, subject and key words. Some secondary journals also provide a detailed abstract (or summary) of each paper. These abstracting and indexing services are expensive to purchase and will generally only be found in university and specialist libraries. Readers wishing to conduct an exhaustive literature search would be strongly advised to use two or more secondary journals. Although this approach will inevitably result in some duplication of references, it ensures that relevant publications are not missed.

The secondary journals most frequently used by entomologists are listed below; all provide varying degrees of coverage to world wide literature.

Biological Abstracts, *published by BIOSIS*

Biological Abstracts started publication in 1926. It now abstracts and indexes nearly 6500 scientific periodicals. One printed issue is published each fortnight and has author, biosystematic, generic and subject indexes. A cumulative index is produced every six months. As *Biological Abstracts* is very bulky in its published format, many researchers prefer to use the on-line or CD-ROM version.

CAB Abstracts, *published by CAB International*

Over 12 000 journals are scanned by CAB International to maintain two large bibliographic databases called CAB Abstracts and CAB Health. From their databases, CABI produce two printed abstracting services that are of particular interest to entomologists: *Review of Agricultural Entomology* and *Review of Medical and Veterinary Entomology*. The publications first appeared in 1913 under the titles *The Review of Applied Entomology Series A: Agricultural* and *The Review of Applied Entomology Series B: Medical and Veterinary*.

Review of Medical and Veterinary Entomology is published on a monthly basis. It provides full references, indexes and abstracts to publications concerned with insects and other arthropods that transmit diseases or are injurious to humans and domesticated or wild animals.

Review of Agricultural Entomology is also published monthly. It covers the world literature of insects and other arthropods that are pests of cultivated crops, forestry and stored foods. It also includes beneficial insects that are parasites or predators of pests.

These two publications benefit from offering very detailed abstracts. This helps entomologists to evaluate a publication before going to the time and expense of obtaining a copy.

Science Citation Index, *published by Institute for Scientific Information*

Science Citation Index is published bi-monthly with an annual cumulation and is based on the contents of over 3000 leading scientific journals. It consists of four indexes, covering author, institution, subject and a citation index. The citation index is unique, as it enables a researcher to trace which authors are being cited in scientific bibliographies. This enables a researcher to answer questions that other secondary journals cannot address. For example, if a researcher has found a useful reference published in 1985, he or she could check which authors have subsequently cited this paper, thereby tracing the development of the topic forwards in time. Science Citation Index coverage for entomology is good but not comprehensive.

Zoological Record, *published by BIOSIS and The Zoological Society of London*

Zoological Record provides a detailed annual index to the zoological literature. Each year, one volume of *Zoological Record* is produced, consisting of 27 separately issued sections. Six sections cover the insects, General Insects and Smaller Orders, Coleoptera, Diptera, Lepidoptera, Hymenoptera and Hemiptera. Each of these sections has five very useful indexes enabling users to locate information by author, subject or geographical area, plus a palaeontological and a systematic index. The full bibliographic reference is provided in the author index. No abstracts are provided.

Entomologists often consider *Zoological Record* as purely taxonomic in content. Although this is a very important aspect of the work, it underrates its value as a broad entomological index.

The first volume of this monumental work (formally known as *The Record of Zoological Literature*) was published in 1865 and covered the world zoological literature for 1864. At present electronic records of the printed version are only available since 1978. Anyone conducting a thorough literature search covering the period before 1978 will need to spend time studying a large number of printed volumes.

On-line services and CD-ROMs

There can be no doubt that computerised information systems have revolutionised the whole process of locating entomological information. High-powered database systems and CD (Compact Disc) technology have combined to enable complex searches to be carried out. The results can leave the researcher deluged with almost too much data.

In the late 1970s printed secondary journals like *Biological Abstracts, CAB Abstracts, Science Citation Index* and *Zoological Record* started to make data available on remote computer databases. These databases operate through commercial host systems. These 'on-line' searches are charged on a combination of time connected to the database and the number of references retrieved. Such systems require specialist training as the search language is not easily understood. Most searches are carried out by trained librarians on behalf of the researcher. On-line searching is rapidly becoming viewed as old technology. Many research and public libraries have now invested heavily in CD-ROM technology. CD-ROMs (Compact Discs Read Only Memory) are very similar to music compact discs except that they hold textual information and sometimes images. A huge amount of data can be stored on one disc, which can be searched using special software on a PC that has a CD-ROM drive.

Unfortunately, specialist CD-ROMs of interest to entomologists are very expensive, usually because the price includes a subscription to future updates, and are generally only found in research or specialist libraries. The great advantage of these systems is that users can quickly search for information and then browse through any interesting references at their leisure. There is no charge for time spent searching on the system or for retrieving information. References that are of interest can be either printed or downloaded onto disc.

A wide range of CD-ROMs is now available based on the information published in secondary journals. The following CD-ROMs are of particular interest to entomologists.

BIOSIS Previews

Based on data from *Biological Abstracts* but with additional references. Covers records since 1985.

CAB Abstracts

Records since 1972 available on CD. Other specialist discs produced by CABI include:

CABPest, crop protection and pest management literature since 1973
E-CD, environmental literature since 1973
Tree, ecology and management of forests since 1939
Soil, soils and land management since 1973

Science Citation Index

CD-ROMs issued on a quarterly basis since 1980 but lack abstracts, also issued monthly since 1991 with abstracts.

Zoological Record

Records available on CD-ROM since 1978.

Internet services and the World Wide Web

The Internet is a global network of computers, connecting millions of users around the world. All one needs to join this network is a PC with appropriate software and a modem that links the computer to the telephone system. Individuals wanting to connect to the Internet also need an account with an Internet Service Provider. These are commercial companies which for a monthly or annual service charge provide access to the Internet. Many public libraries are now offering access to the Internet, and high street Cyber-cafes are now becoming popular.

Using electronic mail (e-mail), entomologists can now discuss projects, submit and edit publications, and arrange meetings with colleagues, around the world. All this can be done at a fraction of the cost and time of using traditional postal or telephone services.

The World Wide Web enables individuals and organisations to make information available to a global Internet audience. Text and images are structured into a series of files called Web pages. These pages then form a Web site. Any user around the world who has the appropriate software (called a browser) can access and read these pages. Web site addresses are known as URLs (Uniform Resource Locators). Once connected, it is easy to browse through the system. By clicking on highlighted words or images, the reader can jump from one web page to another. Special software packages called Search Engines exist, which help to locate relevant sites.

An increasing amount of useful entomological information is available on the World Wide Web. This includes information about entomological societies and institutions, updates on entomological projects, access to databases and library catalogues and details about books for sale from specialist booksellers. The Internet is still in a developmental stage and is an unstable structure. Information that is available one week may change address or disappear the next.

The best way to search for entomological information on the Internet is to use a search engine using key words like 'entomology', 'entomological' or 'insects'. A large number of entomological sites already exist. Two of the most useful entomological URLs for finding more information on the Internet are listed below. If they cannot be found, you should use a search engine to relocate the site.

Iowa State Entomology Index of Internet Resources: a directory and search engine of insect-related resources on the Internet
http://www.ent.iastate.edu/list

Entomology at Colorado State University
http://www.colostate.edu/Depts/Entomology/ent.html

Joining an electronic mailing list or listserver can be an excellent way of getting into contact with people who share your interest, whether they live in the UK or overseas. People send messages to a central e-mail account; mail is then automatically forwarded to all the other list subscribers. Current information on mailing lists can be found on the above URLs.

Key references

Entomology libraries have in their reference collections a few works that answer most of the standard enquiries. Listed below are some of these key works.

General guides to information sources

Colvin, M. & Reavey, D. 1993. *A directory for entomologists* (2nd edn). The Amateur Entomologists' Society, Pamphlet No. 14.
   A very useful small guide to sources of entomological information in the UK. Some details such as telephone numbers are now out of date. Includes information on national and local recording schemes, field courses, research grants, libraries, museums, trade fairs and traders.

Gilbert, P. & Hamilton, C.J. 1990. *Entomology: a guide to information sources* (2nd edn). Mansell, London.
   Although some sections are becoming rather out of date, this is still a good guide to entomological information. It includes sections on naming and identification of insects, specimens and collections, the literature of entomology, searching and locating literature, keeping up with current events, and entomologists and their organisations.

Hamilton, C.J. 1991. *Pest management: a directory of information sources.* Vol. 1, *Crop protection.* CABI, Oxon.
   A useful reference work providing sources of information on crop protection.

Checking early references

Zoological Record is invaluable for verifying and finding references. For checking publications before *Zoological Record* started in 1864, the following references are particularly useful.

Hagen, H.A. 1862–3. *Bibliotheca entomologica.* 2 vols. W. Engelmann, Leipzig.
   An alphabetical index by author from earliest times to 1862. Also includes a useful subject index. A very reliable source of bibliographic references.
Horn, W. & Schenkling, S. 1928–9. *Index litteraturae entomologicae.* Ser. I, *Die Welt-Literatur über die gesamte Entomologie bis inklusive 1863.* 4 pts. W. Horn, Berlin.
   Similar to Hagen but includes additional references. Does not have a subject index. Additional supplements by Gaedike, R. & Smetana, O. Ergänzungen und Berichtigungen zu Walter Horn und Sigmund Schenkling: Index Litteraturae Entomologicae, Serie I, die Welt-Literatur über die gesamte Entomologie bis inklusive 1863. *Beiträge zur Entomologie* (1978) 28: 329–436, and (1984) 34: 167–291.

For checking entomological works published between 1864 and 1900 the following works are useful sources.

Derksen, W. & Scheiding, U. 1963–75. *Index litteraturae entomologicae.* Ser II, *Die Welt-Literatur über die gesamte Entomologie von 1864 bis 1900.* 5 vols. Deutsche Akademie der Landwirtschaftswissenschaften, Berlin.
   Volume 5 includes a list of abbreviated periodical titles and a subject index.
Freeman, R.B. 1980. *British natural history books: 1495–1900 A handlist.* Dawson and Archon Books, Folkestone and Connecticut.
   This contains an alphabetical list by author, a list of titles in date order from 1495 to 1800 and a subject index.

Nomenclators

A common need is to find the reference that contains the original description of an insect genus or species. The following references are key works in this area:

Neave, S.A. 1939–40. *Nomenclator zoologicus. A list of the names of genera and subgenera in zoology from the 10th edition of Linnaeus 1758 to the end of 1935.* 4 vols, plus supplement. Zoological Society of London, London.
   Later volumes have also been published under this title, volumes 5–9 (1950–96) including names listed up to 1994. This work provides the bibliographic reference for every generic and subgeneric zoological name described. Section 20 of the *Zoological Record* also provides a list of new taxonomic names.
Sherborn, C.D. 1902. *Index animalium 1758–1800.* Cambridge University Press, Cambridge.
   Provides a list and reference to all zoological generic and specific names described between these dates.

Sherborn, C.D. 1922–33. *Index animalium 1801–1850.* 33 parts. British Museum (Natural History), London.
Continues the above work up to 1850.

Glossaries

Nichols, S.W. (Compiler) & Schuh, R.T. (Ed.) 1989. *The Torre-Bueno glossary of entomology.* New York Entomological Society and the American Museum of Natural History, New York.
Provides definitions of entomological terms.

Biographical and manuscript resources

Bridson, G.D.R., Phillips, V.C. & Harvey, A.P. 1980. *Natural history manuscript resources in the British Isles.* Mansell and Bowker, London and New York.
An annotated bibliography guiding researchers to sources of original manuscript material. Includes indexes to name, place and subject.
Gilbert, P. 1977. *A compendium of the biographical literature on deceased entomologists.* British Museum (Natural History), London.
A good source of references to biographical information about entomologists.
Harvey, J.M.V., Gilbert, P. & Martin, K. S. 1996. *A catalogue of manuscripts in the Entomology Library of The Natural History Museum, London.* Mansell, London.
Lists entomological manuscripts, collecting notebooks, diaries and correspondence held in The Natural History Museum. Contains biographical information and author and geographical indexes.

Bibliographic references

There are very few guides that explain how to interpret bibliographic references. Modern publishing practice encourages authors to provide full and complete references in their publications. Authors should aim to provide references that give all the information necessary to trace and obtain a copy. However, some references, especially in older works, can seem incomprehensible to many people.

This section has two aims: to explain the format of standard references and to discuss possible causes of confusion.

Understanding references

Many amateur entomologists are surprised at the great variation in the way references are listed in different publications. On occasions it can even be difficult to decide whether a reference describes a book or a journal publication.

References to books

A simple reference referring to a book may appear similar to the following example.

Gauld, I.D. & Bolton, B. (Eds) 1996. *The Hymenoptera*. Oxford University Press, Oxford.

The authors normally appear first in the reference, as surname followed by initials. In most publications a maximum of three authors are listed. If more than three authors are involved in a publication, the first author is normally given followed by the Latin abbreviation *et al.*, meaning 'and others'. In the above example, the authors have edited the work. This is shown by the abbreviation (Eds). The date of publication may appear in a range of positions in the reference, sometimes in round brackets. The title of the book can be indicated by a number of different methods: italics, underlining, boldface or sometimes a different font size. (This may be followed by the number of pages, known as pagination. Roman numerals show the number of preliminary pages, which may include the title page, contents, acknowledgements and a preface.) The pagination will indicate how large the work is, and may help to indicate the level of detail. It is now standard practice to include the publisher and place of publication. This will assist the reader in acquiring a copy from a library or bookseller. The ISBN (International Standard Book Number) does not appear in a reference.

A more complex reference to a book might appear as:

Agassiz, D.J.L. 1985. Douglasiidae. *In* Heath, J. & Emmet, A.M. (Eds) *The moths and butterflies of Great Britain and Ireland* **2**, pp. 408–9. Harley Books, Colchester.

In this example the author, Agassiz, has contributed a section on Douglasiidae in a work edited by Heath and Emmet. The work, *The moths and butterflies of Great Britain and Ireland,* has been published in a series of volumes (known as a monographic series). This reference refers to volume 2, which is indicated in bold type. In such references the pages written by the contributing author (pp. 408–9) are usually noted. The full pagination of the book is sometimes also given.

References to journal papers

A great deal of specialist entomological literature is published as papers in scientific journals. Journals are also known as periodicals or serials. Any serious entomological study will result in the need to refer to papers published in a wide range of journal publications.

A simple reference to a publication in a journal will appear in a format similar to the one shown below:

Verrall, G.H. 1909. The Large Copper butterfly *(Chrysophanus dispar)*. *Entomologist* **42**: 183.

The author's name is followed by the date of publication and the title of the paper. The paper is published in the journal with the full title *The Entomologist* (shown in this example in italics) in volume 42 (shown in bold) on page 183.

Sources of confusion

Some references can appear complex owing to the use of abbreviated words or Latin expressions. Standard abbreviations in a bibliography include the following.

ibid. or *ib.*, an abbreviation for the Latin *ibidem*, meaning in the same place. This indicates that a reference has been published in the same journal as the preceding reference in the bibliography.

idem or *id.*, Latin, meaning the same word. This indicates that the author or title is the same as in the preceding reference.

Both these terms were commonly used in the past and are frequently found in old publications when similar references were cited. The two references below illustrate their use.

Donisthorpe, H. StJ. K. 1907. *Cephalonomia formiciformis. Entomologist's Rec. J. Var.* **19**: 260.
Idem 1929. *Histeromerus mystacinus* Wesmael: a Coleopterous parasite. *Ibid.* **41**: 125.

In this example *Idem* indicates that Donisthorpe is also the author of the second paper and *ibid.* indicates that the paper also appears in the journal *Entomologist's Record and Journal of Variation.*
    Many librarians have been asked for the paper by the author Idem in the journal Ibid.!

[*sic*] This Latin word, meaning so or thus, appears in square brackets. It indicates that a word or spelling was incorrect in the original reference and is always used immediately after the word to which it applies.

[1920] Use of square brackets: this is used to show that information has been inferred. It is most commonly used when dates of publication have been deduced from other sources.

In press: Indicates that the reference has not yet been published, but has been accepted by the journal indicated.

More complex journal references

Entomological literature is filled with many examples of more complex journal references. There are many different reasons why they can be complicated. Some examples are given in this section.

Withycombe, C.L. (1925) Some aspects of the biology and morphology of the Neuroptera. With special reference to the immature stages and their possible phylogenetic significance. *Trans. Ent. Soc. London* (1924): 303–411.

In this example two dates of publication appear in the reference. The first date, 1925, is the actual year of publication. However, the papers were published as a result of being presented at a meeting that was held in 1924. The second date relates to the calendar year that the *Transactions* cover. As these transactions have no volume number, individual volumes are referred to by the calendar year. The title of the journal has been given in abbreviated form.

Many journals are issued as numbered parts of a volume. Authors do not normally refer to these parts in a bibliography unless each part has its own separate pagination, as shown in the following reference. In this example the volume number is shown in bold, the part number in brackets.

Efflatoun, H.C. 1922. A monograph of Egyptian Diptera (Part 1. Fam. Syrphidae). *Mém. Soc. ent. Égypte* **2**(1): 1–123.

This can cause confusion with another type of reference shown below where a journal has been published in different numbered series. Each series is normally indicated by a number in brackets followed by the volume number.

Waterhouse, C.O. 1888. New species of Lucanidae, Cetoniidae, and Bruprestidae in the British Museum. *Ann. Mag. nat. Hist.* (6) **1**, 260–264.

Journal abbreviations

Until recently it was common practice for authors to use journal abbreviations to help save space in the bibliography. Some of the most common abbreviations used in journal titles are listed below.

A.	*Annual*
Abstr.	*Abstracts*
Annls.	*Annales*
Ann.	*Annals*
Arch.	*Archiv*
Beitr.	*Beiträge*

Ber.	*Bericht*
Bolm	*Boletim*
Boln.	*Boletin*
Boll.	*Bollettino*
Bull.	*Bulletin*
Circ.	*Circular*
C.r.	*Compte Rendu*
Conf.	*Conference*
Contr.	*Contributions*
Dep.	*Department*
Dt.	*Deutsche*
Ent.	*Entomological or Entomologische*
Int.	*International*
J.	*Journal*
Mag.	*Magazine*
Mém.	*Mémoires*
Mem.	*Memoirs*
Mitt.	*Mitteilungen*
Mon.	*Monthly*
Nat. Hist.	*Natural History*
Newsl.	*Newsletter*
Occ.	*Occasional*
Pap.	*Papers*
Proc.	*Proceedings*
Publs.	*Publications*
R.	*Royal*
Rec.	*Record*
Rep.	*Report*
Rev.	*Review*
Riv.	*Rivista*
Sber.	*Sitzungsberichte*
Sci.	*Science*
Soc.	*Société or Society*
Tech.	*Technical*
Trans.	*Transactions*
Univ.	*University*
Z.	*Zeitschrift*
Zool.	*Zoological or Zoologische*

When authors abbreviate journal titles all small insignificant words are excluded. This would commonly include words such as a, in, of, and the. So, for example, *The Annals and Magazine of Natural History* becomes abbreviated to *Ann. Mag. nat. Hist.*

Understanding English abbreviated titles can take some practice, but the process becomes more difficult when other languages are involved. For example *Arb. morph. taxon. Ent. Berl.* is *Arbeiten über morphologische und taxonomische Entomologie aus Berlin-Dahlem.*

A number of libraries have published information on their journal holdings, which can help to trace entomological journal references.

The Natural History Museum: serial titles held in the Department of Library and Information Services. 1995 (4th edn). 5 vols. The Natural History Museum, London.

> Over 25 000 natural history journal titles are listed. Unfortunately, abbreviated journal titles, which were provided in earlier editions, are not included.

Leonard, B. 1983. *Serial publications in the library of the Royal Entomological Society of London* (2nd edn). Royal Entomological Society, London.

> A small but very useful work listing the serials held in this important entomological library. Journal abbreviations are listed following the standard in the *World list of scientific periodicals.*

World list of scientific periodicals 1963–5 (4th edn). 3 vols. Butterworths, London.

> Journals that required authors to use abbreviated titles in their references often insisted that standard abbreviations were used as listed in the *World list of scientific periodicals.* Reference is often made to the *World list* in bibliographies and instructions to authors. Originally published in 3 volumes, additional supplements were published up to 1980. This work is still invaluable in interpreting difficult references although details on library holdings, etc., are now very out of date.

Another useful source for interpreting abbreviations is the 5th volume of Derksen, W. and Scheiding, U. 1963–75. *Index litteraturae entomologicae* (see above). This lists journals in their abbreviated form and is particularly useful in finding full titles of journals cited in early literature.

Public, university and specialist entomological libraries

Public libraries can provide an excellent resource for amateurs, providing a wide range of information services. However, serious entomologists will need to refer to specialist publications, which cannot be supplied through the normal Public Lending Library or inter-library loan system. For this material an entomologist will have to make use of one of the Entomological Libraries in the UK.

Public libraries

Although public libraries have suffered reductions in spending budgets, they remain an important source of reference material and are the first port of call for many amateur entomologists.

A small local library is likely to stock only a handful of natural history titles, of which a few may be of entomological interest. The library computerised catalogue will give access to a wider range of stock held in public libraries in the area. It should be fairly easy for the library staff to obtain these titles for you. Some central libraries have excellent reference collections and will stock a larger collection of natural history works, although the emphasis will tend to be modern works on the UK fauna. If you need to refer to a modern entomological work that is in print, it is always worth requesting the item from the library. If the librarian considers the work to be of interest to other readers, the book may be purchased for the library. This will benefit you and other readers.

Inter-library loans and photocopy services

If you need an item which is not available in the library you can request an inter-library loan. You will be required to complete a form, detailing the book or journal reference required. A small reservation fee is often charged for this service. Some public libraries have agreements with other local authorities so that material can be borrowed at a low cost.

If local arrangements cannot produce an item, libraries are likely to use the British Library Document Supply Centre, based at Boston Spa (West Yorkshire). This is the national centre for inter-library loans in Britain. Some public libraries do make an additional charge for using the BLDSC service to cover the fees that are incurred. When ordering such material it is always important to supply full and accurate details of the reference. Books published in the UK within the past 20–30 years should in most cases be supplied in a matter of weeks. Journal references will often be supplied as photocopies within a few days. BLDSC successfully supplies approximately 90% of requests. In general it will not be possible to obtain loans or copies of early or antiquarian material or trace incomplete references through BLDSC. To trace difficult references you will probably need to visit a specialist entomological library.

Some entomology libraries offer a postal photocopy service. For a fee they will supply photocopies from sections of books and journals and post them to you. The amount of copying that can be done will depend on the age and condition of the work and the restrictions of the copyright law. These services aim to cover staff and preservation costs and are therefore more expensive than standard self-service photocopies.

University and research libraries

Students based at universities and institutes of higher education have access to a wide range of library facilities. Many university libraries have an impressive range of computer facilities including CD-ROMs, Internet access and on-line services, as well as standard collections of reference works and journal runs.

Universities that offer life-science courses are likely to have facilities of use to entomologists. If you are not a full-time student but live near to a university campus or an institute of higher education it may be worth contacting the librarian by letter, to discover what facilities may be available to you. Most of these libraries will allow interested individuals access to standard reference facilities.

The Michael Way Library

This research library, managed by CABI, is located on the Imperial College Campus (London University) at Silwood Park, Ascot. It specialises in applied entomology and pest management information, stocks 10000 books and 50000 volumes of periodicals, and receives 66 current journal titles. The library is open to Imperial College staff and students. Members of the public may use facilities on a reference-only basis, at the discretion of the librarian. Payment of an appropriate fee may be required. For further details contact: The Librarian, The Michael Way Library, Silwood Park, Buckhurst Road, Ascot, Berkshire SL5 7TA. Attached to the library is a document delivery service, which can supply copies of many of the items included in CAB Abstracts. For details of this service see the CABI website http://www.cabi.org.

The Hope Library, Oxford University Museum of Natural History

This houses 7000 entomological books and includes many classic British entomological works. One hundred journals are currently subscribed to. The collection also includes 60000 reprints.

The library is open to members of Oxford University and to staff and students from other universities. Other users should apply in writing to the Principal Curator, The Hope Library, Oxford University Museum of Natural History, Parks Road, Oxford OX1 3PW.

Entomological society libraries

Joining an entomological society brings many advantages. Subscription fees often include a free journal, with interesting articles, details of forthcoming events and book reviews. Society events provide a chance to meet other entomologists, exchange ideas and seek advice. Many experienced entomologists are only too happy to recommend useful publications and give advice to those less experienced.

Two entomological societies in the UK also have significant library collections. This is an invaluable resource for members. Extensive use of a library can help to make membership fees seem very reasonable.

The British Entomological and Natural History Society

This society maintains a library at its headquarters in Dinton Pastures, which is open to members on specific days each month. The library includes a good collection of key British entomology books and journal runs. Over 60 current journal titles are subscribed to. Members may borrow items from the library. Further details can be obtained by contacting the Society at: British Entomological and Natural History Society, Dinton Pastures Country Park, Davis Street, Hurst, Reading, Berkshire RG10 0TH.

The Royal Entomological Society Library

This is open to Fellows and Members of the Society. The library is noted for its collection of works on insect taxonomy and general insect biology, with particular reference to the Western Palaearctic Region. The stock includes 11 000 books and 750 journal titles, of which 250 are currently received. Members and Fellows may borrow most library items. Special collections include 1000 rare books, archival material and reprints. The library offers a number of information services including literature searches, scanning journal contents, dealing with subject enquiries and providing photocopies of articles. For further information contact: The Librarian, Royal Entomological Society, 41 Queen's Gate, London SW7 5HR.

National and Museum Reference libraries

The Science Reference and Information Service (SRIS)

This is part of the British Library and is the national library for modern science. The collection is currently divided into two parts in London, at Holborn and Aldwych, the latter stocking life science publications. In the near future the collections will be moving to the new site at St Pancras. Members of the public have access to the library and do not need a pass. For more information contact: The British Library (SRIS), 9 Kean St, Aldwych, London WC2B 4AT

The Entomology Library at the Natural History Museum

This is the most comprehensive entomology library in the world. The collections consists of 90000 volumes, 900 current serials and an extensive collection of manuscripts and drawings. Visitors are requested to make an appointment and will be required to apply for a reader's ticket on their first visit. A postal photocopy service is available. For more information contact: The Librarian, Entomology Library, The Natural History Museum, Cromwell Road, London SW7 5BD

The National Museum of Wales Library

This has good collections of botanical, geological and zoological works. The library is open to the public, but as the entomology collections of journals and books are not housed in the reading room, visitors wishing to use these works should make an appointment in advance. Contact: The Librarian, National Museum of Wales, Cardiff CF1 3NP

The Royal Museum of Scotland Library

This has a good collection of natural history and entomology publications. The library is open to the public for reference only, and visitors are requested to make an appointment in advance. For further information contact: The Librarian, The Royal Museum, Chambers Street, Edinburgh EH1 1JF

Local museums

Some local museums with natural history collections have developed small reference libraries, which may be available to the public.

Library catalogues on-line

Many important library catalogues are now available for searching via the Internet. These catalogues enable entomologists to search for information by using key words (such as author and title words) and subject terms that library staff have applied. Such catalogues can be an excellent way of finding new references and verifying known ones.

Some of the older computer catalogues can appear very confusing when accessed via the Internet. New advanced library systems, which are compatible with the World Wide Web, are much clearer to use and understand. If you should experience problems in using a catalogue it is best to contact the relevant library, as staff should be able to guide you through. Electronic addresses of libraries are still relatively unstable. However, a search on the World Wide Web using a search engine should easily locate important libraries.

The following Web sites are a good source of library information.

The British Library
http://portico.bl.uk

The Natural History Museum Library page
http://library.nhm.ac.uk

NISS (National Information Services and Systems)
http://www.niss.ac.uk

Developing your own entomology library

Many entomologists gradually acquire a collection of key entomology books. Over a period of time this can develop into a small library. Specialist books can appear expensive, but they can bring a great deal of enjoyment to the enthusiast and in some cases can be a good long-term investment. This section looks at the various aspects of buying new and second-hand entomology books.

Buying new books

With so many new entomology books being published it can be very difficult to keep up to date with the range of titles which are available. There are a number of ways to find out about forthcoming and recently published books.

Entomology journals are a good source of information as they often include advertisements for new titles and reviews of recently published books. A good review should provide a rounded appraisal of the work, noting the useful aspects of the work as well as any weaknesses. Such reviews can help readers to decide whether they should consider purchasing a title.

Large bookshops will stock a range of natural history books, although most high street bookshops are likely to display only a handful of entomology titles. All bookshops should be able to check whether UK published material is in print. Such books can be ordered and obtained for you in a matter of days. Bookshops will normally only charge the basic cost of the book for this service, but you should always check first. Most high street shops do not supply overseas publications.

Specialist natural history booksellers will stock a wide range of entomology books including overseas publications. Such specialists should be able to offer advice on which books are available in a given subject area and give guidance on the content of individual titles. Mail-order book services supply detailed catalogues which can help readers to scan recent and forthcoming publications. Some booksellers now have web sites, which can be located by means of a search engine.

Entomological fairs and exhibitions such as the annual exhibition organised by the Amateur Entomologists' Society provide an excellent opportunity to see and purchase new publications. These events are advertised in a number of entomology journals.

Out of print and second-hand books and reprints

Many key entomology works are now out of print. If you are trying to find a copy of a book that has recently gone out of print it is worth checking with specialist entomology booksellers who sometimes still have such titles in stock. Another way to obtain an out of print title is to buy a second-hand copy. Some second-hand titles can be relatively cheap compared with new book prices and can be a good way of starting a book collection.

When authors publish papers in journals they normally receive a number of copies, called reprints (also known as separates or offprints). If you see a recent paper that is of interest to you, a letter to the author requesting a reprint (with a S.A.E.) will often be met with a favourable response. Over a period of time reprint collections on a particular subject can build up to a useful resource.

Specialist booksellers advertise in entomological or other natural history journals, and details can also be sought from entomology libraries or societies.

Collembola: the springtails

(*ca.* 335 species in 11 families)

PETER C. BARNARD

The taxonomic position of this group of small wingless insects has changed many times in recent years. They were once considered, along with the Thysanura, Diplura and Protura, as the subclass Apterygota of the insects. It is now widely recognised that although the Thysanura (with the Archaeognatha also separated off) belong in the true Insecta, the Collembola and Protura are a rather isolated pair of orders, with the Diplura distantly related to them. Some authors place them right outside the insects (but still within the Hexapoda), and others place them as a distinct group, the Parainsecta, which is the sister-group of the rest of the insects, the Euinsecta. This is not the place to discuss this question further, especially as the situation is far from settled, but the springtails are given a place in this book, partly because they have always traditionally been regarded as insects (e.g. in the Royal Entomological Society's handbooks and checklists), and also because the latest monograph places them firmly back in the group Insecta, at least in the broad sense (Hopkin, 1997).

Springtails are tiny animals found mainly in soil and leaf litter, well-known for their springing organ, or furca, which enables them to leap considerable distances when disturbed. They are also commonly found on the seashore, with several species also living on the surface of fresh water, often in large populations. Despite their small size, their occurrence in vast numbers in the soil makes them a highly significant group in soil ecology, with some species even reaching pest proportions. It is therefore unfortunate that, like many soil organisms, they are poorly known and little studied. Hence there are very few useful taxonomic references: Gough (1977) provides a good key to families, but to take identification any further entails the study of a large number of specialist papers, usually covering the whole European fauna, and usually not in English. The best source of these references is Hopkin (1997), which is an essential book for anyone becoming interested in this group of organisms.

The higher classification of the British Collembola, based on Hopkin (1997) but treating the Collembola as an order rather than a class, is:

Suborder Arthropleona
Superfamily Poduroidea: Hypogastruridae, Neanuridae, Onychiuridae, Poduridae
Superfamily Entomobryoidea: Cyphoderidae, Entomobryidae, Isotomidae,
    Oncopoduridae, Tomoceridae
Suborder Neelipleona: Neelidae
Suborder Symphypleona: Sminthuridae (including Sminthurididae)

Fjellberg, A. 1980. *Identification keys to Norwegian Collembola.* Norwegian
    Entomological Society, Oslo.
    Simple, practical keys that cover many British species.
Goto, H.E. & Lawrence, P.N. 1964. Collembola. *In*: Kloet & Hincks: a check list of
    British insects. Part 1, small orders and Hemiptera. *Handbooks for the
    Identification of British Insects* **11**(1): 4–11.
    See Gough (1978) for an update.
Gough, J.J. 1977. A key for the identification of the families of Collembola recorded
    from the British Isles. *Entomologist's Monthly Magazine* **109**: 159–61.
    Useful family key, which treats Sminthurididae as a distinct family.
Gough, H.J. 1978. The British insect fauna: check list: Collembola. *Antenna* 2: 51.
    Updates the Goto & Lawrence (1964) check list.
Hopkin, S.P. 1997. *Biology of the springtails (Insecta: Collembola).* Oxford University
    Press.
    Excellent monograph on all aspects of the group on a world basis, with an extensive bibliography.

Protura: the proturans

(12 species in 3 families)

PETER C. BARNARD

A small group of tiny soil-dwelling animals, related to the Collembola. The taxonomic position of these two groups is discussed in the Collembola chapter. Like the springtails, they are poorly known, yet can be common in soil and leaf litter. Their small size (usually less than 2 mm in length) means they can usually only be seen under the microscope, and many details of their life history and biology are unknown.

There is no identification guide to the British species, but Tuxen (1964a) provides the standard monograph on the group, with Nosek (1973) providing additional information.

The three families in Britain are: Eosentomidae, Protentomidae, Acerentomidae.

Janetschek, H. 1970. Ordnung Protura (Beintastler). *Handbuch der Zoologie* 4(2). 72pp.
   General review of the group, in German.
Nosek, J. (1973). *The European Protura, their taxonomy, ecology and distribution, with keys for determination.* Muséum d'Histoire Naturelle, Geneva.
   Good general biology of the European species, with illustrated keys and distributional information. Continues on from Tuxen (1964a).
Tuxen, S.L. 1964a. *The Protura: a revision of the species of the world, with keys for determination.* Hermann, Paris.
   The standard monograph on this group.
Tuxen, S.L. 1964b. Protura. *In*: Kloet & Hincks: a check list of British insects. Part 1, small orders and Hemiptera. *Handbooks for the Identification of British Insects* 11(1): 3.
   The only British checklist.

Diplura: the two-tailed bristletails

(12 species in 1 family)

PETER C. BARNARD

Although they are small wingless insects, found in damp soil or detritus, often under stones, the Diplura have the distinctive features of two long cerci, which makes them among the more noticeable of the soil insects. They sometimes occur in small colonies, and the British species (all in the genus *Campodea*) seem be mainly herbivorous, feeding on decaying plant material. Some other families of Diplura found in Europe, but not found in Britain, are mainly carnivorous. Delany's (1954) handbook is the only treatment of the British species.

The Campodeidae is the only family represented in Britain.

Delany, M.J. 1954. Thysanura and Diplura. *Handbooks for the Identification of British Insects* 1(2): 7pp.

Very basic key, covering all except one of the 12 British species.

Paclt, J. 1957. Diplura. *Genera Insectorum* 212: 123pp.

A standard monograph on the group.

Steel, W.O. 1964. Diplura. *In*: Kloet & Hincks: a check list of British insects. Part 1, small orders and Hemiptera. *Handbooks for the Identification of British Insects* 11(1): 2.

The only British checklist of this group.

Thysanura: the silverfish and firebrats

(2 species in 1 family)

PETER C. BARNARD

This order, sometimes also known as the Zygentoma, was previously grouped with the Archaeognatha, but as now constituted it contains only two British species, the silverfish (*Lepisma saccharina*) and the firebrat (*Thermobia domestica*). Both orders superficially resemble the Diplura, but have two long cerci and a central epiproct, giving them a 'three-tailed' appearance. The silverfish is commonly seen in cool, damp parts of houses, often where food such as cereals has been spilt. It is an omnivorous species, which even causes damage to books by eating paper and glue. It covering of shiny scales gives it a fish-like appearance, and also makes it difficult to pick up. In contrast, the firebrat lives in much warmer places such as bakeries and large commercial kitchens. It has longer antennae than the silverfish, but is a similar opportunistic feeder on a range of foodstuffs. The two species are easily separated using Delany's (1954) key.

The Lepismatidae is the only family represented in Britain.

Delany, M.J. 1954. Thysanura and Diplura. *Handbooks for the Identification of British Insects* 1(2): 7pp.
  Adequate key, which includes the Archaeognatha in the Thysanura.
Steel, W.O. 1964. Thysanura. *In*: Kloet & Hincks: a check list of British insects. Part 1, small orders and Hemiptera. *Handbooks for the Identification of British Insects* 11(1): 1.

Archaeognatha: the bristletails

(7 species in 1 family)

The Archaeognatha and Thysanura were previously treated as a single order, but are now separated because of fundamental differences in the structure of their mouthparts. They are active insects up to 18 mm long, often found under stones or in leaf litter and detritus. Some species live on the seashore; the largest British species, *Petrobius maritimus*, is frequently seen on rocks near the sea. Their scaly appearance makes them superficially similar to the Thysanuran silverfish, but they are less flattened and can jump when disturbed. The British species can be separated using Delany's (1954) key.

The Machilidae is the only family represented in Britain.

Delany, M.J. 1954. Thysanura and Diplura. *Handbooks for the Identification of British Insects* 1(2): 7pp.

> This key includes the Archaeognatha in the Thysanura. Note that the figures of *Petrobius maritimus* and *P. brevistylis* have been reversed.

Steel, W.O. 1964. Thysanura. *In*: Kloet & Hincks: a check list of British insects. Part 1, small orders and Hemiptera. *Handbooks for the Identification of British Insects* 11(1): 1.

Ephemeroptera: the mayflies or up-winged flies

(50 species in 8 families)

STEPHEN J. BROOKS

There are 50 species of mayfly recorded from Britain, although one of these, *Arthroplea congener*, has only been recorded once in southern England and is apparently extinct (Bratton, 1990). Mayfly larvae occur in a wide range of freshwater habitats and many species have exacting habitat requirements, often related to current flow and levels of dissolved oxygen. For this reason they are useful indicators of freshwater quality, and are particularly sensitive to organic pollution. They are one of the key indicators used in biological monitoring of water quality and are widely sampled by the Environment Agency and water authorities. Despite this there is little accessible information available on the distribution of mayflies in Britain. The Mayfly Recording Scheme is currently inactive and no distribution maps have been published.

The larvae can be distinguished from those of most other freshwater insects by the three tails at the end of the abdomen and the feather-like gills along each side of the abdomen. Most species are confined to flowing water and those that live in torrential streams have flattened, streamlined bodies. They can be readily found by examining the underside of stones and boulders. Those in slow-flowing and standing water either burrow into the silt and fine gravel or are active swimmers. Most species browse algae off the surface of stones and plants but some larvae are carnivorous. Larvae should be collected into 70% alcohol, isopropanol or weak formalin solution. The families are identifiable by their characteristic body shapes. Generic and specific identification relies mainly on the shape of the abdominal gills and mouthparts and can only be achieved with the aid of a ×50 microscope and reference to Elliott *et al.* (1988).

Adult mayflies are small to medium-sized insects with two or three long, trailing tails, short antennae and two pairs of densely veined membranous wings. The hind wings are very small or absent. The wings are held upright above the body when the insect is at rest; this habit has earned them the angler's name of up-winged flies. They are the only group of insect to moult fully functional wings. Sexually immature subimagos, or duns, have dull body colours and milky wings and moult into mature imagos, or spinners, within a few hours of emerging from the water. Males have long fore legs and enlarged colourful eyes. Mayflies are predominantly day-flying insects, although they will fly at night, when they can be seen at lights. During spring and early summer near rivers and ponds, especially in late afternoon, mayflies are often seen in large undulating mating swarms. Because they are an important food source for trout they are mimicked by fly-fishermen who tie artificial mayflies and, as a result, many British species have fisherman's names. Adult mayflies do not have functional mouths and so cannot feed. They rely on the reserves built up during the larval stage. The adults, therefore, are very short-lived and most individuals do not survive for more than a few days. The males of many species have distinctive colour patterns and these can be identified in the field with reference to Goddard (1988, 1991). Several species are also illustrated in colour in Harker (1989). However, family and generic identifications should be confirmed by microscopic examination of the wing venation. Species are best determined by examination of the male claspers and genitalia at the tip of the abdomen with reference to Elliott & Humpesch (1983). The species of some genera are unidentifiable if only females are available. Adults are best preserved in 70% alcohol, isopropanol or weak formalin solution. They are too fragile to last long intact in a pinned collection. Two species were recently added to the British list (Gunn & Blackburn, 1997, 1998) but their adults cannot be readily identified.

The eight families in Britain are: Siphlonuridae, Baetidae, Heptageniidae, Leptophlebiidae, Potamanthidae, Ephemeridae, Ephemerellidae, Caenidae.

References

Bratton, J.H. 1990. A review of the scarcer Ephemeroptera and Plecoptera of Great Britain. *Research & Survey in Nature Conservation* 29: 40pp. Nature Conservancy Council, Peterborough.
Review of the status and distribution in Britain of the five rarest Ephemeroptera.

Brittain, J.E. 1982. Biology of mayflies. *Annual Review of Entomology* 27: 119–47.
Account of mayfly biology on a world basis with full bibliography to other works.

Elliott, J.M. & Humpesch, U.H. 1983. A key to the adults of the British
Ephemeroptera. *Scientific Publications of the Freshwater Biological Association*
47: 101pp.
Key to adult males of British species (most females can only be identified to generic level)
with ecological notes on adult behaviour.

Elliott, J.M., Humpesch, U.H. & Macan, T.T. 1988. Larvae of the British
Ephemeroptera. A key with ecological notes. *Scientific Publications of the
Freshwater Biological Association* 49: 145pp.
Key to larvae of British species with ecological notes on larvae.

Goddard, J. 1988. *John Goddard's waterside guide.* Unwin Hyman Ltd, London.
Pocket guide to common freshwater insects, especially those of interest to fly-fishermen.
Illustrated with colour photographs.

Goddard, J. 1991. *Trout flies of Britain and Europe.* A & C Black, London.
Similar to the guide above but with additional species from Britain and Europe.

Harker, J. 1989. *Mayflies.* Naturalists' Handbooks No. 13. Richmond Publishing Co.
Ltd, Slough.
Account of the biology and ecology of all life-stages of the British mayflies with colour
paintings of selected species with illustrated keys to larvae and adult males to species-
level and females to generic level.

Gunn, R.J.M. & Blackburn, J.H. 1997. *Caenis peudorivulorum* Kieffermüller (Ephem.,
Caenidae), a mayfly new to Britain. *Entomologist's Monthly Magazine* 133:
97–100.
Contains a revised to key to nymphs of British *Caenis* species.

Gunn, R.J.M. & Blackburn, J.H. 1998. *Caenis beskidensis* Sowa (Ephemeroptera,
Caenidae), a mayfly new to Britain. *Entomologist's Monthly Magazine* 134: 94.
New record based on nymphs only, although the taxonomic status of this species needs
clarification.

Macan, T.T. 1982. *The study of stoneflies, mayflies and caddisflies.* Amateur
Entomologists' Society, Hanworth.
Provides a readable introduction to the biology and ecology of mayflies.

Odonata: the dragonflies and damselflies

(45 species in 9 families)

STEPHEN J. BROOKS

The order Odonata includes both dragonflies (suborder Anisoptera) and damselflies (suborder Zygoptera). There are 38 species currently breeding in the British Isles but a further four (*Hemianax ephippiger, Sympetrum vulgatum, Sympetrum fonscolombii* and *Sympetrum flaveolum*) are frequently recorded as non-breeding migrants from continental Europe. In addition three species (*Coenagrion scitulum, Coenagrion armatum* and *Oxygastra curtisii*) that formerly bred in England have become extinct in the past 50 years. One species *(Coenagrion lunulatum)* is currently known only from Ireland, where it was discovered in 1981. The current British list (Merritt *et al.*, 1996) also includes several species that have been recorded as very rare vagrants.

Dragonfly larvae occur in a wide range of freshwater habitats but most species are restricted to standing or slow-flowing water. Only *Calopteryx virgo* and *Cordulegaster boltonii* occur in fast-flowing streams. Dragonfly larvae are carnivorous and feed mostly on insect larvae and other aquatic invertebrates; however, the larger larvae, which may grow up to 5 cm, will take tadpoles and small fish. Prey is captured with an extendible lower lip equipped with sharp spines. Damselfly larvae can be recognised by the three leaf-like appendages (the caudal lamellae) at the tip of the abdomen, which serve as gills and assist as paddles when the larvae are swimming. The shape and patterning of the lamellae can be a useful guide to identifying damselfly larvae. Nevertheless, many species are difficult to separate without microscopical examination of the mouthparts. Dragonfly larvae do not possess caudal lamellae; instead, an array of five short spines are present at the tip of the abdomen. The gills are internal and water is pumped over them through the anus. To escape predators a sudden contraction of the rectal muscles expels a jet of water, propelling the larva to safety. At genus-level, dragonfly larvae have characteristic body shapes but species-level identification requires examination of body spines and mouthparts under a low-powered microscope or hand lens. Only mature odonate larvae and exuviae (the larval skin left after the adult dragonfly has emerged) can be identified by using existing keys and even then the characters used are prone to variation.

Adult dragonflies and damselflies are unmistakable. They are large day-flying insects, capable of fast flight, and have prodigious aerobatic skills. They have large eyes, minute antennae and two pairs of wings with a complex network of veins and a dark pigmented pterostigma near the tip. The abdomen is long, slender and brightly coloured. Many species can be identified in the field, often in flight by experienced observers, by their characteristic body markings; however, the many age-related colour forms can be confusing and are not illustrated in most guides. A pair of close-focusing binoculars can be useful for identifying perched dragonflies in the field. Other species need to be identified in the hand by careful examination of the male claspers at the tip of the abdomen or the shape of the prothorax in female damselflies. Because of their attractive, photogenic appearance and fascinating life histories, dragonflies are the focus of interest of many amateur naturalists. The British Dragonfly Society currently has about 1500 members and there is an active and well-subscribed recording scheme. For these reasons there are several identification guides available to the British species. Hammond (1983) is useful but only includes the breeding species and frequent migrants. Identification of the rare vagrants can be accomplished using Askew (1988), which treats the entire western European fauna. A comprehensive distribution atlas (Merritt *et al.*, 1996) is also available and there are a few books on dragonfly biology, of which the most accessible is Miller (1995). Brooks & Lewington (1997) comprehensively covers identification, biology, ecology and distribution of all species on the British list.

Further details of the British Dragonfly Society, which produces a journal and newsletter twice a year, and the Dragonfly Recording Scheme can be obtained from The Secretary, Dr W.H. Wain, The Haywain, Hollywater Road, Borden, Hants GU35 0AD.

The 9 families in Britain are:

Suborder Zygoptera: Calopterygidae, Lestidae, Platycnemididae, Coenagrionidae
Suborder Anisoptera: Gomphidae, Cordulegastridae, Aeshnidae, Corduliidae,
    Libellulidae

References

Askew, R.R. 1988. *The dragonflies of Europe*. Harley Books, Colchester.
  Includes keys, distribution maps, brief biological notes and large colour paintings of
  adults and larval keys to all species occurring in western Europe. Large format.
British Dragonfly Society. [1992]. *Dig a pond for dragonflies*. British Dragonfly Society
  Publication.
  Contains advice on pond creation and maintenance to attract dragonflies.
Brooks, S.J. & Lewington, R. 1997. *A field guide to the dragonflies of Britain and
  Ireland*. British Wildlife Publishing, Hook.
  Colour paintings of adults of all species on the British list, including most of the age-
  related colour forms. Illustrated keys to larvae, distribution maps, regional accounts,
  descriptions and detailed account of ecology of the breeding (and extinct) species.
Corbet, P.S. 1962. *A biology of dragonflies*. Witherby, London (reprinted by E.W.
  Classey in 1983).
  Seminal work on dragonfly biology including examples from Britain and the rest of the
  world. A sequel to this book, covering the intervening years, is due to be published in
  1999.
Corbet, P.S., Longfield, C. & Moore, N.W. 1960. *Dragonflies*. Collins New Naturalist,
  London (reprinted in 1985).
  Classic work on the biology of British dragonflies.
Gibbons, B. 1986. *Dragonflies and damselflies of Britain and northern Europe*.
  Country Life Guides.
  Concise pocket guide to northern European dragonflies, including all the British species
  and many of the migrants, illustrated with colour photographs.
Hammond, C.O. 1983. *The dragonflies of Great Britain and Ireland*, (2nd edn, revised
  by R. Merritt). Harley Books, Colchester.
  Keys to adults and larvae, distribution maps, very brief biological notes and large format
  colour paintings of adults.
Merritt, R., Moore, N.W. & Eversham, B.C. 1996. *Atlas of the dragonflies of Britain and
  Ireland*. HMSO, London.
  Comprehensive distribution maps including records up to 1990. High-quality colour
  photographs of many species as adults and larvae, notes for each species on identifica-
  tion, ecology, flight period and status.
Miller, P.L. 1995. *Dragonflies*. Naturalists' Handbooks No. 7. Richmond Publishing
  Co. Ltd, Slough.
  Concise, informative and stimulating account of biology and ecology of British species,
  also includes colour paintings of adults of selected British species and keys to all adults
  and larvae.

Plecoptera – the stoneflies

(36 species in 7 families)

STEPHEN J. BROOKS

This small group of insects with aquatic larvae includes 36 species recorded from Britain. Many species are widespread but some, such as *Protonemura montana, Capnia atra, Capnia vidua* and *Brachyptera putata*, are largely restricted to northern uplands. *Nemoura dubitans* occurs only in East Anglia and southern England. Several of the British species have not been recorded from Ireland but this may reflect a lack of recording effort rather than a real biological absence. Of the 36 species one, *Isoperla obscura,* is thought to be extinct in Britain; three taxa, *Taeniopteryx nebulosa brittanica, Brachyptera putata* and *Capnia vidua anglica,* are endemic. The distribution and status of stoneflies in Britain is imperfectly known but has been reviewed by Bird (1983), Costello (1988) and Bratton (1990). Stoneflies occur in a wide range of aquatic habitats but most species are restricted to fast-flowing streams or the wave-washed shores of large lakes, particularly in upland regions. The larvae of Perlidae, Perlodidae and Chloroperlidae are carnivorous; the others are herbivores. Although most larvae grow to no more than 10 mm, some are among the largest insects in fast-flowing streams, with *Perla bipunctata* reaching up to 35 mm. These larger larvae, living in nutrient-poor streams, may take three years to complete development but smaller species, living in southern England in nutrient-rich chalk streams or in the organic detritus in the bottom of lakes, complete development within a year. Most stonefly larvae are very sensitive to organic pollution, although they are less sensitive to acidification or heavy metal pollution. They are an important group in biotic indices and are routinely used in the assessment of water quality. Stonefly larvae can be distinguished from most other aquatic insects by the pair of long tails at the tip of the abdomen. They have long antennae and most species also have external gills, which are sausage-shaped or filiform and are positioned below the thorax. The larvae can be accurately identified only by use of a stereo microscope to examine details of setation and mouth-part morphology.

Adult stoneflies emerge during summer. They closely resemble larvae in appearance, particularly so in those species in which the males are brachypterous, that is their wings are short and they are incapable of flight. Stoneflies have two pairs of membranous wings, which in most species are held flat over the body but in Leuctridae are wrapped around the abdomen to form a tube. The wings have a dense network of veins forming a characteristic ladder-shape in the middle of the wing. Adult stoneflies stay close to water, where they can be found among fringing vegetation and scurrying around on stones at the water's edge, or in mid-stream. Most species do not feed, although some scrape algae and lichens from the surfaces of boulders, and others have been recorded as feeding on nectar or pollen or predating other insects. Artificial flies, tied to resemble both larval and adult stoneflies, are used by trout fishermen; many species are known by anglers' names (Goddard, 1988, 1991). Stoneflies can be identified to family by differences in wing venation. Species-level identification is dependent on examination of the genitalia under a low-powered stereo microscope. Both larvae and adults can be identified by using Hynes (1977). Information on the biology of stoneflies is summarised by Macan (1982).

The seven families in Britain are: Taeniopterygidae, Nemouridae, Leuctridae, Capniidae, Perlidae, Perlodidae, Chloroperlidae.

References

Bird, L.M. 1983. Records of stoneflies (Plecoptera) from rivers in Great Britain. *Entomologist's Gazette* 34: 101–11.
Distributional records of riverine species.

Bratton, J.H. 1990. A review of the scarcer Ephemeroptera and Plecoptera of Great Britain. *Research & Survey in Nature Conservation* 29: 40pp. Nature Conservancy Council, Peterborough.
Review of the status and distribution in Britain of the rarer Plecoptera.

Costello, M.J. 1988. A review of the distribution of stoneflies (Insecta, Plecoptera) in Ireland. *Proceedings of the Royal Irish Academy* 88B(1): 1–22.
Distributional records of Irish species.

Goddard, J. 1988. *John Goddard's waterside guide.* Unwin Hyman Ltd, London.
Pocket guide to common freshwater insects, especially those of interest to fly-fishermen. Illustrated with colour photographs.

1991. *Trout flies of Britain and Europe.* A & C Black, London.

Similar to the guide above but with additional species from Britain and Europe.

Hynes, H.B.N. 1977. Adults and nymphs of British stoneflies (Plecoptera). *Scientific Publications of the Freshwater Biological Association* **17**: 92pp.

Definitive key to adults and larvae with brief ecological notes.

Macan, T.T. 1982. *The study of stoneflies, mayflies and caddisflies.* Amateur Entomologists' Society, Hanworth.

Provides a readable introduction to the biology and ecology of stoneflies.

Phasmida: the stick-insects[†]

(3 species in 1 family)

JUDITH A. MARSHALL

Although there are several species of stick-insect in Europe, most of the group are tropical. However, during this century, and particularly in recent years, many species of tropical stick-insect have been brought to Europe and successfully reared indoors. Specimens may often escape (or be released), and the laboratory or Indian stick-insect (*Carausius morosus*) in particular is often found on privet hedges in summer. This is the only species that readily feeds on privet; most other tropical species will prefer bramble and other Rosaceae. The three species that have established breeding colonies in Britain are of New Zealand origin, having been imported with plants, probably as eggs, during the early 1900s. They have become well established in the south-west of England — Devon, Cornwall and Isles of Scilly — and Eire. They feed, and are found, on a variety of bushes and trees including Japanese Cedar.

Stick-insects in Britain are slender, wingless, stick-like and usually green, up to 125 mm in length when adult, and are parthenogenetically reproducing females. Adults may easily be identified, using Brock (1987) or Marshall & Haes (1988); nymphs are less easy to identify with certainty, so should, if possible, be reared to adulthood. Large tropical species, which may have wings

[†] Note: Phasmida is the earliest (and simplest) name proposed for this order. Although the names of family-group taxa are subject to the rules of the International Commission on Zoological Nomenclature there are no such rules for the formation of names above this level, so that for the Order Phasmida, the names Phasmatodea, Phasmatoptera and Cheleutoptera are also in use. A family-group name is formed by adding a Latinised suffix to the stem of the name of the type genus, the stem being obtained from the genitive singular of the name; for *Phasma* this means *Phasmat-*, hence family Phasmatidae, but 'Phasmatodea' is an unnecessary complication.

or be very spiny, may be successfully reared in captivity but if released out-of-doors are unlikely to survive through the winter.

The Phasmid Study Group was formed in 1980 to foster the study of stick-insects; membership is world-wide though UK-based; a quarterly newsletter is produced, and scientific articles are published in the journal *Phasmid Studies*. Details of membership may be obtained from the Treasurer and Membership Secretary, Paul Brock: 'Papillon', 40 Thorndike Road, Slough, Berks. SL2 1SR, UK.

Brock (1986) described the three species breeding here as *Acanthoxyla geisovii, A. inermis* and *Clitarchus hookeri,* although Salmon (1991) placed both *A. geisovii* and *A. inermis* as subspecies of *A. prasina.* All are placed in the family Phasmatidae.

References

Brock, P.D. 1987. A third New Zealand stick insect (Phasmatodea) established in the British Isles, with notes on the other species, including a correction. *In* Mazzini, M. & Scali, V. (Eds) *1st International Symposium on Stick Insects.* University of Siena (1986): pp. 125–32.
  Includes a simplified key to separate the three species and a full taxonomic bibliography.

Brock, P.D. 1991. *Stick insects of Britain, Europe and the Mediterranean.* Fitzgerald Publishing, London.
  Contains identification keys, photographs, drawings of eggs, body parts and food plants.

Haes, E.C.M. & Harding, P.T. 1997. *Atlas of grasshoppers, crickets and allied insects in Britain and Ireland.* ITE Research Publication No. 11. The Stationery Office, London.
  Distribution maps; no keys, but includes discussion of introductions, both casual and deliberate, and their potential survival.

Marshall, J.A. & Haes, E.C.M. 1988. *Grasshoppers and allied insects of Great Britain and Ireland.* Harley Books, Colchester.
  Descriptions and keys to the adults of the three New Zealand species and the Laboratory Stick-insect, with colour plate.

Salmon, J.T. 1991. *The stick insects of New Zealand.* Reed Books, Auckland.
  Keys to all New Zealand species, with colour drawings and colour photographs of eggs.

Orthoptera: the grasshoppers, crickets and bush-crickets[†]

(33 species in 6 families)

JUDITH A. MARSHALL

The saltatorial Orthoptera are those with enlarged hindlegs with powerful muscles for jumping. This includes the Acridoidea (grasshoppers, locusts and groundhoppers), which have short thick antennae, short ovipositors, abdominal hearing organs, and stridulate using hind legs and wings; the Grylloidea, or true crickets and mole-crickets, and Tettigonioidea, the bush-crickets, camel-crickets and others, also called katydids (and previously, 'long-horned grasshoppers'), having long, slender antennae, sword-shaped ovipositors, fore-tibial hearing organs and stridulation by fore wings only.

Of the 33 species, 29 are natives, two are established aliens and two are regarded as probable natives. The majority of native species occur in southern areas with only the more cold-tolerant having reached northern Scotland, and the brachypterous meadow grasshopper (*Chorthippus parallelus*) apparently not having reached Ireland. The lesser mottled grasshopper (*Stenobothrus stigmaticus*) occurs only on the Isle of Man and is regarded as a probable native, as also is the scaly cricket (*Pseudomogoplistes squamiger*), found in Dorset, Devon and Sark. Most orthopteroids live in a range of grassland habitats but the few wetland species have become much restricted, with the mole-cricket (*Gryllotalpa gryllotalpa*) now very rare and local. Recent hot summers have increased the incidence of macropterous forms of normally brachypterous species, particularly Roesel's bush-cricket (*Metrioptera roeselii*), with resultant spread of the species.

[†] The name Orthoptera has been used to refer to two related groups of insects, the Orthoptera Saltatoria (those that jump i.e. the grasshoppers etc.), and the Orthoptera Cursoria (those that run, which includes the earwigs, cockroaches, mantids and stick-insects). See note under Phasmida, and also Marshall & Haes, 1988: 16–19.

The house-cricket (*Acheta domesticus*) has been established in Britain for centuries, normally as a pest in protected environments, but warmer summers have seen the species spread far from human habitation. The southern field-cricket (*Gryllus bimaculatus*) has also been recorded out of doors in hot summers, resulting from casual introductions or escapes from the many cultures available in Britain. Many other species are occasionally introduced with bananas from the West Indies or with tropical plants, or as stowaways with a variety of goods; few survive long or could become established in the UK.

The six families in Britain are: Rhaphidophoridae, Tettigoniidae, Gryllidae, Gryllotalpidae, Acrididae, Tetrigidae.

References

Bellmann, H. 1988. *A field guide to the grasshoppers and crickets of Britain and northern Europe*. Collins, London.
  Illustrated key to species and songs, colour photographs of adults and of mating, oviposition, developmental stages and ecdysis.

Burton, J.F. & Ragge, D.R. 1987. *Sound guide to the grasshoppers and allied insects of Great Britain and Ireland*. Cassette. Harley Books, Colchester.
  Recordings of the songs, with announcements by David Ragge.

Haes, E.C.M. & Harding, P.T. 1997. *Atlas of grasshoppers, crickets and allied insects in Britain and Ireland*. ITE Research Publication No. 11. The Stationery Office, London.
  Brief species accounts of current status with cross-references to Marshall & Haes (1988).

Marshall, J.A. & Haes, E.C.M. 1988. *Grasshoppers and allied insects of Great Britain and Ireland*. Harley Books, Colchester.
  Illustrated key to all species, detailed discussions of habitats, and distribution maps with vice-county records and offshore island records to 1988.

Ragge, D.R. 1965. *Grasshoppers, crickets and cockroaches of the British Isles*. Warne, London.
  Illustrated keys to species (excluding Channel Islands) and to the named colour varieties of grasshoppers.

Ragge, D.R. & Reynolds, W.J. 1998. *The songs of the grasshoppers and crickets of western Europe*. Harley Books, Colchester.
  Key to 170 European species based on their readily identifiable songs.

Ragge, D.R. & Reynolds, W.J. 1998. *A sound guide to the grasshoppers and crickets of western Europe*. Two compact discs. Harley Books, Colchester.
  Over 1200 recordings, including special courtship songs.

Dermaptera: the earwigs

(7 species in 4 families)

JUDITH A. MARSHALL

The common earwig, *Forficula auricularia*, is a well-known insect, Palaearctic in origin but now of world-wide distribution. Although almost ubiquitous in Britain it is poorly recorded.

Earwigs are easily recognised by the abdominal pincers or forceps, straight in females but curved and often elaborately enlarged in adult males. Many earwigs are fully winged and capable of flight; this is rarely seen in *F. auricularia* but the lesser earwig (*Labia minor*) flies readily. At rest the delicate hind wings are almost totally concealed under the short, tough, protective fore wings, which may be held firmly closed by a 'zip' structure formed by rows of hook-like spines on the fore wings and thorax.

The degree of maternal care seen in *F. auricularia* is assumed to occur in all earwigs. The eggs are laid in underground chambers and carefully tended by the mother; without such care they will not hatch. After hatching the nymphs may stay with the mother until the second or third instar. As with other insects having incomplete metamorphosis, the newly moulted insect is pure white until the cuticle darkens and hardens, resulting in many queries about albino earwigs.

There are four native species and three established aliens, and other occasional introductions, but the species causing problems is invariably the common earwig. They are omnivorous and may cause serious damage to young plants and flowers, and sometimes occur in very large numbers. Earwigs are occasionally host to a small roundworm parasite, *Mermis nigrescens*, picked up by eating eggs on plant-leaves. The worms may be seen when earwigs are disposed of after invading homes or other premises, but are perfectly harmless to people, plants and domestic animals.

The four families in Britain are: Forficulidae, Anisolabiidae, Labiduridae, Labiidae.

References

Haes, E.C.M. & Harding, P.T. 1997. *Atlas of grasshoppers, crickets and allied insects in Britain and Ireland.* ITE Research Publication No. 11. The Stationery Office, London.

Brief species accounts of current status with cross-references to Marshall & Haes (1988).

Hincks, W.D. 1956. Orthoptera, Dermaptera (2nd edn). *Handbooks for the Identification of British Insects* **1**(5): 24pp.

Key to species still usable, but other books listed here are better for current distribution and biology.

Marshall, J.A. & Haes, E.C.M. 1988. *Grasshoppers and allied insects of Great Britain and Ireland.* Harley Books, Colchester.

Introduction to the order and keys to species, with illustrations.

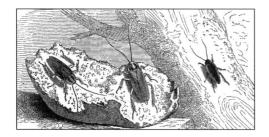

Blattodea: the cockroaches[†]

(9 species in 3 families)

JUDITH A. MARSHALL

Of the 4000 or so described species of cockroach in the world, less than 1% are recognised pests. However, these pests may occur in such large and unpleasant numbers in domestic environments that they have earned a bad reputation for the whole group. The majority of cockroach species are wild-living, and in northern Europe are small and inconspicuous; there are three native cockroaches, *Ectobius* spp., in the British Isles, confined to Wales and southern England and the Channel Isles. There are also a number of established pest species, some long-standing, others more recent introductions. The common or Oriental cockroach (*Blatta orientalis*) has been established in Britain for centuries, possibly crossing Europe from the eastern Mediterranean area with the returning crusaders.

Native cockroaches are largely diurnal, present in a wide range of habitats. *E. lapponicus* and *E. pallidus* may be found in woodland rides and clearings and on scrubby heathland; *E. lapponicus* will fly readily in warm weather if disturbed from vegetation. *E. panzeri* is a mainly coastal species, found in low vegetation on sand-dunes, shingle beaches etc. All three species lay oothecae, which over-winter, *E. lapponicus* and *E. pallidus* having a two-year life cycle, over-wintering as nymphs during the second year; adults appear from May onwards, surviving until September or October.

The majority of pest species require the protection of artificially heated surroundings and a ready supply of food to scavenge from; blocks of flats, restaurants, warehouses, rubbish tips, are all ideal for the *Periplaneta* spp., *Blattella* and *Supella*. These species have a continuous life cycle, laying

---

[†] Note: The Blattodea have previously been considered, with the Mantodea, as a suborder of Order Dictyoptera: the names Blattaria and Blattoptera are also in use. See note under Phasmida.

oothecae, which may take 2–3 months to hatch; development is faster at higher temperatures. *Pycnoscelus surinamensis* is a burrowing root feeder which may become a nuisance in commercial glasshouses, infesting potted plants; this species reproduces parthenogenetically, and retracts the ootheca into a brood pouch to protect the eggs during development (ovoviviparity). Many other species are introduced with foods, plants and other goods; some are cosmotropical pests, other are casual introductions, such as those imported with bananas from the West Indies. Other tropical species are held in culture: the Blattodea Culture Group (formed in 1987) fosters the study of this diverse and fascinating group of insects. Details are available from Adrian D. Durkin, 8, Foley Road, Pedmore, Stourbridge DY9 0RT.

The three families in Britain are: Blattidae, Blattellidae, Blaberidae.

References

Brown, V.K. 1973. A key to the nymphal instars of the British species of *Ectobius* Stephens (Dictyoptera: Blattidae). *The Entomologist* **106**: 202–9.
   An illustrated key to species and all instars.
Haes, E.C.M. & Harding, P.T. 1997. *Atlas of grasshoppers, crickets and allied insects in Britain and Ireland.* ITE Research Publication No. 11. The Stationery Office, London
   Brief species accounts of current status with cross-references to Marshall & Haes (1988).
Marshall, J.A. & Haes, E.C.M. 1988. *Grasshoppers and allied insects of Great Britain and Ireland.* Harley Books, Colchester.
   Key to native species and established aliens, with full descriptions and illustrations.
Ragge, D.R. 1965. *Grasshoppers, crickets and cockroaches of the British Isles.* Warne, London.
   Descriptions and illustrations of some introduced species.

Psocoptera: the booklice and barklice

(*ca.* 90 species in 18 families)[†]

JON H. MARTIN

Members of the Psocoptera are all generally referred to as 'psocids', although the family Psocidae is only one of 18 in the British Isles. The colloquial names 'booklice' or 'dustlice' are often used, particularly for members of the Liposcelidae, but these common names are both misnomers. Although Psocoptera are closely related to true lice (Phthiraptera), their appearance, feeding methods and ecology are quite different.

Superficially, psocids resemble psyllids (Hemiptera: Sternorrhyncha), whose wing venation and body shape are quite similar. However, psocids have chewing mouthparts and do not jump as psyllids do. Most adult psocids have two pairs of well-developed wings, but brachyptery and aptery are quite common; adulthood in such cases is most commonly indicated by the presence of three-segmented tarsi (at least in the British fauna), whereas nymphs have two. Most macropterous species have six nymphal instars, but the number of instars may be reduced in cases of aptery or brachyptery. Reproduction may be either sexual or parthenogenetic, but most species are oviparous. New (1987) gives a good general account of psocid biology.

Psocids are most often encountered as the indoor inhabitants referred to as booklice, but the majority of British species live out of doors. They are grazing feeders, specialising in the exploitation of microepiphytes (terminology of Smithers & Lienhard, 1992) such as algae, lichens, mildews and other fungi. The species that feed on mildews and moulds are associated with damp conditions, and the name 'booklice' arose from the association of these insects with damp newspapers, books and wallpaper inside dwellings, where feeding on damp glues can lead to books falling apart and wallpaper separating from walls. In domestic situations they are not, in themselves, often pests but when present in very large numbers they may cause indirect

---

[†] New (1974) stated that the British fauna comprises around 50 species that are known to occur naturally, but that around 40 more have been recorded sporadically. Chinery (1993) noted the same total number but considered that several occurred only in artificial conditions. As with the scale insects (Hemiptera: Coccoidea), it is somewhat arbitrary whether certain species are regarded as 'British' or as mere casual introductions. The world fauna stands at over 3500 described species (Smithers & Lienhard, 1992).

feeding damage, faecal contamination and staining. In the stored products industry psocids are regarded as serious pests, although their presence in large numbers is actually indicative of environmental problems, notably dampness. The best control is keeping houses, warehouses and holds dry enough for moulds and mildews not to develop. Broadhead (1954) recorded 30 species of psocid as pests of stored products in Britain and considered that *Trogium pulsatorium* was then the commonest in domestic situations; *Liposcelis bostrychophila* now has that status. Mound (1989) regarded only 8 of the species discussed by Broadhead as being commonly encountered.

Interestingly, some domestic psocids reveal their presence by the sounds they make. *Trogium pulsatorium* makes a sound like a ticking watch when on a substrate that adequately amplifies its tapping; *Lepinotus* species emit a faint chirping sound (Mound, 1989).

Psocids shrivel when dried, and collections generally rely on spirit storage and slide mounts. As with psyllids, these insects need a degree of dissection to enable a range of characters to be seen on slides.

The Psocoptera in Britain may be identified by using New's (1974) *Handbook*. This book includes large sections on biology and morphology as well as an account of preservation, preparation and study techniques. Separate keys are provided to suborders, families, genera and the species within each genus. Mound (1989) gave a short key to the genera most commonly found as pests of stored products. Little work has been done on the psocid fauna in Britain since these works were published.

The eighteen families in Britain are placed in three suborders:

Suborder Trogiomorpha: Lepidopsocidae, Trogiidae, Psoquillidae, Psyllipsocidae
Suborder Troctomorpha: Liposcelidae, Pachytroctidae, Sphaeropsocidae
Suborder Psocomorpha: Epipsocidae, Caeciliidae, Stenopsocidae, Lachesillidae, Ectopsocidae, Peripsocidae, Trichopsocidae, Elipsocidae, Philotarsidae, Mesopsocidae, Psocidae

References

Broadhead, E. 1954. The infestation of warehouses and ships' holds by psocids in Britain. *Entomologist's Monthly Magazine* **90**: 103–105.
  The author lists 30 species of psocid found in ships' holds and warehouses around Britain during a period of about three years. No guide to recognition is given.
Chinery, M. 1993. *A field guide to the insects of Britain and Northern Europe.* Collins, London.
  So little has been written about the systematics of British psocids that the short chapter in this book is a useful guide when New's more comprehensive work is unavailable.

Mound, L.A. (Ed.) 1989. *Common insect pests of stored products.* British Museum (Natural History), London.

This useful little guide to stored-product pests contains a short section on psocids and a key to the four genera most often encountered in this context.

New, T.R. 1974. Psocoptera. *Handbooks for the Identification of British Insects* **1**(7): 102pp.

This is the only work that covers the British psocid fauna in detail, and is discussed above.

New, T.R. 1977. Notes on the identification of nymphs of the British Psocoptera. *Entomologist's Gazette* **28**: 61–71.

Keys to families, and tabular summary of characteristics of nymphs of each genus.

New, T.R. 1987. Biology of the Psocoptera. *Oriental Insects* **21**: 1–109.

This work gives an updated account of reproduction, feeding, natural enemies, ecology and biodiversity of Psocoptera.

Smithers, C.N. & Lienhard, C. 1992. A revised bibliography of the Psocoptera (Arthropoda: Insecta). *Technical Reports of the Australian Museum* **6**: 86pp. Australian Museum, Sydney.

This work is an essential guide to the literature for any worker contemplating a deeper interest in the Psocoptera.

Phthiraptera: the lice

(545 species in 16 families)

CHRISTOPHER H.C. LYAL

Lice are a common group, but most species are rarely encountered, because of where they live. They are ectoparasites, spending all their lives on the bodies of birds or mammals. Even the eggs are found only on the host, cemented to the plumage or fur. Lice feed on the feathers, hair or blood of the host. In order to move from host to host, lice must wait until the two animals are in physical contact, since they can neither fly nor jump. The lice most frequently encountered are the human-associated species, particularly the head louse.

Most species of lice are found on only a single host species, for example the three species found on humans: *Pediculus humanus, Pediculus capitis* and *Pthirus pubis*. Other species, however, can be found on more than one host, a good example being one of the lice of gulls, *Saemundssonia lari*, which is parasitic on every species of *Larus* in Britain. A species of mammal or bird may, however, be parasitised by more than one type of louse, some having three or four. Often these different species of louse live on different parts of the host's body; for example, the short, slow, round-bodied *Craspedorhynchus platystomus* lives on the head and neck of the golden eagle (*Aquila chrysaetos*) whereas the slender, rapidly moving *Falcolipeurus suturalis* lives among the wing feathers of the same species.

Not all mammals have associated lice; for example bats, insectivores and whales lack them. However, there are almost certainly lice on all species of birds in Britain, although not all have been recorded. Because lice are so intimately associated with their hosts, there is a natural expectation that if a known host is found in the UK, all of its louse species will be too. This, however, is not necessarily so, since louse distribution is not always the same as that of the host. Nevertheless, a very useful starting point for louse identification is a host–parasite list, so the likely possibilities can be determined.

There are few general treatments of lice allowing identification of a full fauna. Certainly there are no faunistic treatments for the UK, and not all of the British species have been described. The concept of a 'UK fauna' is anyway rather a vague one. There are well-established populations of lice on such exotic imports as guinea pig (which are on the British List), coypu and ostrich, not to mention birds and mammals in zoos and wildlife parks. Therefore, a list of some of the more comprehensive works from around the world will provide a useful starting point, even though most of these will be difficult to obtain. For accurate species identification specialist literature will often be needed. In addition, the lice may need to be cleared and mounted on a microscope slide for examination of setae and genitalia. Thus it will prove difficult to identify the majority of lice to species.

There are different opinions about the classification of lice. Many books and scientific papers divide the lice into two orders: Anoplura ('sucking lice', all blood-feeding, living on mammals) and Mallophaga ('chewing lice' or 'bird lice', lice feeding on blood, feathers, hair or sebaceous material, found on both birds and mammals, and including the Amblycera, Ischnocera and Rhynchophthirina). Modern classification puts all of the lice in a single Order, Phthiraptera, with four suborders: Amblycera (on birds and mammals), Ischnocera (on birds and mammals), Rhynchophthirina (on elephants and wart-hogs) and Anoplura (on mammals). A key to these can be found in Ledger (1980), perhaps one of the most comprehensive works listed below. There are also different views on the number and extent of genera in the Amblycera and Ischnocera, with British and American workers favouring fewer than some Eastern Europeans. The references given below are fairly consistent in their treatment.

The sixteen families in Britain are placed in three suborders:

Suborder Amblycera: Menoponidae, Laemobothriidae, Ricinidae, Trimenoponidae, Gyropidae
Suborder Ischnocera: Trichodectidae, Philopteridae, Goniodidae
Suborder Anoplura: Echinophthiriidae, Haematopinidae, Hoplopleuridae, Linognathidae, Pediculidae, Enderleinellidae, Polyplacidae, Pthiridae

References

Blagoveshchenskii, D. I. 1964. 16. Order Mallophaga – Biting lice. *In* Bei-Bienko. G.Y. (Ed.) Keys to the insects of the European USSR. Vol. 1. *Apterygota, Palaeoptera, Hemimetabola.* (Keys to the Fauna of the USSR no. 84) pp. 385–403. Israel program for scientific translations, Jerusalem. [English translation of the original Russian.]
    Illustrated keys to families, genera and some species of Ischnocera, Amblycera and Anoplura.
Borror, D.J., Triplehorn, C.A. & Johnson, N.F. 1989. *An introduction to the study of insects.* (6th edn). Saunders College. Publishing, Philadelphia.
    Key to families of lice.

Clay, T. 1969. A key to the genera of Menoponidae (Amblycera: Mallophaga: Insecta). *Bulletin of the British Museum (Natural History)* (Entomology) **24**: 1–26.
   Illustrated key covering most of the British Amblycera. The families can be keyed using Clay (1970).

Clay, T. 1970. The Amblycera (Phthiraptera: Insects). *Bulletin of the British Museum (Natural History)* (Entomology) **25**: 75–98.
   Keys to suborders of Phthiraptera, families of Amblycera and genera of Boopiidae; useful discussion of amblyceran morphology.

Clay, T. & Moreby, C. 1967. Mallophaga (biting lice) and Anoplura (sucking lice). Part 11: Keys and locality lists of Mallophaga and Anoplura. *Antarctic Research Series* **10**: 157–95.
   Useful illustrated keys and host list, covering some of the UK fauna, particularly seabirds.

Durden, L.A. & Musser, G.G. 1994. The sucking lice (Insecta, Anoplura) of the world: a taxonomic checklist with records of mammalian hosts and geographical distributions. *Bulletin of the American Museum of Natural History* **218**: 1–90.
   Useful list, including synonyms and references to original descriptions.

Emerson, K.C. 1956. Mallophaga (chewing lice) occurring on the domestic chicken. *Journal of the Kansas Entomological Society* **29**: 63–79.
   Key and illustrations of the Amblycera and Ischnocera found on the domestic chicken.

Emerson, K.C. & Price, R.D. 1981. A host-parasite list of the Mallophaga on mammals. *Miscellaneous publications of the Entomological Society of America* **12**(1): 1–72.
   List of Amblycera and Ischnocera of mammals, arranged by host.

Ferris, G.F. 1951. The Sucking Lice. *Memoirs of the Pacific Coast Entomological Society* **1**: 1–320.
   Keys to Anoplura of the world, with descriptions of genera and some species, well illustrated with good line drawings.

Hopkins, G.H.E. 1949. The host-associations of the lice of mammals. *Proceedings of the Zoological Society of London* **119**: 387–604.
   An older list of mammals and their lice than others mentioned below, but perhaps easier to obtain for British users.

Hopkins, G.H.E. & Clay, T. 1952. *A checklist of the genera and species of Mallophaga.* British Museum (Natural History), London.
   For additions and corrections see *Annals and Magazine of Natural History* (12) **6**: 434–48 and **8**: 177–90. Lists all Amblycera, Ischnocera and Rhynchophthirina known at the time, giving also the type hosts (although not the complete host ranges).

Kéler, S. von 1960. Bibliographie der Mallophagen. *Mitteilungen aus dem Zoologischen Museum in Berlin* **36** (2): 145–403.
   A bibliography covering all works on Amblycera and Ischnocera up to 1959.

Kéler, S. von 1971. A revision of the Australasian Boopiidae (Insecta: Phthiraptera) with notes on the Trimenoponidae. *Australian Journal of Zoology* (suppl.) **6**: 1–126.
   Covers a number of amblyceran lice occurring in the UK on introduced mammals.

Kim, K.C. & Ludwig, H.W. 1978. The family classification of the Anoplura. *Systematic Entomology* **3**: 249–84.
   Key to families of Anoplura.

Kim, K.C., Pratt, H.D. & Stojanovich, C.J. 1986. *The sucking lice of North America: an illustrated manual for identification*. Pennsylvania State University Press, University Park.
  Illustrated keys to the families, genera and species of North American Anoplura, with a number of species descriptions. Covers much of the UK fauna. Also contains information on collecting, morphology and biology. Lists of parasites and their hosts (and hosts and their parasites) given.

Ledger, J.A. 1980. The arthropod parasites of vertebrates in Africa South of the Sahara. Vol. IV, Phthiraptera (Insecta). *Publications of the South African Institute for Medical Research* **56**: 1–327.
  Good introduction to the group; keys to suborders, families, genera; host-parasite list.

Lyal, C.H.C. 1985. A cladistic analysis and classification of trichodectid mammal lice (Phthiraptera: Ischnocera). *Bulletin of the British Museum (Natural History) (Entomology)* **51**: 187–346.
  Keys to subfamilies, genera, and subgenera of Trichodectidae of the world, with numerous illustrations.

Smith, V. & Page, R. 1997. Tree of Life: Phthiraptera. World Wide Web pages at
  http://taxonomy.zoology.gla.ac.uk/tol/phthiraptera/phthiraptera.html
  Useful references, introduction to the group, and images.

Timmermann, G. 1965. Die Federlingsfauna der Sturmvögel und die Phylogenese des procellariiformen Vogelstammes. *Abhandlungen und Verhandlungen des Naturwissenschaftlichen Vereins in Hamburg (N.F.)* **8** (suppl.): 1–249.
  Keys to genera, species groups and species of lice parasitic on Procellariiformes (in German).

Werneck, F.L. 1948. *Os Malóphagos de mamíferos*. Parte I, *Amblycera e Ischnocera (Philopteridae e parte de Trichodectidae)*. Revista Brasileira de Biologia, Rio de Janeiro
  Descriptions and illustrations of mammal lice (Amblycera and Trichodectidae) of the world.

Werneck, F.L. 1950. *Os Malóphagos de mamíferos*. Parte II, *Ischnocera (continuacao de Trichodectidae) e Rhyncophthirina*. Instituto Oswaldo Cruz, Rio de Janeiro.
  Descriptions and illustrations of mammal lice (Trichodectidae) of the world.

Thysanoptera: the thrips

(160 species in 3 families)

JON H. MARTIN

The scientific name Thysanoptera means 'fringed wings'. The adults of the great majority of species are winged and only 1–3 mm long. The largest British species reaches 7 mm, and the largest in the world 14 mm. Members of this order feed either on plant or fungal material, and their method of feeding is most unusual. Thrips have a structure called a mouthcone under the head, and a single needle-like lance (evolved from the left mandible) which pierces the substrate. The maxillary stylets (paired in the usual way) are then used to suck liquid food, which may be either cell sap from living plants or liquid from fungal sources, including spores.

The thrips life cycle is a curious intermediate between holometaboly and hemimetaboly. There are two larval stages, which resemble wingless adults, in a typically hemimetabolous way. These stages are followed by a prepupa and one (Terebrantia) or two (Tubulifera) non-feeding pupal stages prior to the final emergence of the adult. Although the prepupa and pupa stages cannot feed they do have legs and are sometimes mobile to a limited degree. Reproduction is another thrips curiosity: fertilised (diploid) eggs develop into females and unfertilised (haploid) ones into males. Populations of some species are entirely parthenogenetic. Perhaps the best general account of the group was provided by Lewis (1973), and Kirk (1996) also gave a good account of the biology of thrips.

The common name 'thunder fly' is applied to one particular thrips species in Britain, *Limothrips cerealium*. This common name is used because of the propensity of this species for flying in huge numbers in the still, humid conditions that are often associated with thundery weather. In practice such flights usually occur when any warm, still day occurs after a prolonged period of cooler weather, during which the population has built up steadily without conditions being right for flight. When swarms occur these tiny insects get into people's eyes, into their hair, and inside clothes, causing extreme irritation. They come into houses, covering the insides of windows, and windowsills are later littered with dead individuals. These insects are so adept at getting into tiny spaces that the manufacturers of smoke alarms have to take thrips into account when designing their products: earlier designs were plagued by false alarms triggered by thrips getting into the sensors. They also regularly insinuate themselves behind the glass of picture frames.

With the exception of swarms of thunder flies, thrips are rarely noticed and most people with a casual interest in entomology would be surprised to find that there are over 150 species in the British Isles. The order Thysanoptera is divided into the suborders Tubulifera (with the abdomen terminating in a tubular structure) and Terebrantia (with the abdomen apically blunt). The Tubulifera contains the single family Phlaeothripidae, with 3000 species worldwide and 39 in Britain. The Terebrantia includes 7 families of which only the Thripidae and Aeolothripidae are found in Britain. The Thripidae comprises some 1700 species worldwide, 107 of them in Britain, and the Aeolothripidae contains 250 species worldwide of which just 13 occur in the UK (Kirk, 1996).

A comprehensive account of adult thrips in Britain was produced by Mound *et al.* (1976). Kirk's (1996) book is also very valuable, providing a key to families and immature stages and also keys to adult Aeolothripidae, adult Thripidae and adult Phlaeothripidae: it is not comprehensive, however, taking some groups only as far as genus. Speyer & Parr (1941) provided accounts of the larval stages. More of historical value nowadays is a work by Morison (1947–9), who gave an account of the Thysanoptera of the London area.

The three families in Britain are: Phlaeothripidae, Thripidae, Aeolothripidae.

References

Kirk, W.D.J. 1996. *Thrips.* Naturalists' Handbook No. 25. Richmond, Slough.
This guide to British thrips is a mixture of general account and identification guide. Although Kirk acknowledges that his book is not comprehensive in its coverage of the British fauna, the keys cover all but the most obscure species. The whole book is copiously illustrated, and the key sections have line drawings of characters in the margin, alongside the text, making the keys very user-friendly. There are thorough chapters on biology, behaviour and handling techniques, including slide-mounting methodology.
Lewis, T. 1973. *Thrips: their biology, ecology, and economic importance.* Academic Press, London.
This book gives a thorough account of all aspects of thrips, with the exception of providing any form of identification guide. It is in many ways regarded as the 'thysanopterists Bible'.
Morison, G.D. 1947–1949. Thysanoptera of the London Area (London Thrips). *The London Naturalist* 59 (suppl.): 1–131 (in 3 parts).
This work provided a general account of the group, keys to genera and species and a

checklist of British thrips. Unfortunately it was very poorly illustrated, limiting its usefulness for non-specialists.

Mound, L.A., Morison, G.D., Pitkin, B.R. & Palmer, J.M. 1976. Thysanoptera.
Handbooks for the Identification of British Insects 1(11): 79pp.
Following the standard format of the Royal Entomological Society volumes, this provides
a checklist of British Thysanoptera and separate keys to the species included in each of
the three families. The volume is well illustrated. There is only a short section on biology.

Speyer, E.R. & Parr, W.J. 1941. The external structure of some Thysanopterous larvae.
Transactions of the Royal Entomological Society of London **91**: 559–635.
This work gives a good account of the larval stages of thrips and also a key to species, but
only to members of the Terebrantia.

Hemiptera: the true bugs

(*ca.* 1700 species in 61 families)

JON H. MARTIN AND MICK D. WEBB

The Hemiptera are the fifth most speciose order of insects in Britain, behind the Hymenoptera, Diptera, Coleoptera and Lepidoptera. The name True Bugs is nowadays used in order to distinguish them from the colloquially used word 'bug', which is often applied to anything vaguely insect-like. There is controversy concerning the higher classification of the True Bugs, with specialists in North America and in Europe tending to adopt different systems. The traditional European system divided the Hemiptera into two suborders, Heteroptera and Homoptera, with the Homoptera further divided into two Series, Auchenorrhyncha and Sternorrhyncha. The traditional North American system generally regards the Hemiptera as an order that accommodates only the Heteroptera; the Homoptera is also regarded as an order, with the Auchenorrhyncha and Sternorrhyncha being suborders. The system now favoured by most workers regards Heteroptera, Auchenorrhyncha and Sternorrhyncha as three suborders of the Hemiptera (with 'Homoptera' effectively redundant): this is discussed by Gullan & Cranston (1994) and is followed by Carver *et al.* (1991). This controversy may be resolved following DNA studies, and an overview of the current results of such work was given by Schuh & Slater (1995).

Members of the Hemiptera are characterised by their sucking mouthparts, formed by the mandibles and maxillae, which are modified to form stylets, all accommodated within a beak-like labium; the whole structure is termed a rostrum (or, less commonly, proboscis). In addition to the rostrum, most True Bugs have well-developed legs, three or two tarsal segments, two tarsal claws, and four wings with reduced venation: there are many exceptions even to this basic diagnosis. The hemipteran life cycle is technically hemimetabolous, but some Sternorrhyncha, especially whiteflies, have life cycles that strongly resemble holometaboly.

Although only fifth in terms of species numbers, the Hemiptera is probably the most diverse order in Britain, in terms of variety of form and habit. Most members of the Heteroptera and Auchenorrhyncha have fully developed adults, with some displaying a degree of aptery or brachyptery. The Sternorrhyncha, on the other hand, includes species with adult forms that have extreme reductions of wings, antennae and even legs and visible body segmentation. The great majority of True Bugs are phytophagous. The members of most families of Heteroptera are either all predatory (all the aquatic and some terrestrial families) or all phytophagous, but some terrestrial families (e.g. Pentatomidae, Miridae) have predatory groups within predominantly phytophagous assemblages.

There are few books that deal with the whole British hemipterous fauna, but recent general entomological textbooks, such as Gullan & Cranston (1994), give a good general biological account of True Bugs. The most recent checklist of the British Hemiptera was that of Kloet & Hincks (1964), which clearly is now considerably out of date. Dolling (1991) provides a good general account of the Hemiptera, including keys to families and separate chapters on Heteroptera, Auchenorrhyncha and Sternorrhyncha; also included are chapters on field techniques, morphology, natural enemies, symbiosis, dispersal and an overview of the British fauna in a world context. Another recent book of great value to those with a broad interest in Hemiptera, but not requiring a detailed identification tool, is McGavin (1993): this book presents an informative account of general hemipteran biology and basic systematics, and is also the showcase for an exceptional range of habitus colour photographs of all main groups of worldwide Hemiptera, taken by K. Preston-Mafham.

Higher classification of British Hemiptera

Suborder Heteroptera
Infraorder Pentatomomorpha: Aradidae, Acanthosomatidae, Cydnidae, Pentatomidae, Scutelleridae, Piesmatidae, Berytidae, Lygaeidae, Pyrrhocoridae, Stenocephalidae, Coreidae, Alydidae, Rhopalidae
Infraorder Cimicomorpha: Tingidae, Miridae, Microphysidae, Nabidae, Anthocoridae, Cimicidae, Reduviidae
Infraorder Leptopodomorpha: Aepophilidae, Saldidae,
Infraorder Dipsocoromorpha: Dipsocoridae, Ceratocombidae
Infraorder Gerromorpha: Mesoveliidae, Hydrometridae, Hebridae, Veliidae, Gerridae
Infraorder Nepomorpha: Nepidae, Corixidae, Naucoridae, Aphelocheiridae, Notonectidae, Pleidae

Suborder Auchenorrhyncha
Superfamily Cicadoidea: Cicadidae
Superfamily Cercopoidea: Cercopidae, Aphrophoridae
Superfamily Membracoidea: Cicadellidae, Membracidae
Superfamily Fulgoroidea: Cixiidae, Delphacidae, Issidae, Tettigometridae

Suborder Sternorrhyncha
Superfamily Aphidoidea: Adelgidae, Aphididae, Phylloxeridae
Superfamily Aleyrodoidea: Aleyrodidae
Superfamily Psylloidea: Calophyidae, Homotomidae, Psyllidae, Triozidae
Superfamily Coccoidea: Asterolecaniidae, Coccidae, Cryptococcidae, Diaspididae,
    Eriococcidae, Kermesidae, Margarodidae, Ortheziidae, Pseudococcidae

General references to Hemiptera

(see notes above)

Carver, M., Gross, G.F. & Woodward, T.E. 1991. *In* Naumann, I.D. *et al.* (Eds) *The insects of Australia* Vol. I. Melbourne, University Press.
Dolling, W.R. 1991. *The Hemiptera.* Oxford University Press.
Gullan, P.J. & Cranston, P.S. 1994. *The insects: an outline of entomology.* Chapman & Hall, Melbourne.
Kloet, G.S. & Hincks, C. 1964. A check-list of British Insects. (2nd edn (completely revised)). Part 1. *Handbooks for the Identification of British Insects* 11(1): 119pp.
McGavin, G.C. 1993. *Bugs of the world.* Blandford, London
Schuh, R.T. & Slater, J.A. 1995. *True bugs of the world (Hemiptera: Heteroptera).* Cornell University, New York.

Suborder Heteroptera

The Heteroptera are notable for their utilisation of different habitats (bark, soil, mud, water, birds' and ants' nests, stored products and houses) and food sources (foliage, fruits, seeds, fungi, insects, birds and mammals). In parallel with this versatility there is considerable variation in body form and in particular the legs can be saltatorial, raptorial, fossorial or modified for swimming. Probably the most widely known Heteroptera are the shield-bugs (Pentatomoidea) whose common name reflects their enlarged scutellum, which almost entirely covers the wings in some species, and the bed-bug (*Cimex lectularius*), whose flattened body is well suited for hiding in crevices etc. The assassin-bugs (Reduviidae) have a short, robust rostrum, ideal for catching and feeding on their arthropod prey; in addition, the immature of *Reduvius personatus*, which also lives in association with human habitation, camouflages its body with detritus (hence the species' common name of masked assassin-bug) and the three species of *Empicoris* are strangely mosquito-like with mantid-like raptorial front legs. In contrast to the formidable assassin-bugs the Anthocoridae or flower-bugs, which have also been known to bite humans, are small and delicate and feed on helpless soft-bodied prey such as aphids. Also of delicate appearance are the Tingidae or lace-bugs, which have intricate reticulate patterns on their dorsum and are most frequently found feeding on thistles, and members of the largest family, the Miridae (or capsid bugs), which possibly owe their success to the ease with which they can walk on various kinds of leaf surfaces.

Diversity of feeding strategy has been accomplished by bringing the mouthparts forward and separating them from the prosternum by an area of cuticle referred to as the gula. Other morphological features of note are the flattened abdominal dorsum, with lateral connexivum and fore wings divided into two areas of different texture, the apical (membranous) areas of which overlap in repose. Scent glands on the larval abdominal dorsum and adult thorax sides produce acrid-smelling fluid, used mainly in defence, hence the name stinkbugs for the Pentatomoidea; defence of a different kind is shown by the female of *Elasmucha grisea* (Acanthosomatidae) which stands guard over her eggs until they hatch, feeding only on the surrounding mesophyll of the leaf. Aggregational and sexual communication is by chemical means but also by sound-producing organs such as abdominal tymbals (as in the Auchenorrhyncha) and stridulatory structures. The latter methods are favoured by the aquatic bugs of which the Gerridae (pondskaters or waterstriders) on the water surface and the Notonectidae and Corixidae (water boatmen) under the water are most commonly seen. Truly aquatic forms have reduced antennae and breathe either from a captured air bubble or through a siphon as in *Nepa* and *Ranatra* (Nepidae, the water scorpions).

The best source of information (including keys) on the British terrestrial fauna remains Southwood & Leston (1959), although this is now out of print. Savage (1989) covers the aquatic groups. More recent relevant revisions have been published in the *Faune de France* series by various authors, and some parts of the Palaearctic catalogue have been published or are currently being prepared.

References

Aukema, B. & Rieger, C. (Eds.) 1995. *Catalogue of the Heteroptera of the Palaearctic region.* Vol. 1. Netherlands Entomological Society.

Gives recent references to each species and latest classification and nomenclature of Palaearctic Enicocephalomorpha, Dipsocoromorpha, Nepomorpha, Gerromorpha and Leptopodomorpha. In addition recent references to each species are given.

Aukema, B. & Rieger, C. (Eds.) 1966. *Catalogue of the Heteroptera of the Palaearctic region.* Vol. 2. Netherlands Entomological Society.

Includes recent references to each species and latest classification and nomenclature of Palaearctic Cimicomorpha.

British Museum (Natural History) 1973. *The Bed-bug.* Economic series 5: 17pp.

Gives information on morphology, biology and control measures.

Butler, E.A. 1923. *A biology of the British Hemiptera-Heteroptera.* London.

Detailed biological information with figures of the immature stages of many species.

Jessop, L. 1983. The British species of *Anthocoris* (Hem., Anthocoridae). *Entomologist's Monthly Magazine* 119: 221–4.

Gives an update on the nomenclature and a key to species of *Anthocoris*.

Kirby, P. 1992. A review of the scarce and threatened Hemiptera of Great Britain. *UK Nature Conservation* 2. Joint Nature Conservation Committee, Peterborough.

More useful and up-to-date information on distribution than in Southwood & Leston (1959).

Moulet, P. 1995. Hémiptères Coreoidea (Coreidae, Rhopalidae, Alydidae) Pyrrhocoridae, Stenocephalidae Euro-Méditerranéens. *Faune de France* 81: 336pp.

Descriptions and keys in French and figures of European Coreoidea (including British species).

Péricart, J. 1983. Hémiptères Tingidae Euro-Méditerranéens. *Faune de France* 69: 618pp.

Descriptions and keys in French and figures of European Tingidae (including British species).

Péricart, J. 1987. Hémiptères Nabidae d'Europe occidentale et du Maghreb. *Faune de France* 71: 185pp.

Descriptions and keys in French and figures of European Nabidae (including British species).

Péricart, J. 1990. Hémiptères Saldidae et Leptopodidae d'Europe occidentale et du Maghreb. *Faune de France* 77: 238pp.

Gives descriptions and keys in French and figures of European Saldidae and Leptopodidae (including British species).

Poisson, R. 1957. Hétéroptères aquatices. *Faune de France* 61: 263pp.

Descriptions and keys in French and figures of European aquatic bugs (including British species).

Savage, A.A. 1989. Adults of the British aquatic Hemiptera Heteroptera: a key with ecological notes. *Scientific Publications of the Freshwater Biological Association* 50: 173pp.

Provides a more up-to-date key and ecological notes to aquatic adults than in Southwood & Leston (1959).

Southwood, T.R.E. & Leston, D. 1959. *Land and water bugs of the British Isles.* F. Warne & Co., London.

The only recent complete account on British Heteroptera including keys and information on host plants, biology and distribution. Remains the standard reference work.

Wachmann, E. 1988. *Wanzen beobachten – kennenlernen*. Neumann-Neudamm,
Melsungen.

Gives notes (in German) and excellent colour photographs on German Heteroptera
(many British).

Wagner, E. & Weber, H. 1964. Hétéroptères Miridae. *Faune de France* 67: 592pp.

Gives descriptions and keys in French and figures of French Miridae (including British
species).

Suborder Auchenorrhyncha

This entirely phytophagous group comprises the Cicadoidea (cicadas),
Cercopoidea (froghoppers or spittlebugs), Membracoidea (Cicadellidae,
leafhoppers and Membracidae, treehoppers) and Fulgoroidea (planthop-
pers). The group are the most abundant insects of grassland ecosystems, and
leafhoppers are also found on a variety of trees and shrubs. All parts of the
plant are utilised for feeding, including the roots, and cell sap is extracted
from mesophyll cells, phloem and xylem. Feeding from xylem results in the
production of copious quantities of watery excreta, which in the nymphs of
Cercopoidea is used to form 'bubble nests' from which the common name
spittlebugs has been derived. With the exception of the cicadas the group are
known for their ability to jump (which has given rise to many of their
common names, noted above). In this connection they have long back legs
with various setal configurations, which are useful for identifying at the
family level. A reduction in mobility, with a more cryptic appearance, is characteristic of two membracoid groups: the largest British cicadellid and only
British ledrine, *Ledra aurita*, which is inconspicuous on lichen-covered
branches, and the only two British membracids, which have spine-like
extensions of the pronotum. Paucity of numbers in more tropical groups are
also evident in the Fulgoroidea, with only four of the 20 families present in
Britain of which one of the two issids, *Issus muscaeformis*, is known only from
a few female specimens and Tettigometridae is represented only by the
scarce *Tettigometra impressopunctata*. In addition, the single British cicada,
Cicadetta montana, known only from the New Forest, Hampshire, is now
threatened with extinction (Kirby, 1992).

The most recent keys to the British fauna are by LeQuesne (1960, 1965,
1969) and LeQuesne & Payne (1981), the latter also including a checklist;
more detailed figures are to be found in Ribaut (1936, 1952), della Giustina
(1989) and Ossiannilsson (1978, 1981, 1983). Identifications in many cases
require reference to the male genitalia, for which microscopic examination is

required. For the identification of nymphs the work by Vilbaste (1982) on the
North European fauna should be consulted.

References

della Giustina, W. 1989. Homoptères Cicadellidae. Vol. 3, Compléments aux ouvrages
d'Henri Ribaut. *Faune de France* 73: 350pp.
This is an up-date to the volumes by Ribaut.

Kirby, P. 1992. A review of the scarce and threatened Hemiptera of Great Britain. *UK
Nature Conservation* 2. Joint Nature Conservation Committee, Peterborough.
**Gives more useful and up-to-date information on distribution than in LeQuesne's hand-
books.**

LeQuesne, W.J. 1960. Hemiptera (Fulgoromorpha). *Handbooks for the Identification
of British Insects* 2(3). 68pp.
**Keys (which also include details of host plants and distribution) and figures to the British
Fulgoroidea.**

LeQuesne, W.J. 1965. Hemiptera Cicadomorpha (excluding Deltocephalinae and
Typhlocybinae). *Handbooks for the Identification of British Insects* 2(2a): 64pp.
**Keys (which also include details of host plants and distribution) and figures to the British
Membracoidea (in part), Cicadoidea and Cercopoidea.**

LeQuesne, W.J. 1969. Hemiptera Cicadomorpha Deltocephalinae. *Handbooks for the
Identification of British Insects* 2(2b): 148pp.
**Keys (which also include details of host plants and distribution) and figures to the British
Cicadellidae, Deltocephalinae.**

LeQuesne, W.J. & Payne, K.R. 1960. Cicadellidae (Typhlocybinae) with a check list of
the British Auchenorrhyncha (Hemiptera, Homoptera). *Handbooks for the
Identification of British Insects* 2(2c): 95pp.
**Keys (which also include details of host plants and distribution) and figures to the British
Cicadellidae, Typhlocybinae, together with a check list of the Auchenorrhyncha.**

Ossiannilsson, F. 1978, 1981, 1983. The Auchenorrhyncha (Homoptera) of
Fennoscandia and Denmark. *Fauna Entomologica Scandinavica* 7(1–3): 980pp.
**Includes descriptions (in English), keys, biological information and excellent figures of
many British species.**

Remane, R. & Wachmann, E. 1993. *Zikaden kennenlernen beobachten.* Naturbuch
Verlag, Augsburg.
**Gives notes (in German) and excellent colour photographs on German Auchenorrhyncha
(many British).**

Ribaut, H. 1936. Homoptères Auchénorhynques. I (Typhlocybidae). *Faune de France*
31: 321pp.

Ribaut, H. 1952. Homoptères Auchénorhynques. II (Jassidae). *Faune de France* 57:
474pp.
**These give keys and descriptions (in French) which together with Ossiannilsson's revi-
sions cover the British fauna with superb figures. Nomenclature updated by della
Giustina (1989).**

Vilbaste, J. 1968. Preliminary key for the identification of the nymphs of North-

European Homoptera Cicadinea. I. Delphacidae. *Annales Entomologici Fennici*
32: 65–74.
Gives keys and figures to the North European Delphacidae.
Vilbaste, J. 1982. Preliminary key for the identification of the nymphs of North
European Homoptera Cicadinea. II. Cicadelloidea. *Annales Zoologici Fennici* 19:
1–20.
Gives keys and figures to the North European Cicadelloidea (Membracoidea).
Wilson, M.R. 1978. Description and key to genera of the nymphs of British woodland
Typhlocybinae (Homoptera). *Systematic Entomology* 3: 75–90.
**Although limited in coverage, still a useful generic key to tree- and shrub-associated
nymphs of this subfamily.**

Suborder Sternorrhyncha

Along with the Auchenorrhyncha, all members of the Sternorrhyncha are
sap-sucking insects. The four superfamilies Aphidoidea, Aleyrodoidea,
Psylloidea and Coccoidea include many of the most serious pests of agricultural and horticultural plants. All sternorrhynchous pests cause direct
feeding damage and some also transmit viruses and other plant diseases. The
very high fecundity of some of these insects, notably aphids, leads to enormous populations building up very quickly. Some, especially some scale
insects, are protected by waxy or chitinous coverings, which render contact
insecticides almost useless. The speed of reproduction in pest species leads
to rapid development of resistance to chemical controls.

Adult Sternorrhyncha are extremely variable in their degree of physiological development. The jumping plant lice have hard-bodied adults, which
superficially resemble some Auchenorrhyncha: at the opposite extreme, all
female scale insects are neotenic and some are reduced to little more than
sacs of delicate cuticle with stylets plumbed into plant tissue.
Sternorrhyncha that are sufficiently developed differ from Auchenorrhyncha
as follows: tarsi one- or two-segmented; rostrum arising well underneath the
body, apparently between the front legs; antennae very variable but never
with a terminal bristle-like ariston. In addition, sternorrhynchs have much
simpler wing venation than do auchenorrhynchs. Whiteflies and male scales
have life cycles that approach holometaboly, with a distinct pupal stage
instead of the gradual development seen in other bugs.

Superfamily Aphidoidea: aphids, adelgids and phylloxerids

Aphids, as all members of this superfamily are normally called, feed on the
phloem sap of plants. Most are to be found colonising new growth, typically
clustered around shoots. Some species feed underneath more mature leaves,

a few through the bark of their woody hosts, and still others specialise in feeding on roots. Appearance is widely variable. Some are bright and shiny, green, brown and even pink or red; others are matt or mealy as a result of waxy secretions, and some secrete copious amounts of waxy 'wool' which renders the insects themselves almost invisible. Many species are normally found in association with ants, which harvest excreted honeydew and afford the aphids a considerable degree of protection. Most aphids display at least some degree of host preference, with just a few (all of them members of the Aphididae) being truly polyphagous. Many species are restricted to members of particular plant families or genera and host information can thus be a useful identification tool.

Many aphids have complex life cycles that involve alternation of host plants and up to seven distinct morphological forms. All species of the very large family Aphididae are parthenogenetic for all or part of their life cycles but many species (especially those with a host-plant alternation) have a sexual phase, with males and egg-laying sexual females occurring in the autumn, enabling survival through the winter as robust eggs. The vast majority of Aphidoidea are viviparous during their parthenogenetic phase, giving birth to active first-instar larvae, but members of the small families Adelgidae and Phylloxeridae are entirely oviparous and lay eggs even when no sexual fertilisation has taken place: in these two families no overwintering eggs are produced and the species pass the winter as larvae. Dixon (1973, 1985) and Blackman (1974) gave good accounts of aphid biology and ecology, Hille Ris Lambers (1966) dealt specifically with polymorphism and van Emden (1972) provided much information on investigative techniques.

The aphids are a group of bugs with a remarkably high proportion of the world fauna represented in Britain. The British fauna comprises around 500 species in 3 families, more than ten per cent of the world total. This unusual situation has arisen because the group has evolved as a predominantly temperate, northern hemisphere, assemblage which has a poor representation in the tropics. Again unusually, new British aphid species are still being discovered despite the group being very well studied in Britain, leading to the rather tentative total quoted above.

The complexity of life cycles is one reason why there is no recent guide to aid the identification of all British aphids. Further, those works that aim to enable the identification of selected groups sometimes contain a confusing mixture of characteristics displayed by wingless forms (apterae) and those with wings (alatae), and care has to be exercised when interpreting particular characters. The last attempt at a comprehensive systematic account of British aphids was by Theobald (1926, 1927, 1929) but knowledge of the group has advanced tremendously since then. Stroyan provided several piecemeal updates to Theobald's monographs, between 1950 and 1979, and Hille Ris Lambers published the first five volumes of an intended comprehensive monograph of European aphids, from 1938 to 1953: these works were useful for identifying the British species of several genera but have been rendered outdated by more recent publications and are not listed here. Shaposhnikov (*in* Bei Bienko, English translation 1967) provided keys to genera and species of aphids of the western parts of the former USSR, and this is still useful in studies of the British aphid fauna.

Although no comprehensive identification guide exists, several works have recently been published that are of great use to those interested in British aphids. A key to families of British aphids was provided by Dolling (1991), as part of a general account of the Hemiptera. Two groups have been covered by books in the Royal Entomological Society's *Handbooks for the identification of British Insects* series, both by Stroyan: Chaitophoridae and Callaphididae [now regarded as subfamilies](1977) and Pterocommatinae and Aphidinae (Aphidini)(1984). Heie (1980, 1982, 1986, 1992, 1994, 1995) has provided keys to Scandinavian aphids, the fauna being very similar to that of Britain. Blackman (1974) gave a key to genera and an illustrated identification guide to over 100 of the more common British species, as well as diagrams annotated with aphid systematic terms. Blackman & Eastop have given host-plant-based keys to aphids on the world's crops (1984) and trees (1994), with short accounts of each species: although many of the species concerned are not found in Britain, these books contain much valuable information on British aphids as well.

References

Blackman, R.L. 1974. *Aphids*. Ginn & Co., London & Aylesbury.
> A good general account of aphids in Britain, with a simplified generic key and accounts of nearly 100 of the more common British species.

Blackman, R.L. & Eastop, V.F. 1984. *Aphids on the world's crops: an identification and information guide.* Wiley, Chichester.
> This book relies upon the investigator knowing the identity of the host plant, with keys to those aphids found on each agricultural crop in turn. Many non-British species are included, but this is none the less of great use in a British context.

Blackman, R.L. & Eastop, V.F. 1994. *Aphids on the world's trees: an identification and information guide.* CAB International, Wallingford.
> Following the proven formula of the previous book, this enormous work treats all the world's aphids associated with tree hosts, some 1750 species within 270 genera: as before, this has some sections of particular relevance to the British fauna.

Brown, P.A. 1989. Keys to the alate *Aphis* (Homoptera) of northern Europe. *Occasional Papers on Systematic Entomology* 5: 29pp.

This work provides a means to identify the winged individuals of this large and difficult genus. It is particularly useful to those carrying out investigations based on trap catches.

Carter, C.I. 1971. Conifer woolly aphids (Adelgidae) in Britain. *Forestry Commission Bulletin* 42: 51pp. HMSO, London.

This work, combined with the one below, represents the best aid to identification to British adelgids currently available.

Carter, C.I. 1976. A gall-forming Adelgid (*Pineus similis* (Gill)) new to Britain with a key to the Adelgid galls on Sitka Spruce. *Entomologist's Monthly Magazine* 111: 29–32. See above.

Carter, C.I. 1982. Conifer Lachnids. *Forestry Commission Bulletin* 58: 75pp. HMSO, London.

This is an identification guide to the conifer-feeding species of the subfamily Lachninae, then regarded as a family. It includes keys to members of the large genus *Cinara* feeding on particular conifer genera, and is also particularly useful for identifying the small genera *Cedrobium, Schizolachnus* and *Eulachnus*. It is well illustrated with line drawings and photographs.

Dixon, A.F.G. 1973. *Biology of aphids*. Arnold, London.

A good introductory guide to aphid biology, but without identification guidance.

Dixon, A.F.G. 1985. *Aphid ecology*. Blackie, Glasgow & London.

An expanded and updated version of his earlier book. Dixon here gives an excellent account of most aspects of aphid biology and ecology. Again, this is not a systematic account.

Dolling, W.R. 1991. *The Hemiptera*. Oxford University Press, Oxford.

This book provides an general account of the Hemiptera. Keys to aphid families [by J.H. Martin] are provided. Care has been taken to clarify reference to apterous and alate characters, and efforts have been made to incorporate atypical members of each family into the key couplets.

Eastop, V.F. 1972. A taxonomic review of the species of *Cinara* Curtis occurring in Britain. *Bulletin of the British Museum (Natural History)* (Entomology) 27: 103–86.

A comprehensive identification guide to the British members of the most speciose conifer-feeding aphid genus. The species of *Cinara* are notoriously difficult to identify.

Emden, H.F. van (Ed.) 1972. *Aphid technology, with special reference to the study of aphids in the field*. Academic Press, London.

Now somewhat dated, this nonetheless provides a good overview of techniques used in the study of aphidology. A small section on identification provides a classification of the Aphidoidea that makes an interesting comparison with more recent ideas.

Heie, O.E. 1980. The Aphidoidea (Hemiptera) of Fennoscandia and Denmark. I. General Part. The Families Mindaridae, Hormaphididae, Thelaxidae, Anoeciidae, and Pemphigidae. *Fauna Entomologica Scandinavica* 9: 236pp.

This is the first of a series of six volumes and contains a general introduction to aphid biology and classification. The keys are copiously illustrated. Genera included are *Anoecia, Aploneura, Baizongia, Cerataphis, Colopha, Eriosoma, Forda, Geoica, Glyphina, Gootiella, Hamamelistes, Hormaphis, Kaltenbachiella, Melaphis, Mimeuria, Mindarus, Pachypappa, Pachypappella, Paracletus, Pemphigus, Prociphilus, Smynthurodes, Tetraneura, Thecabius, Thelaxes*.

Heie, O.E. 1982. The Aphidoidea (Hemiptera) of Fennoscandia and Denmark. II. The Family Drepanosiphidae. *Fauna Entomologica Scandinavica* 11: 176pp.

This second volume includes the genera *Atheroides, Betulaphis, Calaphis, Callaphis, Callipterinella, Caricosipha, Chaetosiphella, Chaitophorus, Chromaphis, Clethrobius, Ctenocallis, Eucallipterus, Euceraphis, Iziphya, Laingia, Monaphis, Myzocallis, Nevskyella, Periphyllus, Phyllaphis, Pterocallis, Saltusaphis, Sipha, Subsaltusaphis, Symydobius, Therioaphis, Thripsaphis, Tinocallis, Tuberculatus.*

Heie, O.E. 1986. The Aphidoidea (Hemiptera) of Fennoscandia and Denmark. III. Family Aphididae: Subfamily Pterocommatinae and Tribe Aphidini of Subfamily Aphidinae. *Fauna Entomologica Scandinavica* 17: 314pp.

The third volume in the series deals with the genera *Aphis, Brachysiphum, Cryptosiphum, Hyalopterus, Melanaphis, Neopterocomma, Plocamaphis, Pterocomma, Rhopalosiphum, Schizaphis, Toxopterina.*

Heie, O.E. 1992. The Aphidoidea (Hemiptera) of Fennoscandia and Denmark. IV. Family Aphididae: Part 1 of Tribe Macrosiphini of Subfamily Aphidinae. *Fauna Entomologica Scandinavica* 25: 190pp.

This volume includes the genera *Acaudinum, Anuraphis, Aspidaphis, Brachycaudus, Brachycolus, Brachycorynella, Brevicoryne, Cavariella, Ceruraphis, Coloradoa, Decorosiphon, Diuraphis, Dysaphis, Elatobium, Ericaphis, Hayhurstia, Hyadaphis, Liosomaphis, Lipaphis, Longicaudus, Muscaphis, Myzaphis, Myzodium, Nearctaphis, Semiaphis, Vesiculaphis.*

Heie, O.E. 1994. The Aphidoidea (Hemiptera) of Fennoscandia and Denmark. V. Family Aphididae: Part 2 of Tribe Macrosiphini of Subfamily Aphidinae. *Fauna Entomologica Scandinavica* 28: 242pp.

This volume includes the genera *Acyrthosiphon, Anthracosiphon, Aulacorthum, Capitophorus, Chaetosiphon, Corylobium, Cryptaphis, Cryptomyzus, Hyalopteroides, Hyperomyzus, Idiopterus, Impatientinum, Jacksonia, Linosiphon, Macrosiphum, Metopolophium, Microlophium, Myzus, Nasonovia, Ovatomyzus, Ovatus, Paramyzus, Pentalonia, Phorodon, Pleotrichophorus, Rhodobium, Rhopalomyzus, Rhopalosiphoninus, Sitobion, Subacyrthosiphon, Tubaphis, Utamphorophora, Volutaphis.*

Heie, O.E. 1995. The Aphidoidea (Hemiptera) of Fennoscandia and Denmark. VI. Family Aphididae: Part 3 of Tribe Macrosiphini of Subfamily Aphidinae, and Family Lachnidae. *Fauna Entomologica Scandinavica* 31: 222pp.

This volume includes the genera *Amphorophora, Cinara, Delphinobium, Eulachnus, Galiaphis, Illinoia, Lachnus, Macrosiphoniella, Maculolachnus, Megoura, Megourella, Metopeurum, Microsiphum, Neoamphorophora, Neotrama, Protrama, Schizolachnus, Sitomyzus, Staticobium, Stomaphis, Titanosiphon, Trama, Tuberolachnus, Uroleucon, Wahlgreniella.* This last volume in the series includes host-plant lists and an index to all six volumes.

Hille Ris Lambers, D. 1966. Polymorphism in Aphididae. *Annual Review of Entomology* 11: 47–78.

This paper gives an overview of the phenomenon of polymorphism in the aphids, with a glossary of terms and sections describing the various morphs and their functions and is thus an interesting insight to the diversity of aphid life cycles.

Martin, J.H. 1983. The identification of common aphid pests of tropical agriculture. *Tropical Pest Management* 29(4): 395–411.

This work is of use in a British context primarily because it presents a detailed method of slide preparation. It will be useful for identifying some pest species in Britain, but should be used with caution because the British aphid fauna contains many other species very similar to those found in tropical agriculture.

Shaposhnikov, G.K. 1967. Aphidinea. *In* Bei-Bienko, G.Y. (Ed.) Keys to the Insects of the European USSR. Vol I, Apterygota, Palaeoptera, Hemimetabola. (Keys to the Fauna of the USSR no. 84), pp. 616–799. Israel Program for Scientific Translations, Jerusalem. [English translation of the original Russian.]

This work is still useful in the study of some British aphids, but the terminology is sometimes non-standard and the work is poorly illustrated.

Stroyan, H.L.G. 1977. Aphidoidea – Chaitophoridae and Callaphididae. *Handbooks for the Identification of British Insects* 2(4a): 130pp.

This volume, with the one below, follows the standard format of this series. This volume has the preferred system of incorporating illustrations close to the relevant discursive text.

Stroyan, H.L.G. 1984. Aphids – Pterocommatinae and Aphidinae (Aphidini). *Handbooks for the Identification of British Insects* 2(6): 232pp.

This book provides conventional keys to most genera, but the huge genus *Aphis* is treated by a series of mini keys to those species found on each host genus.

Theobald, F.V. 1926, 1927, 1929. *The plant lice or Aphididae of Great Britain.* Vols 1–3. Headley Brothers, Ashford and London.

Now very much out-of-date, these three volumes are none the less of great historical value, representing the first attempt at providing a comprehensive guide to the aphids in Britain.

Superfamily Aleyrodoidea: whiteflies

The whiteflies (the word 'whitefly' is also acceptable when used as a plural) are a group whose biology is somewhat intermediate between that of the jumping plant lice and the scale insect families. Adult whiteflies are tiny insects, mostly only a millimetre, or so, in length. They are fully winged and almost always have their bodies and wings coated with white waxy meal. Reproduction is usually sexual, occasionally parthenogenetic, but eggs are always laid. The first-instar larva is mobile and can walk sufficiently to locate a suitable feeding site. Once the first moult has taken place, however, the remaining three larval instars are sessile and unable to relocate themselves if feeding conditions deteriorate. The final larval stage is usually termed a 'puparium'. This is a technical misnomer but reflects the extreme morphological difference between this stage and the winged adult.

Unusually among insects, the taxonomy of whiteflies is based on the puparial stage, and adults in isolation can rarely be identified. This situation has one distinct advantage: because specimens can only be obtained by collection directly from their host plants, achieving accurate host information is easy. Unfortunately, whitefly puparia are notorious for displaying variation induced by their physical environment. Such variation seems to result from factors such as the degree of hairiness of leaf lamellae, waxiness of leaves and whether the puparium develops on the lower or (more rarely) upper leaf surface. There is thus a situation where robust characters, such as the presence of stout setae or tubercles, may be taxonomically irrelevant. Fortunately, the fauna of the British Isles is so small that this does not present an identification problem in Britain.

The representation of whiteflies in the British Isles is the complete opposite to the situation with aphids. Worldwide the whiteflies comprise around 1200 described species within a single family, the Aleyrodidae. The great majority of whiteflies are tropical species, a very high proportion of the world fauna remaining undescribed. Only a few species are found in the temperate regions. The British fauna comprises only 17 species, and even in the warmer climate of the Mediterranean basin only around 60 species are known. Just 13 of the British species are found in the open air, the remainder being restricted to glasshouses. Fortunately for British agriculture the notorious pest *Bemisia tabaci* has only rarely been found in the British Isles, except in glasshouses. Three other species, *Aleuropteridis ficicola, Aleurotulus nephro-lepidis* and *Filicaleyrodes williamsi* are only known in the British Isles from glasshouses: the latter has only ever been found under glass, in Britain and Hungary, and its geographical source remains unknown.

With the sparse whitefly fauna in Britain, a revision of the family by Mound (1966) remains the only substantial systematic work on the group in this country. *Aleurochiton aceris*, which feeds on Norway Maple (*Acer plata-noides*) and was only tentatively included by Mound, has since been confirmed as occurring in Britain (Martin, 1978) and is now common in southern England on this host plant. The only completely new addition to the British whitefly fauna has been the discovery of one colony of *Aleurochiton acerinus*, which feeds on Field Maple (*Acer campestre*), in Kent (Dolling & Martin, 1985). Tropical species are occasionally intercepted by quarantine agencies: although none has become established, these temporary introductions often involve known pest species, which may be identified using Martin (1987).

References

Dolling, W.R. & Martin, J.H. 1985. *Aleurochiton acerinus* Haupt, a maple-feeding whitefly (Hom., Aleyrodidae) new to Britain. *Entomologist's Monthly Magazine* **121**: 143–4.

This short communication formally records this new British introduction and discusses the occurrence of *Aleurochiton* in Europe. The differences between *A. acerinus* and *A. aceris* are too subtle for a simple identification key couplet, but each of these two white-flies appears to be restricted to its own host species.

Martin, J.H. 1978. *Aleurochiton complanatus* (Baerensprung) [= *aceris* (Modeer)] (Homoptera, Aleyrodidae) – confirmation of occurrence in Britain. *Entomologist's Monthly Magazine* 113: 7.

This short note confirmed that this whitefly does indeed breed in Britain, on the Norway Maple, *Acer platanoides*.

Martin, J.H. 1987. An identification guide to common whitefly pests of the world (Homoptera, Aleyrodidae). *Tropical Pest Management* 33: 298–322.

This comprises a key to 46 whitefly species known as pests around the world, with illus-trations of each one. A diagram annotated with puparial terminology is given, along with a method for preparing microscope slide mounts. A few British species are included, but this work is most useful for identification of species imported with plant material.

Mound, L.A. 1966. A revision of the British Aleyrodidae (Hemiptera: Homoptera). *Bulletin of the British Museum (Natural History)* (Entomology) 17(9): 399–428.

The work contains a general introduction to the family, a key to genera in the British Isles and mini keys to those genera represented in Britain by more than one species. The puparium of each species is illustrated, but only the vasiform orifice in some cases.

Superfamily Psylloidea: jumping plant lice

Of all the groups of Sternorrhyncha in the British Isles, the psyllids (as all members of the superfamily are generally called) are probably the group least known to general entomologists, often being casually mistaken for leaf-hoppers (Auchenorrhyncha). Adult psyllids are similar to leafhoppers in general appearance and, most notably, in the way in which they jump strongly when disturbed. Their two-segmented tarsi, multi-segmented antennae and simplified wing venation at once distinguish them from members of the Auchenorrhyncha, and the habits and appearance of most psyllid larvae are also very different. Reproduction is sexual and eggs are always laid. Each of the five larval stages is fully mobile, even though many species develop within leaf galls. Leaf-edge roll-galls are formed in response to the feeding of larvae of several British psyllids, notably *Trioza alacris* on bay laurel (*Laurus nobilis*), *Psyllopsis* spp. on ash (*Fraxinus excelsior*) trees and *Trichochermes walkeri* on buckthorn (*Rhamnus cathartica*).

Identification of Psylloidea in the British Isles is comprehensively covered by two key works, both of them parts of the Royal Entomological Society's *Handbooks for the Identification of British Insects* series. Hodkinson & White (1979) dealt with the adults, and White & Hodkinson (1982) provided keys to the fifth-instar larvae (nymphs). The volume on nymphal identification is particularly useful because nymphs are often found without adults being present, depending on the time of year.

Since the volume on adults was published two additional psyllid species have become established in Britain, and the higher classification has changed such that the British fauna, of 83 species, is now divided among only four families, Psyllidae, Homotomidae, Calophyidae and Triozidae, rather than the six recognised by Hodkinson & White. One further work is particularly useful in studies of the psyllid fauna of the British Isles: Ossiannilsson (1992) completed an excellent and comprehensive work on the Psylloidea of Scandinavia, Finland and Denmark, and this fauna is very similar to that of Britain.

References

Burckhardt, D. & Lauterer, P. 1997. A taxonomic reassessment of the triozid genus *Bactericera* (Hemiptera: Psylloidea). *Journal of Natural History* 31: 99–153.
  This paper lists all those species of *Bactericera* that were included in *Trioza* in Hodkinson & White's adult handbook, and thus updates the British list substantially.
Hodkinson, I.D. & Hollis, D. 1980. *Floria* [=*Livilla*] *variegata* Löw (Homoptera: Psylloidea) in Britain. *Entomologist's Gazette* 31: 171–2.
  A second note on the introduction into Britain of the Laburnum psyllid, this includes an update of the adults handbook, enabling the identification of this species.
Hodkinson, I.D. & White, I.M. 1979. Homoptera: Psylloidea. *Handbooks for the Identification of British Insects* 11 (5a): 98pp.
  This handbook covers the adults of the British psyllid fauna. Diagrams annotated with psyllid terminology are provided, along with keys to families, genera within each family and species within each genus. Confirmatory descriptions are given for the species of the large genera *Psylla* and *Trioza*. A psyllid species checklist is given, along with a list correlated with host data and references to biology, and an index to host plants. The book is copiously illustrated.
Hollis, D. 1978. *Floria variegata* Löw (Homoptera: Psylloidea) on *Laburnum* in Britain. *Plant Pathology* 27: 149.
  A preliminary note on this new introduction to the British psyllid fauna.
Martin, J.H. & Malumphy, C.P. 1995. *Trioza vitreoradiata*, a New Zealand jumping plant louse (Homoptera: Psylloidea), causing damage to *Pittosporum* spp. in Britain. *Bulletin of Entomological Research* 85: 253–8.
  This paper records the second new psyllid introduction to Britain since publication of the identification guides, updates Hodkinson & White's (1979) key to adults and discusses the recognition of the nymphs.
Ossiannilsson, F. 1992. The Psylloidea (Homoptera) of Fennoscandia and Denmark. *Fauna Entomologica Scandinavica* 26: 346pp.
  This book covers the identification to both adults and fifth-instar nymphs, and is copiously illustrated by drawings of very high quality. In nomenclatural terms it provides an update to the two Royal Entomological Society handbooks.

White, I.M. & Hodkinson, I.D. 1982. Psylloidea (Nymphal Stages): Hemiptera, Homoptera. *Handbooks for the Identification of British Insects* 11(5b): 50pp.
This work is similar in structure to the handbook to adult psyllids, but lacks the general sections covered by that book. *Livilla variegata* is included here, under the genus *Floria*, which is now placed in synonymy with *Livilla*. Like the adult volume, it is abundantly illustrated.

Superfamily Coccoidea: scale insects

The superfamily Coccoidea is often referred to as comprising the 'scale insects', but this term is a great oversimplification. Families such as the armoured scales (Diaspididae) have sessile, scale-like, adult females but others, such as the mealybugs (Pseudococcidae) and ensign scales (Ortheziidae), have females that are largely or entirely macropodous and usually mobile. All female coccoids are technically neotenic, reproducing while larviform. Most coccoids are thought to be biparental in their reproduction, but the males of many species are unknown. Adult male coccoids of all families are fragile insects, which usually have well-developed legs and a single pair of wings. Their hind wings are vestigial, known as hamulohalteres, and male coccoids in trap samples are thus often mistaken for small flies. Some adult males are degenerate, lacking wings, and all adult males lack feeding mouthparts and are thus very short-lived. In most species it is the first two male instars that feed, these stages being immediately followed by the non-feeding pre-pupal and pupal stages (which are unique to the Coccoidea) before eventual emergence of the adult. Adult male scales, like adult whiteflies, are rarely used taxonomically; all the discussions below refer to adult females.

The sheer variety of form and biology is such that it is almost impossible completely to define the superfamily in simple terms. Legs are always present in the first instar: here, and where legs are present in later instars, they usually have only one tarsal segment and always have but a single claw on each tarsus. The single tarsal claw at once sets the coccoids aside from all other Hemiptera, which have paired tarsal claws. Insects can also usually be readily recognised as coccoids by the presence of two pairs of large, ventral, spiracles and various glandular structures that are unique to members of this superfamily.

Worldwide there are around 7000 described species of Coccoidea, placed in families that number up to 25, depending on the classification followed (C.P. Malumphy, pers. comm.). Following the classification of Boratynski & Williams (1964), the indigenous British coccoid fauna of 107 species is distributed between 9 families: Ortheziidae, Margarodidae, Pseudococcidae, Coccidae, Kermesidae, Cryptococcidae, Eriococcidae, Asterolecaniidae and Diaspididae. However, the status of several of the smaller coccoid families is the source of controversy amongst systematists: few accept the Cryptococcidae, and the included species are usually regarded as members of the Eriococcidae. Kosztarab & Kozár (1988), and others, even regard the scale insect groups as belonging to a suborder, Coccinea, separate from the Sternorrhyncha. As with whiteflies this is a predominantly tropical group and the fauna of the British Isles is relatively sparse. The British fauna comprises a mixture of native species, which are to be found in the open, and 'exotics', which are mostly known from glasshouses and nurseries. Some exotic species, notably *Pulvinaria regalis* and *Chloropulvinaria floccifera*, have become well established in glasshouses and also out of doors but the inclusion of many others in the British fauna is necessarily somewhat arbitrary. The crypsis of coccoids, particularly when present in small numbers, means they are easily overlooked by those involved in the plant trade. Sharp-eyed quarantine officials, trained to recognise them, intercept more species of coccoids on contaminated plants than any other superfamily of invertebrates. The recent phenomenon of holiday parks with tropical 'leisure domes' has provided many coccoids with ideal environments where they flourish if introduced on amenity plants.

There are no comprehensive works available on the systematics of the entire British coccoid fauna. The only extensive works on British species were provided by Newstead, in the early years of the twentieth century, and by Green in the 1920s: both are long out of date and difficult to obtain. Only the families Pseudococcidae and Eriococcidae have been the subject of relatively recent identification guides to the species found in the British Isles (see below). A key to families was provided by Dolling (1991). The most recent and comprehensive work on European Coccoidea is that by Kosztarab & Kozár (1988). Danzig (English translation, 1967) provided identification keys to Coccoidea of western parts of the former USSR, and this work can still be useful in the British context. More up to date is Danzig's (1993) account of the Diaspididae of Russia and neighbouring countries, well illustrated but with text in Russian. Although dealing with coccoids of the South Pacific Region, three volumes by Williams & Watson (1988a, b, 1990) are extremely useful for the identification of imported species, because many of the taxa included are cosmopolitan pests.

The four most important families of British coccoids are briefly discussed below.

Family Pseudococcidae: mealybugs

Mealybugs are so called because the insects are usually coated in a rather granular, or sometimes woolly, wax secretion and their true body coloration is thus difficult to determine without removing the wax. The majority of species have well-developed walking legs, but some have their legs reduced or even absent: the apodous species are usually those living underground or developing within the protection of leaf sheaths. Many mealybugs feed in a similar fashion to aphids, aerially on new growth of their hosts. However, a large proportion of the native British species habitually live in sheltered situations, particularly under stones or logs where they feed on roots or etiolated stems: despite the tropical bias of the coccoid fauna, mealybugs that have evolved to live in such sheltered niches can also be found in countries such as Iceland and Greenland.

Mealybugs are often encountered as serious pests in warmer parts of the world, where some cause severe distortion of new growth and may transmit virus diseases. Introduced species, surviving in glasshouses, periodically bring these problems to Britain.

Williams (1962) provided the most recent identification guide and account of British mealybugs, with 42 species included. Of those, 13 were recognised as 'greenhouse species', more-or-less established in artificial environments. Subsequently over 65 more species have been recorded as quarantine interceptions or other casual introductions (C.P. Malumphy, pers. comm.). The most common native British mealybug is *Phenacoccus aceris*, whose preferred hosts are hawthorn (*Crataegus*), gorse (*Ulex*) and birch (*Betula*). Williams's work remains essentially complete, albeit with a few name combinations having now been revised, but about five further species are tentatively regarded as established in Britain.

Family Eriococcidae: felt scales

The eriococcids are similar in appearance to pseudococcids; the free-living members of the family are essentially another group of mealybugs. A combination of gland and pore characters set them aside from the pseudococcids. The biology of this family is little known in Britain, but the species are thought to be univoltine in the UK. The British species are all oviparous; so far as is known the females have three instars and the males five (Williams, 1985).

Most British eriococcids feed on grasses. *Eriococcus devoniensis* feeds exclusively on heather (*Erica*), where its feeding causes characteristic bending or spiralling of stems. *Pseudochermes fraxini* and *Cryptococcus fagisuga* (placed in the family Cryptococcidae by Kosztarab & Kozár, 1988) feed in bark crevices on ash (*Fraxinus*) and beech (*Fagus*), respectively. *C. fagisuga* is the most important British eriococcid, sometimes building up huge colonies that are rendered highly visible by their white woolly wax secretions. Cochineal insects, *Dactylopius* spp., periodically turn up in Britain, on potted cactus plants. According to some classifications of the Coccoidea *Dactylopius* is regarded as an eriococcid, but is generally assigned to its own family, Dactylopiidae.

The indigenous British eriococcid fauna comprises only 12 species. An account of this family was provided by Williams (1985), who also discussed four introduced species; a few more species have since been intercepted by quarantine authorities.

Family Coccidae: soft scales

These are the scale insects that most people think of when the name is mentioned. Beyond the first-instar crawler stage they may or may not have legs. Even when legs are present they are completely hidden beneath the flat, scale-like body, and most adult coccids rarely walk. When soft scales do walk it is a laborious process. Most species are cryptic, being yellowish on leaves and brown when they mature, usually on bark. Just a few soft scales, mostly exotics, become brown as they mature on leaves and are thus easily seen. Ant-attendance often gives away the presence of otherwise well-disguised scales. Some soft scales die with their eggs massed underneath the scale, for protection, but others produce a highly visible white-waxy ovisac, which extends posteriorly from the insect.

Worldwide there are in excess of 1100 described species of soft scales: the number of species recorded in Britain, including many exotics, is around 67. There is currently no published account of the British species, and works such as Kosztarab & Kozár (1988) and Danzig (1967) remain the best currently available to assist in species identification. Hodgson (1994) provided a manual for the identification of soft scale genera, and this may be used to provide generic names for British soft scales.

Family Diaspididae: armoured scales

Armoured scales are so called because the feeding insect is protected by a secreted scale, which incorporates the exuviae of the earlier instars. In this they differ from the soft scales, which are always protected by waxy secretions, although these may not be easy to see. With some diaspids the scale protecting the adult female is actually the entire second-instar exuvium, with the adult inside it: such species are described as 'pupillarial'. Diaspid females have only three instars, differing from the majority of other coccoid families, whose life cycles have four stages. The posterior abdominal segments of female diaspids are fused, forming a pygidium, although this structure is sometimes rather membranous in pupillarial species. Of all the coccoid groups the armoured scales are perhaps the least insect-like of all, and their presence may easily be mistaken for a plant disease or other cause of blemishes, especially on fruits.

The Diaspididae is by far the largest coccoid family worldwide, with approaching 2000 described species. In Britain only 10 indigenous species occur, with around 20 exotic species established (mostly indoors) and about 120 having been quarantine-intercepted from time to time. Among the native British diaspids *Carulaspis carueli* and *C. juniperi* are found on members of the Cupressaceae whereas the two commonest species, *Chionaspis salicis* and *Lepidosaphes ulmi*, feed on broadleaved trees and shrubs. As with soft scales there is no published key to British species, and anyone seeking to fill this gap will have to address the question of precisely which introduced species can be regarded as 'naturalised'. Williams & Watson (1988a) is a useful volume to aid identification exotic species under glass, but Kosztarab & Kozár (1988) or Danzig (1967) remain the most appropriate for indigenous species.

References

Boratynski, K.L. & Williams, D.J. 1964. Coccoidea. *In* Kloet, G.S. & Hincks, W.D. (Eds.) A checklist of British Insects, (2nd edn (revised)). Part 1, small orders and Hemiptera. *Handbooks for the Identification of British Insects* 11(1): 87–94.
Not an identification guide, but still a useful list of the coccoids present in the British Isles.

Danzig, E.M. 1967. Coccinea. *In* Bei-Bienko, G.Y. (Ed.) Keys to the Insects of the European USSR. Vol. I, Apterygota, Palaeoptera, Hemimetabola. (Keys to the Fauna of the USSR no. 84), pp. 800–850. Israel Program for Scientific Translations, Jerusalem. [English translation of the original Russain.]
This work is still useful in the study of some British coccoids, but the terminology is often non-standard and the work is inadequately illustrated.

Danzig, E.M. 1993. *[Fauna of Russia and Neighbouring Countries. Rhynchota.* Vol. X, *Scale Insects (Coccinea), families Phoenicococcidae and Diaspididae.]* Nauka, St. Petersburg.
Currently this is perhaps the only available work that deals with many of the diaspids found in Britain. The book is very well illustrated but the text is in Russian, making use of keys difficult.

Dolling, W.R. 1991. *The Hemiptera.* Oxford University Press, Oxford.
This book provides a general account of the Hemiptera. A key to coccoid families [by J.H. Martin] is included.

Hodgson, C.J. 1994. *The scale insect family Coccidae: an identification manual to genera.* CAB International, Wallingford.

This book presents a very good general account of the morphology of the soft scales, history of coccid classification and keys to subfamilies, tribes and genera. With the small British fauna, use of this book is perhaps a case of using a sledgehammer to crack a nut, but it does provide a wealth of information on the group as a whole, including terminology.

Kosztarab, M. & Kozár, F. 1988. *Scale insects of central Europe.* Akadémiai Kiadó, Budapest.

This book provides an informative introduction to coccoid biology, phylogeny, economic importance, field and laboratory techniques, as well as identification keys. The systematic section is organised under the 12 European families (9 of them found in Britain). There are keys to families, genera within families and species keys are provided where needed. The level of illustration is useful but a little disappointing.

Williams, D.J. 1962. The British Pseudococcidae (Homoptera: Coccoidea). *Bulletin of the British Museum (Natural History)*(Entomology) **12**(1): 1–79.

This work contains a key to genera and mini keys to those genera represented by more than single species. All species are illustrated, but the reader is clearly expected to be familiar with pseudococcid terminology. A table records distribution between England, Wales, Scotland, Ireland and Channel Islands.

Williams, D.J. 1985. The British and some other European Eriococcidae (Homoptera: Coccoidea). *Bulletin of the British Museum (Natural History)*(Entomology) **51**(4): 347–93.

This work is similar in format to Williams's pseudococcid work. It, too, assumes a knowledge of the terminology used in the keys. Each species is illustrated with a high-quality drawing of the adult female.

Williams, D.J. & Watson, G.W. 1988a. *The scale insects of the tropical south Pacific region.* Part 1, *The Armoured Scales (Diaspididae).* CAB International, Wallingford.

Williams, D.J. & Watson, G.W. 1988b. *The scale insects of the tropical south Pacific region.* Part 2, *The Mealybugs (Pseudococcidae).* CAB International, Wallingford.

Williams, D.J. & Watson, G.W. 1990. *The scale insects of the tropical south Pacific region.* Part 3, *The Soft Scales (Coccidae) and Other Families.* CAB International, Wallingford.

This trilogy deals with all the known species of the tropical south Pacific, but a high proportion of the fauna comprises well-known and often cosmopolitan pest species, making these works particularly useful for anyone with an interest in scales on glasshouse and other indoor plants in the British Isles. Each volume has its own introductory chapter. A key to families is provided in Vol. 3, which deals with the smaller families in addition to the soft scales. These books are particularly well illustrated and would form an invaluable addition to the library of anyone with more than a passing interest in scale insects.

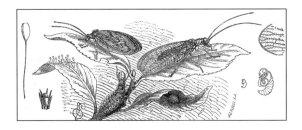

Neuroptera: the lacewings

(65 species in 6 families)

PETER C. BARNARD

Interest in this fairly small group of insects has always been quite strong in Britain, partly because many species are quite large, conspicuous and easy to find, and partly because of the stimulation provided by Killington's (1936–37) exemplary monograph. The latter is still inspirational, but very out of date. Interest in the lacewings has been re-kindled by a new Recording Scheme (see Plant, 1994), as well as by the recent confirmation of populations of the antlion *Euroleon nostras* in East Anglia (Mendel, 1996).

The Neuroptera in the broad sense were always a very diverse group but, even though the Megaloptera and Raphidioptera are now treated as distinct orders, the remaining Neuroptera *sensu stricto* (sometimes known as the Planipennia) still present a highly diverse range of sizes and habits. The tiny Coniopterygidae, or wax flies, resemble hemipteran whiteflies; the giant lacewing *Osmylus* has large strongly patterned wings and a semi-aquatic larva; the Sisyridae, or sponge-flies, have aquatic larvae, which feed on freshwater sponges; the large antlion can be mistaken for a dragonfly, and its larva builds a characteristic conical pit in sandy soil; and both the brown lacewings (Hemerobiidae) and green lacewings (Chrysopidae) are familiar visitors to gardens and houses, especially at night. All the British Neuroptera have predatory larvae, and many adults are also predators on small insects; the Chrysopidae in particular are regarded as beneficial insects because of the large numbers of aphids they consume.

Fraser's (1959) handbook can no longer be recommended for identifying British lacewings, partly because it contained many errors even at the time of publication, and also because the taxonomy has changed dramatically in recent years, with several species being added to the British list. Plant's (1997) key will work for adults of all groups, but Elliott's (1996) book is recommended for the freshwater families, not least because it also covers the known larvae.

The 6 families in Britain are: Coniopterygidae, Osmylidae, Sisyridae, Myrmeleontidae, Hemerobiidae and Chrysopidae.

Aspöck, H., Aspöck, U. & Hölzel, H. 1980. *Die Neuropteren Europas*. 2 vols. Goecke & Evers, Krefeld.

A comprehensive treatment of the European fauna (with an update volume in preparation) intended for the specialist.

Elliott, J.M. 1996. British freshwater Megaloptera and Neuroptera: a key with ecological notes. *Scientific Publications of the Freshwater Biological Association* 54: 69pp.

Very good key to the adults, and all known larvae, of the aquatic families, with excellent biological information; no information on distribution.

Fraser, F.C. 1959. Mecoptera, Megaloptera and Neuroptera. *Handbooks for the Identification of British Insects* 1(12, 13): 40pp.

No longer recommended; see notes above.

Killington, F.J. 1936–37. *A monograph of the British Neuroptera*. 2 vols. Ray Society, London.

Out of date, but still useful.

Mendel, H. 1996. *Euroleon nostras* (Fourcroy, 1785) a British species and notes on ant-lions (Neuroptera Myrmeleontidae) in Britain. *Entomologist's Record and Journal of Variation* 108: 1–5.

The first confirmation of populations of this family in Britain.

Plant, C.W. 1994. *Provisional atlas of the lacewings and allied insects (Neuroptera, Megaloptera, Raphidioptera and Mecoptera) of Britain and Ireland*. Biological Records Centre, Huntingdon.

Good maps showing known distributions based on confirmed records.

Plant, C.W. 1997. A key to the adults of British lacewings and their allies (Neuroptera, Megaloptera, Raphidioptera and Mecoptera). *Field Studies* 9: 179–269.

Useful because it is up-to-date, but the keys are slightly idiosyncratic.

Megaloptera: the alderflies

(3 species in 1 family)

PETER C. BARNARD

The Megaloptera were previously included with the Neuroptera *sensu lato*, along with the Raphidioptera, but have long been recognised as a distinct order of insects. Their larvae are aquatic, found in ponds, lakes, streams and larger rivers, and are active predators on other insect larvae, oligochaete worms and crustaceans. They pass through 10 larval instars, usually over a 2–year life cycle. The adults fly readily in sunny weather and take their common name of alderflies from their habit of settling on trees overhanging the water's edge, although there is no biological connection with alders. Early texts on the British species mention only two species, but a third (*Sialis nigripes*) was first reported in 1977, and was subsequently found to have been overlooked in Britain for over a century.

All three British species are in the family Sialidae.

Elliott, J.M. 1996. British freshwater Megaloptera and Neuroptera: a key with ecological notes. *Scientific Publications of the Freshwater Biological Association* **54**: 69pp.

An excellent key to adults and larvae, with extensive biological information, but no indication of distribution.

Plant, C.W. 1994. *Provisional atlas of the lacewings and allied insects (Neuroptera, Megaloptera, Raphidioptera and Mecoptera) of Britain and Ireland*. Biological Records Centre, Huntingdon.

Useful summary of confirmed distribution records.

Plant, C.W. 1997. A key to the adults of British lacewings and their allies (Neuroptera, Megaloptera, Raphidioptera and Mecoptera). *Field Studies* **9**: 179–269.

Usable for adults, but does not cover the larvae, so Elliott's (1996) key is recommended.

Raphidioptera: the snakeflies

(4 species in 1 family)

PETER C. BARNARD

The snakeflies take their common name from the very elongate pronotum, which is held upright, resembling a snake about to strike its prey. Both the adults and the larvae are predatory, the larvae feeding on other insect larvae on the bark of trees, and the adults on small insects such as aphids high in the canopy. Each of the four British species has been placed in a separate genus by some recent authors, e.g. Aspöck, *et al.* (1991), but such 'inflated' genera appear to have little biological reality, and are most sensibly treated as sub-genera within *Raphidia*. The Raphidioptera have been grouped with the Megaloptera by some authors, but the two groups are not particularly closely related and should be considered as separate orders.

All the British species are in the family Raphidiidae.

Aspöck, H., Aspöck, U. & Hölzel, H. 1980. *Die Neuropteren Europas.* 2 vols. Goecke & Evers, Krefeld.
Excellent keys and illustrations to the European species; in German.

Aspöck, H., Aspöck, U. & Rausch, H. 1991. *Die Raphidiopteren der Erde.* 2 vols. Goecke & Evers, Krefeld.
An expensive monograph of the world species.

Fraser, F.C. 1959. Mecoptera, Megaloptera and Neuroptera. *Handbooks for the Identification of British Insects* 1(12, 13): 40pp.
Fraser's key still works if the venational characters are examined carefully; *R. cognata* is now known as *R. confinis.*

Plant, C.W. 1994. *Provisional atlas of the lacewings and allied insects (Neuroptera, Megaloptera, Raphidioptera and Mecoptera) of Britain and Ireland.* Biological Records Centre, Huntingdon.
Useful distribution maps summarising confirmed records.

Plant, C.W. 1997. A key to the adults of British lacewings and their allies (Neuroptera, Megaloptera, Raphidioptera and Mecoptera). *Field Studies* 9: 179–269.
A more reliable key than Fraser's (1959).

Coleoptera: the beetles

(*ca.* 4000 species in 92 families)

PETER M. HAMMOND AND STUART J. HINE

Among the larger insect groups, the Coleoptera (beetles) rank second only to the Lepidoptera in terms of popular interest with naturalists and collectors. The diversity, ubiquity, and wide range of feeding habits and habitat affiliations exhibited by British beetle species have made them an important focus of ecological, environmental and conservation-related research. Compared with the parasitic Hymenoptera or smaller Diptera, taxonomic understanding of British beetles is good but, unlike the Lepidoptera for example, data on the group remain relatively poorly collated. The Coleoptera are also not well served by comprehensive monographs or keyworks, with the inevitable consequence that their accurate identification requires reference to numerous sources, including many individual papers in journals, as will be seen from the list of references below. Not surprisingly, their immature stages are much less well known than the adults, with even some extremely common British species still completely unknown as larvae.

Like other endopterygote insects with a distinct pupal stage intervening between life as a larva and as a sexually mature adult, the main feeding stage of beetles is the larva. However, unlike the adults of many other endopterygote groups, those of beetles tend to be relatively long-lived. In a number of significant instances adults and larvae occupy the same habitat, and often they occur together consuming the same or approximately the same food. Adult longevity in beetles is likely to be associated with their well-protected and highly 'integral' structure. Typically, the cuticle is thick and hard, and their flight wings are covered by modified fore wings in the form of wing-cases or elytra. This provides protection not only against potential predators and parasites, and against desiccation, but also against incidental damage to the wings and other vulnerable body-parts. Thus, while retaining the advantages of flight, adult beetles are able to burrow, hide in and exploit the soil, wood and many other substrates where other adult endopterygotes are rarely seen.

The general biology, including physiology and behaviour, of beetles is extremely varied. Their feeding biology, in particular, covers an extremely wide range, and there is little in the form of potential nourishment that is not consumed by one or another beetle species. Most, however, have mouth-parts clearly adapted for chewing, with only a few (e.g. Cerylonidae) modified for sucking. Their main impact is through four types of feeding: on vascular plant tissues, on fungi, on decaying animal or plant matter, and as predators. However, some beetles are specialist feeders on, for example, algal cells or slime-mould spores, and several superfamilies include species that are parasitoids.

Serious study of British Coleoptera began relatively early in comparison with that of many other major insect groups. Following the publication in the early nineteenth century of the earliest works devoted specifically to beetles by Thomas Marsham and James Stephens, the study of beetles in the British Isles saw a period of intense activity in the middle and latter half of the nine-teenth century. This culminated in the production of relatively complete checklists of the fauna, and the first really comprehensive and 'modern' account of British beetles, by Canon W.W. Fowler. This magisterial work, pub-lished in five extensively illustrated volumes between 1886 and 1891, was supplemented by a later volume prepared by Fowler together with H. Donisthorpe in 1913.

Although works published prior to that of Fowler retain considerable his-torical interest, most have little value today for practical identification pur-poses. The more useful of them (e.g. Cox's two-volume handbook published in 1874) represent less full or reliable treatments than that of Fowler, and the latter's work represents the real starting point for any guide to literature useful for identifying British beetles.

The next landmark after Fowler's *magnum opus* was the bold attempt made by Norman Joy (1932) to produce a more up-to-date and also more practical handbook. Like Fowler's work, this also soon became out-of-print, with copies fetching a high price in the second-hand book market. Unlike Fowler's work, however, a facsimile edition of Joy's two-volume handbook (one volume is plates) was reprinted in a slightly smaller format in 1976, and again in 1997.

Despite many advances in knowledge of the British beetle fauna since 1932, the works of both Fowler and Joy remain extremely useful if not essen-tial as reference for the serious student of the British beetle fauna. No more recent and complete account of the fauna is available, although Coleoptera parts in the Royal Entomological Society *Handbooks* series now cover nearly half of the species. With the notable exception of *Die Käfer Mitteleuropas* (see under general references, below) published in parts from 1964 onwards (in German), complete modern treatments of the beetle faunas of any parts of mainland Europe are also lacking.

The classification used in this chapter principally follows Lawrence & Newton (1995) (see general references, below).

Higher classification of British Coleoptera

Suborder Adephaga
Carabidae (including Cicindelidae), Haliplidae, Noteridae, Dytiscidae, Hygrobiidae, Gyrinidae

Suborder Myxophaga
Family Microsporidae (= Sphaeriidae)

Suborder Polyphaga
Series Staphyliniformia
Superfamily Hydrophiloidea (including Histeroidea): Hydrophilidae (including Georissidae, Hydrochidae etc.), Histeridae, Sphaeritidae
Superfamily Staphylinoidea: Hydraenidae, Ptiliidae, Leptinidae, Leiodidae, Silphidae, Scydmaenidae, Staphylinidae (including Scaphidiidae), Pselaphidae
Series Scarabaeiformia
Superfamily Scarabaeoidea: Lucanidae, Trogidae, Geotrupidae, Scarabaeidae
Series Elateriformia
Superfamily Scirtoidea (= Eucinetoidea): Clambidae, Eucinetidae, Scirtidae
Superfamily Dascilloidea: Dascillidae
Superfamily Byrrhoidea (including Dryopoidea): Byrrhidae, Psephenidae, Heteroceridae, Limnichidae, Dryopidae, Elmidae
Superfamily Buprestoidea: Buprestidae
Superfamily Elateroidea (including Cantharoidea): Elateridae, Throscidae, Eucnemidae, Drilidae, Cantharidae, Lampyridae, Lycidae
Series Bostrichiformia
Superfamily Dermestoidea: Derodontidae, Dermestidae
Superfamily Bostrichoidea: Anobiidae (including Ptinidae), Bostrichidae (including Lyctidae)
Series Cucujiformia
Superfamily Cleroidea: Phloiophilidae, Trogossitidae (including Peltidae), Cleridae, Melyridae
Superfamily Lymexyloidea: Lymexylidae
Superfamily Cucujoidea: Nitidulidae (including Brachypteridae (=Kateretidae)), Monotomidae (=Rhizophagidae), Sphindidae, Cucujidae, Laemophloeidae,

Silvanidae, Cryptophagidae (including Hypocopridae), Biphyllidae, Byturidae,
Erotylidae, Phalacridae, Cerylonidae, Bothrideridae, Corylophidae,
Coccinellidae, Alexiidae, Endomychidae (including Merophysiidae), Latridiidae
(= Corticariidae)
Tenebrionoidea: Mycetophagidae, Colydiidae, Ciidae, Tenebrionidae (including
Lagriidae), Melandryidae, Tetratomidae, Salpingidae, Mycteridae, Pythidae,
Pyrochroidae, Scraptiidae, Mordellidae, Rhipiphoridae, Oedemeridae,
Meloidae, Anthicidae, Aderidae
Chrysomeloidea: Cerambycidae, Megalopodidae, Chrysomelidae (including
Bruchidae)
Curculionoidea: Nemonychidae, Anthribidae, Attelabidae, Brentidae (including
Apioninae and Nanophyinae), Curculionidae (including Scolytidae,
Platypodidae and Raymondionymidae)

This chapter begins with information about relevant clubs, societies, newsletters and recording schemes. This is followed by sections on: distribution
atlases and checklists; Coleoptera in general; Coleopteran larvae, species of
economic importance; and aquatic Coleoptera. The bulk of the references
are then listed by superfamily in the systematic order given above. In a few
cases separate sections are given for individual families (e.g. Staphylinidae,
Ptiliidae), and in a few instances the references to early stages are given separately.

Throughout the comments in this chapter, 'standard' works means Fowler
(1886–91), Joy (1932), and the Royal Entomological Society *Handbooks.*

Clubs, societies, newsletters and recording schemes

Only one U.K. based society for coleopterists currently exists — the Balfour-
Browne Club — although there is a loose-knit 'interest group' for British
beetles as a whole, centred around the various beetle Recording Schemes
and *The Coleopterist.*

The Balfour-Browne Club. Founded in 1976 to inform those interested in
aquatic British beetles. Publishes *Latissimus* twice a year (superseding
The Balfour-Browne Club Newsletter). Membership details available
from Dr G.N. Foster, 3 Eglinton Terrace, Ayr, Ayrshire KA7 1JJ, Scotland,
UK.

The Coleopterist's Newsletter. 1980–92. Aimed at informing British coleopterists of new literature, recent noteworthy discoveries, etc.; superseded by
a formal publication, *The Coleopterist.* Back-numbers available from P.J.
Hodge, 8 Harvard Road, Ringmer, Lewes, East Sussex BN8 5HJ.

A number of societies and clubs devoted to the study of Coleoptera exist in other countries, notably in Europe and in Japan. Those with an international aspect and of most general interest to UK coleopterists are The Coleopterists' Society (based in the USA), and the European Association of Coleopterists (founded in 1986 and based in Spain).

Fourteen recording schemes devoted to gathering distributional and other data with respect to the British beetle fauna are currently in operation. Some of these schemes have produced newsletters or published reports from time to time, and a few have already generated distribution atlases (see below). The schemes are listed and discussed in a recent publication by Eversham & Harding (1996). The Biological Records Centre acts as an 'umbrella' for the recording schemes, and further information about them may be obtained from BRC, ITE Monks Wood, Abbots Ripton, Huntingdon, Cambridgeshire PE17 2LS.

Eversham, B.C. & Harding, P.T. 1996. Coleoptera Recording Schemes Update. *Coleopterist* 5(1): 13–18.

Distribution atlases and checklists

The most recently published checklist of British Coleoptera is that of Pope (1977), and this remains the basic standard in terms of classification and nomenclature. However, several informal and generally computerised checklists of more recent date are now also in use. A number of amendments to the 1977 list are incorporated in the recently published checklist of Irish Coleoptera by Anderson *et al.* (1997). There are also recently published checklists for several North and West European countries that are useful for studies of the British beetle fauna. The most immediately relevant to the UK are the catalogues incorporated in Volumes 12–14 of *Die Käfer Mitteleuropas* (see below), and the checklists of Hansen (1996) for Denmark and Silfverberg (1992) for Fennoscandia. Also among the most recently completed is that for Italy — *Checklist delle specie della Fauna Italiana* — with the beetle parts (44–61) published between 1993 and 1995.

Distribution atlases are available for a few of the groups covered by recording schemes (e.g. Elateroidea, see below) and others are in preparation. Some of the more modern county lists, e.g. that for Somerset by Duff (1993), also contain much useful information on distribution.

Anderson, R., Nash, R. & O'Connor, J.P. 1997. Irish Coleoptera, a revised and annotated list. *Irish Naturalists' Journal* (Special Entomological Supplement): 81pp.
Duff, A. 1993. *Beetles of Somerset: their status and distribution.* Somerset Archaeological and Natural History Society.

Hansen, M. 1996. Catalogue of the Coleoptera of Denmark. *Entomologiske Meddelelser* 64: 1–231.
Up-to-date list in taxonomic arrangement and nomenclature. Parallel Danish and English texts.

Johnson, C. 1993. *Provisional atlas of the Cryptophagidae: Atomariinae (Coleoptera) of Britain and Ireland.* Biological Records Centre, Huntingdon.

Lindroth, C.H. (ed.) 1960. *Catalogus Coleopterorum Fennoscandiae et Daniae.* Entomological Society, Lund.
Provides distribution by province in tabular form.

Lott, D. 1995. Changes to the British list published in 1994. *Coleopterist* 4: 3–5.

Lott, D. 1996. Changes to the British List published in 1995. *Coleopterist* 5: 1–2.

Luff, M.L. 1998. *Provisional atlas of British Carabidae (Coleoptera).* Biological Records Centre, Abbots Ripton, Huntingdon.

Mendel, H. & Clarke, R.E. 1996. *Provisional atlas of the click beetles (Coleoptera: Elateroidea) of Britain and Ireland.* Ipswich Borough Council Museums, Ipswich.
Revised check list and bibliography as well as maps and notes on scarce species.

Owen, J.A. 1993. An annotated list of recent additions and deletions affecting the recorded beetle fauna of the British Isles. *Coleopterist* 2: 1–18.

Owen, J.A. 1994. Corrections to an annotated list of recent additions and deletions affecting the recorded beetle fauna of the British Isles. *Coleopterist* 2: 67.

Pope, R.D. 1977. (Revised) A check list of British Insects. 2nd edn. *Handbooks for the Identification of British Insects* 11(3): 105pp.

Silfverberg, H. 1992. *Enumeratio Coleopterorum Fennoscandiae, Daniae et Baltiae (A check list of Northern European Coleoptera).* Helsingin Hyonteisvaihtoyhdistys, Helsinki.

Coleoptera: general references

Auber, L. 1960. *Atlas des Coléoptères de France, Belgique, Suisse. Nouvel atlas d'entomologie,* (2nd edn). 2 vols. Boubée, Paris.
Keys to major groups; brief descriptions of each species; excellent illustrations; in French.

Blair, K.G. 1948. Some recent additions to the British insect fauna. *Entomologist's Monthly Magazine* 84: 51–7.
Includes colour plate with figures of five beetle species.

Blair, K.G. 1948. Some alien Coleoptera occasionally found in Britain. *Entomologist's Monthly Magazine* 84: 123–4.
Includes colour plate, with figures of ten beetle species, two of them now established in Britain.

Cooter, J. 1991. *A coleopterist's handbook* (3rd edn). Amateur Entomologist's Society, London.
A useful guide to collecting methods, field work techniques, etc.

Crowson, R.A. 1956. Coleoptera: introduction and key to families. *Handbooks for the Identification of British Insects* 4(1): 59pp.

Good keys, enabling accurate family placement, but not easy to use for the beginner. The family arrangement employed is now a little out-of-date, and a few British family-group taxa (e.g. Derodontidae) are not included.

Crowson, R.A. 1981. *The biology of the Coleoptera.* Academic Press, London.
The most useful general work concerning beetle biology, including an extensive bibliography.

Dibb, J.R. 1948. *Field book of beetles.* Brown, London and Hull.
Adopts the unusual approach of using habitat as primary divisions in its keys, but this works well for many species. Inevitably now out-of-date and not to be recommended for 'difficult' groups.

Fowler, W.W. 1886–91. *The Coleoptera of the British Islands.* 5 vols (in large paper edn). Reeve, London.
A 'standard' work; still useful, but out-of-date in many respects (see 'Introduction' and comments under individual superfamilies and families below).

Fowler, W. W. & Donisthorpe, H. St J.K. 1913. *The Coleoptera of the British Islands.* Vol. 6 (suppl.). Reeve, London.
Keys, descriptions of adults; frequency and distribution; some bionomic data.

Freude, H., Harde, K.W. & Lohse, G.A. (Eds.) 1964–. *Die Käfer Mitteleuropas.* Goecke & Evers, Krefeld.
A 'standard' work; in German. See detailed comments below.

Freude, H., Harde, K.W. & Lohse, G.A. (Eds.) 1965. Einführung in die Käferkunde. *Die Käfer Mitteleuropas.* Vol. 1. Goecke & Evers, Krefeld. 214pp.
General introduction and key to families; in German.

Göllner-Scheiding, U. 1970–71. Bibliographie der Bestimmungstabellen europäischer Insekten (1880–1963). Teil III: Coleoptera und Strepsiptera. *Deutsche Entomologische Zeitschrift* **17**: 33–118, 433–76; **18**: 1–84, 287–360.
A very useful guide to key works on European Coleoptera; arranged by taxon with cross-reference to author; in German.

Hammond, P.M. 1974. Changes in the British coleopterous fauna. *In* Hawksworth, D.L. (Ed.) *The changing flora and fauna of Britain,* pp. 323–69. Academic Press, London.
Summarises changes in British beetle fauna over the past 10 000 years.

Hammond, P.M. 1996. *A taxonomic review of possibly endemic British non-marine invertebrates.* Unpublished report to English Nature. The Natural History Museum, London.
Contains an appendix listing all Coleoptera described from the British Isles since 1900, with notes and full bibliography. To be published by English Nature.

Hammond, P.M., Smith, K.G.V., Else, G.R. & Allen, G.W. 1989. Some recent additions to the British insect fauna. *Entomologist's Monthly Magazine* **125**: 95–102.
Colour photographs and diagnostic notes on six notable British species not included in the standard works.

Harde, K.W. & Hammond, P.M. 1984. *A field guide in colour to beetles.* Octopus Books, London. English edn edited and with additional introductory material by P.M. Hammond. [Original German title (1981): *Die Kosmos-Käferführer.* Kosmos-Verlag, Stuttgart.] [Reprinted in 1998.]

1080 species illustrated in colour, including most of the larger species found in NW Europe.

Hodge, P.J. & Jones, R.A. 1995. *New British beetles: species not in Joy's Practical Handbook*. British Entomological and Natural History Society, Hurst.

A compendium of references to species found to be additional to the British beetle fauna since *ca*. 1930, with useful diagnostic notes on many of them; an essential companion for users of Joy's handbook.

Horion, A. 1941–. *Faunistik der Mitteleuropäischen Käfer*. Several volumes; various publishers.

Still the most comprehensive source of data concerning distribution, habitats, and general biology of European Coleoptera, although partly superseded by the 'Ecology' volumes of *Die Käfer Mitteleuropas* and more specialist works; in German.

Hyman, P.S. & Parsons, M.S. 1992. Review of the Scarce and Threatened Coleoptera of Great Britain. 1. *UK Nature Conservation* **12**: 484pp. Joint Nature Conservation Committee, Peterborough.

Treats species (of all but ten families) categorised as scarce or threatened in Britain, with summaries of known distribution, habitat, ecology, and conservation status; extensive bibliography.

Hyman, P.S. & Parsons, M.S. 1994. Review of the Scarce and Threatened Coleoptera of Great Britain. 2. *UK Nature Conservation* **12**: 248pp. Joint Nature Conservation Committee, Peterborough.

Provides similar coverage to Volume 1 (above) but for the ten 'difficult' families (including Staphylinidae) not there included.

Johnson, C. 1992. Additions and corrections to the British list of Coleoptera. *Entomologist's Record and Journal of Variation* **104**: 305–10.

Taxonomic and nomenclatural notes on species of seven different families.

Joy, N.H. 1932. *A practical handbook of British beetles*. 2 vols. Witherby, London. Reprinted (1976, 1997) by Classey, Farringdon.

A 'standard' work, and the most recent comprehensive account of British Coleoptera in English. Now out-of-date in many parts but still useful (see 'Introduction' and comments under individual superfamilies and families).

Lawrence, J.F. & Newton, A.F. 1995. Families and subfamilies of Coleoptera (with selected genera, notes, references and data on family-group names. *In* Pakaluk, J. & Slipinski, S.A. (Eds) *Biology, phylogeny and classification of Coleoptera: papers celebrating the 80th birthday of Roy A. Crowson*, pp. 779–1006. Muzeum i Instytut Zoologii PAN, Warzawa.

Contains the most up-to-date complete classificatory system for Coleoptera on a world basis; extensive bibliography.

Linssen, E.F. 1959. *Beetles of the British Isles*. 2 vols. Frederick Warne & Co.

A popular guide, but apt to be misleading; not recommended for species identification.

Paulian, R. 1988. *Biologie des Coléoptères*. Lechevalier, Paris.

A good general account, rather more 'popular' in style, and with more emphasis on ecology and behaviour than Crowson's book; includes many intriguing observations, if of uncertain reliability, as sources are not always clearly indicated; in French.

Unwin, D.M. 1984. A key to the families of British beetles (and Strepsiptera). *Field Studies* 6(4):48pp.

A well-illustrated 'AIDGAP' key. Useful for the non-specialist but, like other attempts to provide short and simple keys to beetle families, cannot be relied upon.

Zahradnik, J. 1985. *Käfer Mittel- und Nordwesteuropas. Ein Bestimmungsbuch für Biologen und Naturfreunde.* Paul Parey, Hamburg & Berlin.

A well produced 'popular' guide, with coloured illustrations of some 900 Central and Northwestern European beetle species; in German.

Die Käfer Mitteleuropas

This important series, in German, is a standard reference work for the central European fauna. The contents of each volume are listed below.

Die Käfer Mitteleuropas Vols. 1–11, by Freude, H., Harde, K.W. & Lohse, G.A. Published by Goecke & Evers, Krefeld.

Vol. 1 General introduction to Coleoptera [includes a well-illustrated key to beetle families] '

Vol. 2 Adephaga 1 [Carabidae]

Vol. 3 Adephaga 2, Palpicornia, Histeroidea, Staphylinoidea 1 [16 families, from Hygrobiidae to Scaphidiidae]

Vol. 4 Staphylinidae 1 [all but Aleocharinae]

Vol. 5 Staphylinidae 2 [plus Pselaphidae]

Vol. 6 Diversicornia [22 families, from Lycidae to Byrrhidae]

Vol. 7 Clavicornia [20 families, from Trogossitidae to Ciidae]

Vol. 8 Teredilia, Heteromera, Lamellicornia [18 families, from Bostrichidae to Lucanidae, and Strepsiptera]

Vol. 9 Cerambycidae, Chrysomelidae [includes Megalopodidae]

Vol. 10 Bruchidae, Anthribidae, Scolytidae, Platypodidae, Curculionidae [includes Attelabidae, Brentidae, Nemonychidae and 'broad-nosed' Curculionidae]

Vol. 11 Rhynchophora [covers most Curculionidae *sensu stricto*]

Die Käfer Mitteleuropas Vols. 12–14 by Lohse, G.A. & Lucht, W.H.

Vol. 12 Supplement 1 [1989 supplement to Vols 1–5 and to catalogue]

Vol. 13 Supplement 2 [1992 supplement to Vols 6–8 and to catalogue]

Vol. 14 Supplement 3 [1994 supplement to Vols 9–11 and to catalogue]

Die Käfer Mitteleuropas Vols. E1–E8 by Koch, K.C.

Vol. E1–Vol. E8 Ecology of Coleoptera [contain information on habitat/microhabitat affiliations and feeding biology of individual species, and lists of species associated with various habitats]

Die Käfer Mitteleuropas Vols. L1–L3 by Klausnitzer, B.

Vol. L1 Larvae 1: Adephaga

Vol. L2 Larvae 2: Myxophaga, Polyphaga 1 [17 families, including all Hydrophiloidea *s. str.*, Scirtoidea, Elateroidea and Chrysomeloidea]

Vol. L3 Larvae 3: Polyphaga 2 [23 families, including all Scarabaeoidea and Cleroidea, and many Byrrhoidea and Tenebrionoidea]

Die Käfer Mitteleuropas Vol. K by Lucht, W.H.

Vol. K Catalogue [full checklist as of 1987 with occurrence of species in ten areas of Central areas tabulated]

Coleoptera larvae (general)

Works relating to the early stages of particular beetle groups are referred to under the appropriate superfamily or family headings below.

Emden, F.I. van 1942. Larvae of British beetles. III. Keys to families. *Entomologist's Monthly Magazine* 78: 206–26, 253–72.
Concise key to most families, illustrated by 54 figures.

Emden, F.I. van 1943. Larvae of British beetles. IV. Various small families. Cicindelidae, Hygrobiidae. *Entomologist's Monthly Magazine* 79: 209–23, 259–70.
Covers some 18 families, with keys to all genera and/or species of some; illustrated by 36 figures.

Klausnitzer, B. 1978. *Ordnung Coleoptera (Larven). Bestimmungsbücher zur Bodenfauna Europas.* Junk, The Hague.
Keys to families and genera of soil and surface-substrate groups; in German.

Klausnitzer, B. 1991, 1994, 1996. *Die Käfer Mitteleuropas, Larven* 1, 2 & 3. Goecke & Evers, Krefeld.
Well-illustrated keys in German to a wide range of families including all Adephaga, Elateroidea, Hydrophiloidea *s. str.* and Scarabaeoidea. The major groups *not* covered include Curculionidae and Staphylinidae.

Larsson, S.G. 1968. Løbebillernes larver. *Danmarks Fauna* 76: 282–433.
In Danish.

Lawrence, J., Hastings, A., Dallwitz, M. & Paine, T. 1993. *Beetle larvae of the world.* CSIRO, Australia. (CD-ROM)

Nikitskii, N.B. 1976. [Morphology of larvae of beetles predaceous on and associated with bark beetles in the north-west Caucasus.] *In* Mamaev, B.M. (Ed.) *[Evolutionary morphology of the larvae of insects]*, pp. 175–201. Science Publishers, Moscow.
In Russian.

Viedma, M.G. de 1962–63. Larvas de coleópteros. *Boletin del Servicio de Plagas Forestales* 5: 87–91; 6: 103–21. (Reprinted 1964, *Graellsia* 20: 245–75).
Keys to 31 families affecting forestry; in Spanish.

Species of economic importance

Booth, R.G., Cox, M.L. & Madge, R.B. 1990. *Coleoptera.* IIE guides to insects of importance to man. 3. CAB International, Wallingford, UK.
A very useful general introduction to Coleoptera as a whole, with good illustrated keys enabling identification to all major families.

Balachowsky, A.S. 1962–63. *Coléoptères. Entomologie appliquée à l'agriculture.* Masson et Cie, Paris. (in 2 pts).
    A good general text on beetles of importance to agriculture; includes many useful details on the biology of species, although the sources (and veracity) of some observations are left unclear; in French.

Hickin, N.E. 1963. *The insect factor in wood decay.* Hutchinson, London.
    Good coverage of species affecting worked timber, with the focus on Britain.

Hickin, N.E. 1941. The Lathridiidae of economic importance. *Bulletin of Entomological Research* **32**: 191–247.
    Well illustrated, but now out-of-date for some groups of species in genera such as *Lathridius* and *Enicmus*.

Hickin, N.E. 1945. *A monograph of the beetles associated with stored products.* Vol. I. British Museum (Natural History), London. [reprinted 1963, Johnson Reprint Company].
    Profusely illustrated keys to families, species (adults and larvae). Covers Carabidae, Staphylinidae, Nitidulidae, Lathridiidae, Mycetophagidae, Colydiidae, Murmidiidae (Cerylonidae), Endomychidae, Erotylidae, Anthicidae, Crytophagidae and Dermestidae. [Note that the intended Volume 2 was never published.]

Hickin, N.E. 1945. The Histeridae associated with stored products. *Bulletin of Entomological Research* **35**: 309–40.
    Illustrated keys to adults and larvae.

Nikitskii, N. B. 1976. [Morphology of larvae of beetles predaceous on and associated with bark beetles in the north-west Caucasus.] *In* Mamaev, B.M. (Ed.) *[Evolutionary morphology of the larvae of insects]*, pp. 175–201. Science Publishers, Moscow.
    A useful account of wood-associated beetle larvae, containing information not available elsewhere; in Russian.

Schimitschek, E. 1955. *Die Bestimmung von Insektenschäden im Walde.* P. Parey, Hamburg & Berlin.
    Includes keys based on damage caused; in German.

Viedma, M.G. de 1962–63. Larvas de coleópteros. *Boletin del Servicio de Plagas Forestales* **5**: 87–91; **6**: 103–21. (Reprinted 1964, *Graellsia* **20**: 245–75.)
    Keys to 31 families affecting forestry; in Spanish.

Water beetles: general

This section provides information on works that deal with aquatic beetles as a 'habitat' group, regardless of their systematic position within the Order. The groups covered by such works are generally Dytiscidae, Noteridae, Gyrinidae, Hydrophilidae (including Georissidae, Hydrochidae, Spercheidae, etc.) and Hydraenidae (including Limnebiidae), with the addition of all or some of the following: Elmidae, Dryopidae, Psephenidae, Scirtidae (= Helodidae), Heteroceridae, Limnichidae, and the aquatic or semi-aquatic species of certain other families such as Chrysomelidae (e.g.

Donaciinae), Curculionidae (e.g. Bagoini), Staphylinidae, etc. Note that all species known to be British up to 1988 are included by Friday (1988), and this work is generally a reliable means of identifying aquatic Adephaga (but see additional works listed under Haliplidae, Noteridae, Dytiscidae and Gyrinidae, below).

Bertrand, H.P.I. 1972. *Larves et nymphes des Coléoptères aquatiques du globe.* Paris.
Keys to genera; in French.

Balfour-Browne, F. 1940. *British water beetles.* Ray Society, London. **1**.
Covers Haliplidae, Hygrobiidae and Dytiscidae in part.

Balfour-Browne, F. 1950. *British water beetles.* Ray Society, London. **2**. (Facsimile reprint 1964.)
Covers Dytiscidae in part and Gyrinidae.

Balfour-Browne, F. 1953. Coleoptera: Hydradephaga. *Handbooks for the Identification of British Insects* **4**(3): 33pp.
A rather sparsely illustrated handbook that is difficult to use and now somewhat out-of-date.

Balfour-Browne, F. 1958. *British water beetles.* Ray Society, London. **3**.
Covers Hydrophilidae and Hydraenidae.

Chiesa, A. 1959. *Hydrophilidae Europae (Coleoptera, Palpicornia).* Arnaldo Forni, Bologna.
Keys to genera and species; in Italian.

Franciscolo, M. 1978. Coleoptera Haliplidae, Hygrobiidae, Gyrinidae, Dytiscidae. *Fauna d'Italia* **14**: 804pp.
Keys, illustrated by over 2000 figures, to all Italian species; in Italian.

Freude, H., Harde, K.W. & Lohse, G.A. 1971. Hygrobiidae, Haliplidae, Dytiscidae, Gyrinidae, Rhysodidae. *Die Käfer Mitteleuropas.* Vol. 3.
In German.

Friday, L.E. 1988. *Key to the adults of British water beetles.* (AIDGAP) Field Studies Council.
An excellent key to British water beetles.

Holmen, M. 1987. The aquatic Adephaga (Coleoptera) of Fennoscandia and Denmark. 1. Gyrinidae, Haliplidae, Hygrobiidae and Noteridae. *Fauna Entomologica Scandinavica* **20**: 1–168.
Well-illustrated guide covering all British species.

Lohse, G.A. & Vogt, H. 1971. Hydraenidae, Spercheidae, Hydrophilidae. *Die Käfer Mitteleuropas.* Vol. 3, pp. 95–156.
In German.

Suborder Adephaga

Most species of this well-demarcated 'primitive' suborder are active, and often relatively free-ranging predators. Members of five adephagan families represented in Britain (Haliplidae, Hygrobiidae, Noteridae, Dytiscidae and

Gyrinidae) are all aquatic as larva and adult, and species of these groups are
to be found in almost all types of freshwater habitat, from brackish estuarine
waters to mountain streams and tarns. The prey of aquatic adephagans
ranges from tubificid worms to small fish. Haliplid adults and larvae are
grazers, and their food may include or consist largely of algae as well as
sessile colonial animals such as bryozoans. Adult Gyrinidae, well-known as
whirligig beetles, often swim at the surface in schools, and prey largely on
insects trapped in the surface film; their larvae are bottom-feeding predators. Members of the remaining adephagan family represented in the British
Isles, the Carabidae, are generally referred to as ground beetles. Most British
species are ground-dwelling and cursorial types as adults, although a
number are burrowers (fossorial) and some forage for prey on herbaceous
plants or in trees. As well as many polyphagous predators and omnivores, the
family includes specialist predators of, for example, molluscs or springtails,
parasitoids of other beetles, and seed-feeders.

Family Carabidae

The standard works, particularly Lindroth's (1974) handbook, should suffice
for accurate identification of Carabidae, except for those few additional
species (of *Agonum, Asaphidion, Bembidion, Calathus, Cymindis* and
Pterostichus) first shown to occur in the British Isles over the past two
decades or so, and perhaps also for some of the most difficult species (e.g. of
Amara). Forsythe's (1987) keys will also serve to identify species in all but the
most difficult groups. Virtually all larvae are also identifiable, at least to
genus, using the 1993 Scandinavian handbook and other papers by M.L. Luff.

Allen, A.A. 1985. *Brachinus sclopeta* (Fabricius). Two captures in the present century.
   Entomologist's Record and Journal of Variation 97: 137–9.
Allen, A.A. 1991. Notes on distinction between *Patrobus atrorufus* Ström (*P. excavatus*
   in Joy) and *P. assimilis* Chaudoir. *Entomologist's Record and Journal of Variation*
   103: 71–2.
Anderson, R. 1985. *Agonum lugens* (Duftschmidt) (Coleoptera: Carabidae) new to the
   British Isles. *Entomologist's Monthly Magazine* 121: 133–5.
Anderson, R & Luff, M.L. 1994. *Calathus cinctus* Motschulsky: a species of the
   Calathus melanocephalus/mollis complex (Coleoptera: Carabidae) in the British
   Isles. *Entomologist's Monthly Magazine* 130: 131–5.
   **Good key and male genitalia figures for three British species of the *Calathus melano-
   cephalus* group.**
Bonadona, P. 1971. Catalogue des Coléoptères carabiques de France. *Nouvelle Revue
   d'Entomologie* (suppl.) 1: 1–177.

Crossley, R. & Norris, A. 1976. *Bembidion humerale* Sturm (Col., Carabidae) new to Britain. *Entomologist's Monthly Magazine* **111**(1975): 59.

Emden, F.I. van 1942. A key to the genera of larval Carabidae. *Transactions of the Royal Entomological Society of London* **92**: 1–99.
Keys to 31 tribes and many genera (including most of those represented in Britain), with 100 figures.

Emden, F.I. van 1943. Larvae of British beetles. IV. Various small families. Cicindelidae, Hygrobiidae. *Entomologist's Monthly Magazine* **79**: 209–13.

Forsythe, T. 1987. *Common ground beetles.* Naturalist's handbooks No. 8. Richmond Publishing Co., Slough.
A useful key work for most purposes. Attempts to cover all British species (despite the title), but groups some uncommon and 'difficult' species together under one half of a couplet.

Freude, H., Harde, K.W. & Lohse, G.A. 1976. Carabidae. *Die Käfer Mitteleuropas.* Vol. 2.
In German.

Hammond, P.M. 1982. *Cymindis macularis* (Fischer & Waldheim) (Coleoptera: Carabidae) – apparently a British species. *Entomologist's Monthly Magazine* **118**: 37–8.

Houston, W.W.K. & Luff, M.L. 1975. The larvae of the British Carabidae (Coleoptera). III. Patrobini. *Entomologist's Gazette* **26**: 59–64.
Key to all three British species, with 15 figures.

Houston, W.W.K. & Luff, M.L. 1983. The identification and distribution of the three species of *Patrobus* (Coleoptera: Carabidae) found in Britain. *Entomologist's Gazette* **34**: 283.

Jeannel, R. 1941, 1942, 1949. Coléoptères carabiques. *Faune de France* **39**: 1–571; **40**: 1–601; **51**: 1–51.
In French.

Kevan, D.K. 1949. The sexual and other characteristics of British Nebriini, Notiophilini & Elaphrini, with special reference to the genus *Notiophilus* Dum. (Coleoptera: Carabidae). *Entomologist s Monthly Magazine* **85**: 1–18.

Larsson, S.G. 1968. Løbebillernes larver. *Danmarks Fauna* **76**: 282–433.
In Danish.

Lindroth, C.H. 1972. Taxonomic notes on certain British ground-beetles (Col., Carabidae). *Entomologist's Monthly Magazine* **107**: 209–23.
Distinguishing characters of species not then known from Britain, but likely to occur; well illustrated.

Lindroth, C.H. 1974. Coleoptera: Carabidae. *Handbooks for the Identification of British Insects* **4**(2): 148pp.
See corrections to key in *Antenna* **1**: 25.

Lindroth, C.H. 1985–86. The Carabidae (Coleoptera) of Fennoscandia and Denmark. *Fauna Entomologica Scandinavica* **15**(1); 1–236; **15**(2): 237–497.
Well-illustrated keys, with good information on biology, etc.; in English.

Lohse, G. A. 1983. Die *Asaphidion*-Arten aus der Verwandtschaft des *A. flavipes* L. *Entomologische Blätter für Biologie und Systematik der Käfer* **79**: 33–6.
Key to the three British species of this group.

Luff, M.L. 1969. The larvae of British Carabidae (Coleoptera). I. Carabini and Cychrini. *The Entomologist* 102: 245–63.
Key to all 12 British species, with 75 figures.

Luff, M.L. 1972. The larvae of the British Carabidae (Coleoptera). II. Nebriini. *The Entomologist* 105: 161–79.
Key to 12 out of 13 British species, with 61 figures.

Luff, M.L. 1976. The larvae of the British Carabidae (Coleoptera). IV. Notiophilini and Elaphrini. *Entomologist's Gazette* 27: 51–67.
Key to 11 out of 13 British species, with 47 figures.

Luff, M.L. 1978. The larvae of the British Carabidae (Coleoptera). V. Omophronini, Loricerini, Scaritini and Broscini. *Entomologist's Gazette* 29: 265–87.
Illustrated key to all genera and 13 of the British species.

Luff, M.L. 1980. The larvae of the British Carabidae (Coleoptera). VI. *Entomologist's Gazette* 31: 177–94.

Luff, M.L. 1985. The larvae of the British Carabidae (Coleoptera) VII. Trechini and Pogonini. *Entomologist's Gazette* 36: 301–14.
Well-illustrated keys to ten species.

Luff, M.L. 1990. *Pterostichus rhaeticus* Heer. *Entomologist's Monthly Magazine* 126: 245–9.
Description and figures.

Luff, M.L. 1993. The Carabidae (Coleoptera) larvae of Fennoscandia and Denmark. *Fauna Entomologica Scandinavica* 27: 1–182.
Well-illustrated guide, with excellent habitus figures, covering most British species.

Moore, B.P. 1957. The British Carabidae (Coleoptera) part I: a checklist of the species. *Entomologist's Gazette* 8: 129–37.
Provides a summary of recorded distribution, by county, for all British Carabidae.

Plant, C.W. & Drane, A.B. 1988. Notes on *Acupalpus elegans* (Dejean) and separation from *A. dorsalis* (Fabricius). *Entomologist's Gazette* 39: 227–37.
Detailed discussion of separation of the two species, with male genitalia figures.

Speight, M.C.D., Martinez, M. & Luff, M.L. 1986. The *Asaphidion* (Col.: Carabidae) species occurring in Great Britain and Ireland. *Transactions of the British Entomological and Natural History Society* 19: 17–21.
Key to the North European species of *Asaphidion*; well illustrated.

Trautner, J. & Geigenmüller, K. 1987. *Tiger beetles, ground beetles: illustrated key to the Cicindelidae and Carabidae of Europe.* J. Margraf, Aichtal, Germany.
A rather uneven treatment, with the species keyed out in some genera only; many illustrations and some maps; in English.

Wachmann, E., Platen, R. & Barndt, D. 1995. *Laufkäfer. Beobachtung – Lebensweise.* Naturbuch Verlag, Augsburg.
Good quality colour photographic guide to more than 200 Central European ground-beetle species, most of them occurring in Britain; in German.

Welch, R.C. 1981. *Nebria nivalis* (Paykull). Brief note and description. *Entomologist's Monthly Magazine* 116: 166.

Family Haliplidae (see also water beetles)

Parry, J. 1983. *Haliplus varius* Nicolai (Col., Haliplidae) new to Britain. *Entomologist's Monthly Magazine* **119**: 13–16.

Distinguishing characters; well illustrated.

Vondel, B.J. van 1997. Insecta: Coleoptera: Haliplidae. *Süsswasserfauna von Mitteleuropa* **20**(2): 1–95.

The most thorough modern account of European Haliplidae available; extensively illustrated; keys to larvae of many species; in English.

Family Noteridae (see also water beetles)

Dettner, K. 1997. Insecta: Coleoptera: Noteridae. *Süsswasserfauna von Mitteleuropa* **20**(2): 97–126.

A thorough account of the four Central European species; in English.

Family Dytiscidae (see also water beetles)

Carr, R. 1984. A *Coelambus* species new to Britain (Coleoptera: Dytiscidae. *Entomologist's Gazette* **35**: 181–4.

Distinguishing characters of *C. nigrolineatus*; colour patterns and male genitalia of various species of *Coelambus* illustrated.

Foster, G.N. & Angus, R.B. 1985. Key and descriptions of British *Hydroporus* (Coleoptera; Dytiscidae. *Balfour-Browne Club Newsletter* **33**: 1–19.

Excellent illustrated key to all British species with notes on other species from NW Europe.

Galewski, K. 1973. Some notes on the generic characters of the larvae of the subfamily Colymbetinae (Dytiscidae, Coleoptera) with a key for the identification of the European genera. *Polskie Pismo Entomologiczne* **43**: 215–24.

Galewski, K. 1973. Generic characters of the larvae of the subfamily Dytiscinae (Dytiscidae) with a key to the central European genera. *Polskie Pismo Entomologiczne* **43**: 491–8.

Galewski, K. 1974. Diagnostic characters of larvae of European species of *Graphoderus* Dejean (Coleoptera, Dytiscidae) with an identification key and some notes on their biology. *Bulletin de l'Académie Polonaise des Sciences. Série des Sciences* **22**: 485–94.

Nilsson, A.N. 1981. The Fennoscandian species of the genus *Hydaticus* Leach (Coleoptera: Dytiscidae). *Entomologica Scandinavica* **12**: 103–8.

Nilsson, A.N. 1982. A key to the larvae of the Fennoscandian Dytiscidae (Coleoptera). *Fauna Norrlandica* **2**: 1–45.

Nilsson, N. & Holmen, M. 1993. The aquatic Adephaga (Coleoptera) of Fennoscandia and Denmark. II. Dytiscidae. *Fauna Entomologica Scandinavica* **32**: 1–192.

Very well-illustrated guide, covering almost all British species.

Owen, J.A., Lyszkowski, R.M., Proctor, R. & Taylor, S. 1992. *Agabus wasastjernae* (Sahlberg) (Col.: Dytiscidae) new to Scotland. *Entomologist's Record and Journal of Variation* **104**: 225–30.

Distinguishing characters, with elytral microsculpture figured.

Family Gyrinidae

Foster, G.N. 1981. The British species of *Gyrinus*. *Balfour-Browne Club Newsletter* **20**: 3–7.

Well illustrated.

Ochs, G. 1967. Zur Kenntnis der europäischen *Gyrinus*-Arten. *Entomologische Blätter für Biologie und Systematik der Käfer* **63**: 174–86; (Corrections) **64**: 64.

Suborder Myxophaga

Only one (rare) British species, *Microsporus acaroides* (formerly known as *Sphaerius acaroides*) in the family Microsporidae (= Sphaeriidae), identifiable using the standard works.

Suborder Polyphaga

Superfamily Hydrophiloidea (including Histeroidea)

Although often referred to two distinct superfamilies (Hydrophiloidea and Histeroidea), members of this group have a number of significant morphological and biological characteristics in common. All larvae are predacious but only consume fluids from their prey. The favoured food is generally the larvae of other insects, especially Diptera. Many histerid larvae feed more or less exclusively on fly larvae, although some are predators of other insects, including the larvae or eggs of wood-boring beetles. Adult Histeridae have feeding habits largely similar to those of the larvae, whereas adult Hydrophilidae, unlike their larvae, appear to be exclusively saprophagous or phytophagous. All Histeridae are terrestrial, many being associated with carrion, dung and decaying vegetable matter; others are found in decaying wood, under the bark of dead or dying trees, in birds' nests, in association with ants and in leaf litter. Many Hydrophilidae are either strictly aquatic or semi-aquatic. Many exhibit morphological and behavioural modifications for an aquatic life, including plastron respiration and, in some larvae, gills. The subfamily Sphaeridiinae contains mostly terrestrial species, many associated with dung and decaying vegetable matter.

Family Hydrophilidae (including Georissidae, Hydrochidae, etc.) (see also water beetles)

Most aquatic and semi-aquatic species may be satisfactorily identified using Friday's (1988) key although the fuller accounts available in recent continental works such as that of Hansen (1987) may provide useful confirmation. For the difficult species of *Helophorus* the works of Angus (1992, etc.) are also extremely useful. Hansen's (1987) work is also recommended for the terrestrial Sphaeridiinae.

Allen, A.A. 1969. *Cercyon laminatus* Sharp (Col., Hydrophilidae) new to Britain; with corrections to our list of species, and further notes. *Entomologist's Record and Journal of Variation* **81**: 211–16.

Distinguishing characters; key to '*tristis*' group; no figures.

Angus, R.B. 1973. The habits, life histories and immature stages of *Helophorus* F. (Coleoptera: Hydrophilidae). *Transactions of the Royal Entomological Society of London* **125**(1): 1–26.

Includes key to larvae of British species.

Angus, R.B. 1977. A re-evaluation of the taxonomy and distribution of some European species of *Hydrochus* Leach (Coleoptera: Hydrophilidae). *Entomologist's Monthly Magazine* **112**: 177–201.

Illustrated. Key includes the six British recorded species.

Angus, R.B. 1978. The British species of *Helophorus. Balfour-Browne Club Newsletter* **11**: 2–15.

Well illustrated; essential for identification of British members of this genus.

Angus, R.B. 1992. Insecta: Coleoptera: Hydrophilidae: Helophorinae. *Süsswasserfauna Mitteleuropas* **20**: 1–144.

The most thorough account available of the Central European species of *Helophorus*; profusely illustrated keys to adults and larvae, and figures of many egg cocoons.

Balfour-Browne, F. 1958. *British water beetles.* Ray Society, London. **3**.

Chiesa, A. 1959. *Hydrophilidae Europae (Coleoptera, Palpicornia).* Arnaldo Forni, Bologna.

Keys to genera and species: in Italian.

Foster, G.N. 1983. *Laccobius simulator* d'Orchymont (Coleoptera: Hydrophilidae) confirmed as British. *Entomologist's Gazette* **34**: 265–6.

Distinguishing characters; male genitalia figured.

Foster, G.N. 1984. Notes on *Enochrus* subgenus *Methydrus* Rey (Coleoptera: Hydrophilidae), including a species new to Britain. *Entomologist's Gazette* **35**: 25–9.

Illustrated tabular key to the three British species.

Gentili, E. & Chiesa, A. 1975. Revisione dei *Laccobius* paleartici (Coleoptera, Hydrophilidae. *Memorie della Società Entomologica* **54**: 5–187.

Figures of male genitalia and distribution maps for most species; in Italian, English key pp. 179–84. Key reproduced in *Balfour-Browne Club Newsletter* 4: 8–10 (1977). Addenda by Gentili (1979) in *Bolletino della Società Entomologica Italiana* 111: 43–50.

Hansen, M. 1982. Revisional notes on some European *Helochares* Muls. (Coleoptera, Hydrophilidae. *Entomologica Scandinavica* **13**: 201–11.

English summary with key to the three British species in *Balfour-Browne Club Newsletter* **25**: 1–3.

Hansen, M. 1987. The Hydrophiloidea (Coleoptera) of Fennoscandia and Denmark. *Fauna Entomologica Scandinavica* **18**: 1–254.

A well-illustrated guide, covering almost all British species; includes Hydraenidae (see below).

Hebauer, F. 1981. *Enochrus (Methydrus) isotae* sp.n. – eine neue Hydrophiliden-Art aus Jugoslawien. *Entomologische Blätter für Biologie und Systematik der Käfer* **77**: 137–9.

In German; summary in English; with figures for separation of the three British species of the *E. coarctatus* group in *Balfour-Browne Club Newsletter* **24**: 1–22 (1982).

Henegouwen, A. van Berge 1986. Revision of the European species of *Anacaena* Thomson (Coleoptera: Hydrophilidae). *Entomologica Scandinavica* **17**: 393–407.

Well illustrated key to the five European species.

Henegouwen, A. van Berge 1989. *Sphaeridium marginatum* reinstated as a species distinct from *S. bipustulatum* (Coleoptera: Hydrophilidae). *Entomologische Berichten* **49**: 168–70.

Figures and key to separate all five Palaearctic species of *Sphaeridium*.

Johnson, C. 1967. *Cryptopleurum subtile* Sharp (Col., Hydrophilidae): an expected addition to the British list. *The Entomologist* **100**: 172–3.

Distinguishing characters; no figures.

Lohse, G.A. & Vogt, H. 1971. Hydraenidae, Spercheidae, Hydrophilidae. *Die Käfer Mitteleuropas.* Vol. 3, pp. 95–156.

In German.

Vogt, H. 1968. *Cercyon*-Studien. *Entomologische Blätter für Biologie und Systematik der Käfer* **64**: 172–91.

This key in German to the Central European species of *Cercyon* is translated into English in *Balfour-Browne Club Newsletter* **7**: 4–12 (1978).

Families Histeridae and Sphaeritidae

Members of these two families, sometimes referred to a distinct superfamily (Histeroidea) may be identified satisfactorily with Halstead's (1963) handbook, except for species later recognised as British (see below).

Allen, A.A. 1968. *Plegaderus vulneratus* Panz., a histerid beetle new to Britain. *Entomologist's Monthly Magazine* **104**: 110–12.

Distinguishing characters of the two British species of *Plegaderus* tabled and illustrated.

Halstead, D.G.H. 1963. Coleoptera Histeroidea. Sphaeritidae and Histeridae. *Handbooks for the Identification of British Insects* **4**(10): 16pp.

Hinton, H.E. 1945. The Histeridae associated with stored products. *Bulletin of Entomological Research* **35**: 309–40.

Keys to adults and larvae.

Nash, D.R. 1982. *Epierus comptus* (Erichson) (Col., Histeridae) new to Britain. *Entomologist's Record and Journal of Variation* **94**: 165–7.

    Description of *E. comptus*, with key to European species of the tribe Tribalini.

Vienna, P. 1980. Coleoptera Histeridae. *Fauna d'Italia* **16**: 386pp.

    Keys, well illustrated by habitus and other figures, to the 160 or so Italian species. In Italian.

Witzgall, K. 1971. Histeridae, Sphaeritidae. *Die Käfer Mitteleuropas.* Vol. 3.

    In German.

Superfamily Staphylinoidea

A group of very varied general biology, but usually with adults and larvae occupying the same habitats, and all life stages somewhat hygrophilous (moisture-demanding). Few staphylinoid beetles are strictly phytophagous (some omaliine Staphylinidae feed exclusively on flower buds, flowers or pollen), but most other possible foodstuffs are exploited extensively by members of this superfamily. Adult Hydraenidae are aquatic or semi-aquatic, utilising plastron respiration, and their adaptations resemble those of aquatic Hydrophilidae. Their larvae, on the other hand, have few if any aquatic adaptations and are semi-aquatic or riparian in habit. Both life stages appear to feed more or less exclusively on algae. The mostly minute Ptiliidae, or featherwing beetles, inhabit a wide range of terrestrial habitats, including dung, leaf litter, fungi and decaying wood, where their major food source appears to be fungal spores and hyphae. Leiodidae also occupy a wide range of habitats, including leaf litter and nests, where species of Catopinae in particular are general scavengers. Most species, however, are strictly mycophagous and associated with the fruiting bodies of various fungi including hypogeal (underground) forms, such as truffles. Others feed on slime-mould fruiting bodies. The remaining staphylinoid groups (Scydmaenidae, Silphidae, Pselaphidae and Staphylinidae) are mostly predators, with some exceptions. A few Silphidae are phytophagous, and the sexton or burying beetles (*Nicrophorus*) feed at least in part on carcasses. The Staphylinidae includes many saprophagous species (e.g. many Oxytelinae), algophilous and mycophagous species, as well as slime-mould feeders and parasitoids of dipterous pupae (*Aleochara*), and are found in virtually all types of habitat.

Family Hydraenidae (see also entries under Hydrophiloidea and water beetles)

Carr, R. 1984. *Limnebius crinifer* Rey new to Britain, with a revised key to the British *Limnebius* species (Coleoptera: Hydraenidae). *Entomologist's Gazette* **35**: 99–102.

    Good illustrated key to the five British species.

Foster, G.N. 1979. *Hydraena* aedeagophores. *Balfour-Browne Club Newsletter* **13**: 2–4.
Figures genitalia of all British species of *Hydraena*.

Family Ptiliidae

The standard key works of Fowler and Joy are now particularly out-of-date and now scarcely serviceable for most of the small to extremely tiny feather-wing beetles of the family Ptiliidae. Unfortunately, no modern key work in English is available that covers all or most of the British species. The keys provided by Besuchet & Sundt (1971), supplemented by the brief comments provided by Lohse & Lucht (1989: 118–20), will suffice for the identification of many species, although these keys do not include a few species now known to occur in the British Isles. In truth, identification of many ptiliids, especially in the genus *Acrotrichis*, remains a rather specialist matter, but those wishing to attempt this should begin with the *Käfer Mitteleuropas* keys, and supplement these by reference to the papers by C. Johnson listed below.

Besuchet, C. & Sundt, E. 1971. Familie: Ptiliidae. *Die Käfer Mitteleuropas* Vol. 3: 311–42.
In German.

Easton, A.M. & Johnson, C. 1966. *Acrotrichis rugulosa* Rossk. (Col., Ptiliidae) an addition to the British list. *Entomologist's Monthly Magazine* **102**: 239–40.
Figures of spermathecae.

Johnson, C. 1966. Two species of *Acrotrichis* new to Britain (Col., Ptiliidae). *The Entomologist* **99**: 152–4.
Figures of male genitalia and of spermathecae.

Johnson, C. 1966. Taxonomic notes on British Coleoptera No. 5. Notes on the genus *Acrotrichis* Motsch. (Ptiliidae). *The Entomologist* **99**: 196–9.
Useful figures and key to separate two *Acrotrichis* species.

Johnson, C. 1968. Six species of Coleoptera new to the British list. *The Entomologist* **101**: 28–34.
Key to species of *Ptiliolum spencei* group, with spermathecae illustrated.

Johnson, C. 1968. Two new British species of Coleoptera from Yorkshire. *The Entomologist* **101**: 64–6.
Distinguishing characters of an additional British species of *Ptiliolum*.

Johnson, C. 1975. Five species of Ptiliidae (Coleoptera) new to Britain, and corrections to the British list of the family. *Entomologist's Gazette* **26**: 211–23.
Includes short keys to some species of *Ptenidium* and *Ptilium* and descriptions of new *Ptinella* species.

Johnson, C. 1975. Nine species of Coleoptera new to Britain. *Entomologist's Monthly Magazine* **111**: 177–84.
Distinguishing characters of two additional British species of Ptiliidae.

Johnson, C. 1977. A third immigrant species of *Ptinella* Motschulsky (Col., Ptiliidae) new to the British fauna. *Entomologist's Gazette* **28**: 43–4.

> Distinguishing characters and figures of spermathecae of an additional British species of *Ptinella*.

Johnson, C. 1987. A revised check list of British *Acrotrichis* Motschulsky (Coleoptera: Ptiliidae). *Entomologist's Gazette* **38**: 229–42.

> Lists the 25 known British species with full synonymy, and references to key works. Supersedes a previous list by Johnson published in 1967.

Johnson, C. 1987. Additions and corrections to the British list of Ptiliidae (Coleoptera). *Entomologist's Gazette* **28**: 117–22.

> Good keys and figures to distinguish three additional British species from their close relatives.

Sundt, E. 1958. Revision of the Fenno-Scandian species of the genus *Acrotrichis* Motsch., 1848. *Norsk Entomologisk Tidsskrift* **10**: 241–77.

> Now somewhat out-of-date, but the figures of spermathecae and male genitalia are still useful.

Family Leiodidae

Most identifications in this 'difficult' family may now be tackled with some confidence using Cooter's excellent keys to Leiodinae and the various works on Catopinae listed below, if these are used with care and male genitalia are examined. For the Coloninae, however, some uncertainties may remain even when the most up-to-date keys (e.g. Freude *et al.*, 1971) are used.

Cooter, J. 1978. The British species of *Agathidium* Panzer (Col., Leiodidae). *Entomologist's Monthly Magazine* **113**: 125–35.

> Well illustrated; good practical key.

Cooter, J. 1996. Annotated keys to the British Leiodinae (Col., Leiodidae). *Entomologist's Monthly Magazine* **132**: 205–32, 233–72.

> Excellent, profusely illustrated keys to all British species.

Daffner, H. 1983. Revision der paläarktischen Arten der Tribus Leiodini Leach (Col., Leiodidae). *Folia Entomologica Hungarica* **44**: 9–163.

> Keys to genera and species; good figures of male genitalia, etc.; in German.

Freude, H., Harde, K.W. & Lohse, G.A. 1971. Silphidae, Leptinidae, Catopidae, Colonidae, Liodidae, Clambidae, Scydmaenidae, Orthoperidae (=Corylophidae), Sphaeriidae, Ptiliidae, Scaphidiidae. *Die Käfer Mitteleuropas.* Vol. 3: 190–347.

> Many figures; in German.

Kevan, D.K. 1945a. The aedeagi of the British species of the genus *Catops* Pk. (Col., Cholevidae). *Entomologist's Monthly Magazine* **81**: 69–72.

Kevan, D.K. 1945b. The aedeagi of the British species of the genera *Ptomophagus* Ill., *Nemadus* Th., *Nargus* Th. and *Bathyscia* Sch. *Entomologist's Monthly Magazine* **81**: 121–125.

Kevan, D.K. 1946. The sexual characters of the British species of the genus *Choleva* Lat. including *C. cisteloides* Frohl. new to the British list (Col., Cholevidae). *Entomologist's Monthly Magazine* **82**: 122–30.

Kevan, D.K. 1946. *Catops nigriclavis* Gerh. (Col., Cholevidae) new to the British list. *Entomologist's Monthly Magazine* 82: 155–7.
Good figures comparing sexual and other characters of *C. nigriclavis* and *C. nigrita*.

Kevan, D.K. 1947. A revision of the British species of the genus *Colon* Hbst. (Coleoptera, Cholevidae). *Entomologist's Monthly Magazine* 83: 249–67.

Zwick, P. 1968. Zwei neue Catopiden-Gattungen aus Europa (Auflösung der *nigrita*-Gruppe in der Gattung *Catops*). *Entomologische Blätter für Biologie und Systematik der Käfer* 64: 1–16.
In German.

Family Scydmaenidae

The genera and more distinctive species of this family are identifiable using the standard works, but the more modern keys of Franz & Besuchet (1971) in conjunction with Allen (1969) are recommended for sure identification, especially of *Eutheia, Neuraphes* and *Scydmoraphes* species.

Allen, A.A. 1969. Notes on some British Scydmaenidae (Col.), with corrections to the list. *Entomologist's Record and Journal of Variation* 81: 239–46.
Distinguishing characters of various species; no figures; the two British species of *Microscydmus* not distinguished.

Brown, C. & Crowson, R.A. 1980. Observations on scydmaenid (Col.) larvae with a tentative key to the main British genera. *Entomologist's Monthly Magazine* 115: 49–59.
Good illustrations; key to seven of eight British genera.

Franz, H. & Besuchet, C. 1971. Scydmaenidae. *Die Käfer Mitteleuropas* Vol. 3: 271–303.
Covers all British species except *Euconnus duboisi*. Generally reliable, but note that *Stenichnus* species 14 and 15 are not now regarded as distinct from *S. poweri*.

Last, H.R. 1945. *Euconnus (Napochus) murielae* sp. n., a new British scydmaenid (Col.). *Entomologist's Monthly Magazine* 81: 263–264.
Full description of *E. duboisi* Méquignon (as *E. murielae*).

Family Silphidae

Most species may be reliably identified using the standard British works, but Freude's (1971) key may be referred to if in doubt.

Freude, H. 1971. Silphidae. *Die Käfer Mitteleuropas*. Vol. 3: 190–201.
In German.

Nash, D.R. 1975. *Silpha carinata* Herbst – a remarkable rediscovery in the British Coleoptera. *Entomologist's Record and Journal of Variation* 87: 285–8.
Distinguishing characters from *S. tristis* tabulated; no figures.

Family Staphylinidae (including Scaphidiidae)

With around 1000 British species, this family poses many identification problems to the beginner. Many members of the larger genera such as *Philonthus, Quedius, Stenus* and *Atheta* in particular may prove difficult, even for the more experienced. Other genera such as *Bledius, Carpelimus, Tachyporus* and various Aleocharinae may also present problems not soluble using the standard works. Tottenham's (1954) key remains serviceable for some groups, including most Steninae, but more modern European keys are recommended in most cases. The most useful are the Central European keys (in German) of Lohse *et al.* (1964, 1974), supplemented by Lohse & Lucht (1989) (see *Die Käfer Mitteleuropas* section, above). For figures of male genitalia of Staphylininae the works of Coiffait (1974, 1978) are recommended, and for male genitalia and spermathecae of Aleocharinae figures in various papers by Strand & Vik, all these included also by Palm (1968, 1970, 1972). However, even with these works at hand, reference to others may prove necessary (see the substantial list below).

Allen, A.A. 1969. Notes on some British Staphylinidae (Col.) 1. The genus *Scopaeus* Er., with the addition of *S. laevigatus* Gyll. to our list. *Entomologist's Monthly Magazine* 104: 198–207.

Allen, A.A. 1970. Notes on various little-known, doubtful, or misidentified British Staphylinidae (Col.) *Entomologist's Monthly Magazine* 105: 193–6.
   Taxonomic and nomenclatural notes on 14 species.

Allen, A.A. 1971. Notes on some British Staphylinidae. 2 – three additions to our species of *Philonthus* Curt. *Entomologist's Monthly Magazine* 106: 157–61.
   Distinguishing characters; male genitalia and elytral colour patterns illustrated.

Allen, A.A. 1977. Notes on some British Staphylinidae, 4: *Stenus butrintensis* Smet. new to Britain with brief remarks on a few others of the genus. *Entomologist's Monthly Magazine* 113: 63–9.
   Illustrated key to separate *S. pallitarsis* and *S. butrintensis*.

Allen, A.A. 1994. Notes on some British Staphylinidae (Col.) – 5. the genus *Atheta* Thoms.: three additions to the fauna, a reinstatement, and two new synonymies. *Entomologist's Monthly Magazine* 130: 165–71.
   Full description of *A. verulamii*, with figures to distinguish from *A. puberula*.

Allen, A.A. & Eccles, T.M. 1989. *Hydrosmectina delicatissima* Bernhauer (Col., Staphylinidae), new to Britain. *Entomologist's Monthly Magazine* 124: 215–20.
   Distinguishing characters from *H. subtilissima*; well illustrated, including excellent habitus figure.

Ashe, G.H. 1955. *Atheta (Microdota) alpina* Benick (Col., Staphylinidae) a species new to Britain in Devonshire. *Entomologist's Monthly Magazine* 91: 28.
   Distinguishing features of *Atheta liliputana* (as *A. alpina*), with figures.

Assing, V. 1997. A revision of the Western Palaearctic species of *Myrmecopora* Saulcy, 1864, sensu lato and *Eccoptoglossa* Luze, 1904. *Beiträge zur Entomologie* 47: 69–151.
   In German, with English abstract.

Booth, R.G. 1984. A provisional key to the British species of *Tachyporus* (Col. Staph.) based on elytral chaetotaxy. *Circaea* 2: 15–19.

Booth, R.G. 1988. The identity of *Tachyporus chrysomelinus* (Linnaeus) and the separation of *T. dispar* (Paykull) (Coleoptera; Staphylinidae). *The Entomologist* 107: 127–33.
   Distinguishing characters, with male genitalia and elytral chaetotaxy illustrated.

Bordoni, A. 1980. Studi sui Paederinae III. I *Medon* Steph. paleartici con descrizione di nuove specie mediterranee (Col. Staphylinidae). *Bolletino del Laboratorio di Entomologia Agraria 'Filippo Silvestri'* 37: 73–125.
   In Italian.

Bordoni, A. 1982. Coleoptera Staphylinidae, Generalità – Xantholininae. *Fauna d'Italia* 19: 434pp.
   Subfamily keys to adults and larvae; full and well illustrated treatment of Xantholinini with keys to all Italian species: in Italian.

Brunne, G. 1976. Die Artengruppe des *Philonthus sordidus* Gravenhorst. *Entomologische Blätter für Biologie und Systematik der Käfer* 72: 65–89.
   Provides figures of male genitalia, including *P. pseudoparcus* Brunne; in German.

Coiffait, H. 1972. Coléoptères Staphylinidae de la région paléarctique occidentale. I. Généralités, Sous familles Xantholininae et Leptotyphlinae. *Nouvelle Revue d'Entomologie* (suppl.) 2: 1–651.
   Extensively illustrated account, with keys, covering all British species; in French.

Coiffait, H. 1974. Coléoptères Staphylinidae de la région paléarctique occidentale II. Sous famille Staphylininae Tribus Philonthini et Staphylinini. *Nouvelle Revue d'Entomologie* (suppl.) 4(4): 1–593.
   Extensively illustrated account, with keys, covering all British species; in French.

Coiffait, H. 1978. Coléoptères Staphylinides de la région paléarctique occidentale. III. Sous famille Staphylininae, Tribu Quediini; sous famille Paederinae, Tribu Pinophilini. *Nouvelle Revue d'Entomologie* (suppl.) 8(4): 1–364.
   Extensively illustrated account, with keys, covering all British species; in French.

Coiffait, H. 1984. Coléoptères staphylinides de la région paléarctique occidentale. Sous famille Paederinae, Tribu Paederini 2; Sous famille Euaesthetinae. *Nouvelle Revue d'Entomologie* (suppl.) 8(4): 1–424.
   Extensively illustrated account, with keys, covering all British species; in French.

Cuccodoro, G. & Löbl, I. 1997. Revision of the Palaearctic rove beetles of the genus *Megarthrus* Curtis (Coleoptera: Staphylinidae: Proteininae). *Journal of Natural History* 31: 1347–415.
   Full treatment of all British species, for which some name changes are proposed; well illustrated.

Drane, A.B. 1994. A belated note on *Chloecharis debilicornis* (Wollaston) (Coleoptera: Staphylinidae) new to Britain. *Coleopterist* 3 (1): 2–3.
   See also Duff (1995).

Duff, A.G. 1995. On the genus of *Hypomedon debilicornis* (Wollaston) (Coleoptera: Staphylinidae). *Coleopterist* 4(1): 6.

Easton, A.M. 1970. *Atheta boletophila* Thoms. (Col., Staphylinidae) new to Britain. *Entomologist's Monthly Magazine* 105: 197–8.
Distinguishing features, with sexual characters illustrated.

Easton, A.M. 1971. A new British *Atheta* (Col., Staphylinidae) in a new subgenus. *Entomologist's Monthly Magazine* 107: 24–6.
Full description; well illustrated.

Hammond, P.M. 1971. Notes on British Staphylinidae 1. – The status of *Olophrum nicholsoni* Donisthorpe with notes on the other British species of *Olophrum* (Col., Staphylinidae). *Entomologist's Monthly Magazine* 106: 165–70.
Distinguishing characters of *O. fuscum* and *O. piceum* illustrated.

Hammond, P.M. 1971. Notes on British Staphylinidae 2. On the British species of *Playstethus* Mannerheim, with one species new to Britain. *Entomologist's Monthly Magazine* 107: 93–111.
Illustrated keys, maps, bionomic data.

Hammond, P.M. 1973. Notes on British Staphylinidae 3. The British species of *Sepedophilus* Gistel (*Conosomus* auct.) *Entomologist's Monthly Magazine* 108: 130–65.
Illustrated keys, maps, bionomic data.

Hammond, P.M. 1981. *Aloconota (Aloconota) subgrandis* (Brundin) (Coleoptera: Staphylinidae) new to Britain. *Entomologist's Gazette* 32: 120–2.
Distinguishing characters illustrated.

Hammond, P.M. 1982. On the British species of *Phacophallus* Coiffait (Col., Staphylinidae). *Entomologist's Monthly Magazine* 118: 231.
Distinguishing characters of the two British species; no figures.

Hammond, P.M. & Bacchus, M.E. 1972. *Atheta* (s.str.) *strandiella* Brundin (Col., Staphylinidae) new to the British Isles, with notes on other British species of the subgenus. *Entomologist's Monthly Magazine* 107: 153–7.
Distinguishing characters illustrated.

Johnson, C. 1967. *Hypocyphtus nitidus* Palm (Col., Staphylinidae) in Oxfordshire: an addition to the British list. *The Entomologist* 100: 193–5.
Key and figures to distinguish from related *Cypha* species.

Johnson, C. 1968. Six species of Coleoptera new to the British list. *The Entomologist* 101: 28–34.
Characters to distinguish *Cypha hanseni, Atheta excelsa* and *Ilyobates subopacus* from related species; with figures.

Johnson, C. 1968. Two new British species of Coleoptera from Yorkshire. *The Entomologist* 101: 64–6.
Distinguishing characters, with figures, of *Atheta* species of the *crassicornis* group.

Johnson, C. 1968. *Oxypoda nigricornis* Motschulsky (Col., Staphylinidae) new to Britain. *The Entomologist* 101: 71–2.
Figures sexual characters to distinguish from close relatives.

Kasule, F.K. 1966. The subfamilies of the larvae of Staphylinidae (Coleoptera) with keys to the larvae of British genera of Steninae and Proteininae. *Transactions of the Royal Entomological Society of London* 118: 261–83.

Kasule, F.K. 1968. The larval characters of some subfamilies of British Staphylinidae with keys to the known genera. *Transactions of the Royal Entomological Society of London* **120**: 115–38.

Kasule, F.K. 1970. The larvae of Paederinae and Staphylininae (Coleoptera: Staphylinidae) with keys to the known British genera. *Transactions of the Royal Entomological Society of London* **122**: 49–80.

Kevan, D.K. 1962. *Cypha (Hypocyphtus) imitator* Luze (Col., Staphylinidae), Tachyporinae) new to the British list. *Entomologist's Monthly Magazine* **98**: 51–3.
Distinguishing characters; male genitalia figured.

Kevan, D.K. 1963. *Gabrius exiguus* Nordmann and *Stenus simillimus* Benick (Col., Staphylinidae) new to the British list. *Entomologist's Monthly Magazine* **99**: 80–1.
Distinguishing characters illustrated.

Kevan, D.K. 1965. *Atheta (Acrotona) amplicollis* (Mulsant & Rey) (Col., Staphylinidae) new to the British list. *Entomologist's Monthly Magazine* **101**: 122–4.
Distinguishing characters from *A. fungi*; illustrated.

Kevan, D.K. 1969. Observations on some British species of *Myllaena* Erichson (Col., Staphylinidae). *Entomologist's Monthly Magazine* **104**: 183–7.
Genital and other characters of four little-known species illustrated.

Kevan, D.K. & Allen, A.A. 1962. Notes on some British species of *Stenus* Latreille (Col., Staphylinidae), with additions and amendments to the British list. *Entomologist's Monthly Magazine* **97**: 211–17.
Distinguishing characters of 13 species; male genitalia figured.

Last, H.R. 1948. *Neobisinius cerrutii* Gridelli and var. *rubripennis* Gridelli (Col., Staphylinidae), additions to the British list of Coleoptera. *Entomologist's Monthly Magazine* **84**: 148–50.
Key to the five North European species of *Neobisnius*, with male genitalia illustrated.

Last, H.R. 1950. *Xantholinus meridionalis* (Nordmann) (Col., Staphylinidae) a species new to the British list. *Entomologist's Monthly Magazine* **86**: 138–40.
Distinguishing characters from *X. tricolor*; male genitalia illustrated.

Last, H.R. 1951. *Leptusa norvegica* Strand (Col., Staphylinidae) in Great Britain. *Entomologist's Monthly Magazine* **87**: 182–3.
Distinguishing characters from *L. fumida*; male genitalia illustrated.

Last, H.R. 1952. Taxonomic notes on *Quedius molochinus* Grav. and *Q. pallipes* Lucas (Col., Staphylinidae). *Entomologist's Monthly Magazine* **88**: 148–50.
Distinguishing characters of these two species, with male genitalia illustrated.

Last, H.R. 1955. Two new British species of Staphylinidae (Col.). *Entomologist's Monthly Magazine* **91**: 107.
Distinguishing characters of *Atheta caucasica* and of *A. boreella*; male genitalia and spermathecae of these and related species illustrated.

Last, H.R. 1958. *Atheta* (sub-genus *Microdota*) *spatuloides* G. Benick, a species of Staphylinidae (Col.) confirmed on the British list. *Entomologist's Monthly Magazine* **94**: 143–4.
Distinguishing characters illustrated.

Last, H.R. 1963. Notes on *Quedius molochinus* Gravenhorst (Col., Staphylinidae) with the addition of two species new to the British list. *Entomologist's Monthly Magazine* 99: 43–5.

Distinguishing features; male genitalia figured.

Last, H.R. 1969. *Atheta* (s. str.) *ebenina* Mulsant & Rey (Col., Staphylinidae) new to the British list. *Entomologist's Monthly Magazine* 104: 285–6.

Distinguishing features from *A. euryptera*, with spermathecae and male secondary sexual characters illustrated.

Last, H.R. 1974. *Philonthus mannerheimi* Fauvel (Col., Staphylinidae) and related species. *Entomologist's Monthly Magazine* 109: 85–8.

Löbl, I. 1970. Revision der paläarktischen Arten der Gattungen *Scaphisoma* Leach und *Catyoscapha* Ganglbauer der Tribus Scaphisomini (Coleoptera; Scaphidiidae). *Revue Suisse de Zoologie* 77: 727–99.

In German.

Lohse, G.A. 1985. *Diglotta*-Studien. *Entomologische Blätter für Biologie und Systematik der Käfer* 81: 179–82.

Distinguishing characters of the two much-confused British species, both of them polymorphic. Note that Continental usage of species names is transposed compared with most British usage. In German.

Lohse, G.A. *et al.* 1964, 1974. Staphylinidae I (Micropeplinae bis Tachyporinae); II (Hypocyphtinae und Aleocharinae). *Die Käfer Mitteleuropas.* Vol. 4: 1–264; Vol. 5: 1–304.

In German.

Lott, D.A. 1989. *Hadrognathus longipalpus* (Mulsant & Rey) (Col.: Staphylinidae) new to the British Isles. *Entomologist's Gazette* 40: 221–2.

Distinguishing characters; with habitus figure.

Lott, D. 1993. The British species of the *Thinobius longipennis* (Heer) group (Coleoptera: Staphylinidae). *Entomologist's Gazette* 44: 285–7.

Key to the three British species; male genitalia illustrated.

Lott, D. 1995. *Trichiusa immigrata* Lohse (Coleoptera: Staphylinidae) in Leicestershire. *Coleopterist* 4(1): 14.

MacKechnie-Jarvis, C. 1968. *Lathrobium fennicum* Renk. (Col., Staphylinidae): a species new to Britain. *Entomologist's Monthly Magazine* 104: 123–4.

Male genitalia of *L. fennicum* and *L. quadratum* illustrated.

Muona, J. 1990. The Fennoscandian and Danish species of the genus *Amischa* Thomson (Coleoptera, Staphylinidae). *Entomologisk Tidskrift* 111: 17–24.

Well-illustrated keys, including all British species.

Muona, J. 1991. The North European and British species of the genus *Meotica* Mulsant & Rey. *Deutsche Entomologische Zeitschrift* 38: 225–46.

Well-illustrated key that includes all five British species.

Owen, J.A. 1994. Features to distinguish males as well as females of *Lathrobium fennicum* Renkonen and *L. quadratum* (Paykull) (Col., Staphylinidae). *Entomologist's Monthly Magazine* 130: 67–70.

Illustrated.

Palm, T. 1948–72. Kortvingar, fam. Staphylinidae. *Svensk Insektfauna* 9: 1–985.

Published in nine parts, this standard work remains useful as a reference for its excellent figures, especially those of male genitalia, spermathecae, and secondary sexual characteristics of Aleocharinae (many from the papers of Strand & Vik); in Swedish.

Puthz, V. 1971. Kritische Faunistik der bisher aus Mitteleuropa bekannten *Stenus*-Arten nebst systematischen Bemerkungen und Neubeschreibungen (Coleoptera, Staphylinidae). *Entomologische Blätter für Biologie und Systematik der Käfer* 67: 74–121.
In German.

Reid, C. 1985. External characters for the separation of *Stenus aceris* Stephens and *S. impressus* Germar (Col., Staphylinidae). *Entomologist's Monthly Magazine* 121: 71–2.
Distinguishing characters, with figures, for these two much-confused species.

Steel, W.O. 1948. The British species of *Staphylinus* subgenus *Ocypus* Steph (Col., Staphylinidae). *Entomologist's Monthly Magazine* 84: 271–5.
Good genitalia figures; nomenclature now out of date.

Steel, W.O. 1956. *Bohemiellina paradoxa* Machulka, a staphylinid (Col.) new to Britain. *Entomologist's Monthly Magazine* 91: 296–7.
Distinguishing characters; whole-body figure.

Steel, W.O. 1957. Notes on the Omaliinae (Col., Staphylinidae). *Entomologist's Monthly Magazine* 98: 157–64.
Well-illustrated key to *Acrolocha*.

Steel, W.O. 1961. *Tachinus flavolimbatus* Pandellé, a staphylinid (Coleoptera) new to Britain. *The Entomologist* 94: 77–8.
Distinguishing characters, with male and female secondary sexual characters illustrated.

Steel, W.O. 1966. A revision of the Staphylinid subfamily Proteininae (Coleoptera) 1. *Transactions of the Royal Entomological Society of London* 118: 285–311.
Keys to larvae of the three British genera.

Steel, W.O. 1970. The larvae of the genera of Omaliinae (Coleoptera: Staphylinidae) with particular reference to the British fauna. *Transactions of the Royal Entomological Society of London* 122(1): 1–47.
Key to British genera, with excellent figures.

Strand, A. & Vik, A. 1964. Die Genitalorgane der nordischen Arten der Gattung *Atheta* Thoms. (Col., Staphylinidae). *Norsk Entomologisk Tidsskrift* 12: 327–35.
A set of excellent figures of male genitalia and spermathecae covering most British species of *Atheta* and allied genera; in German.

Tottenham, C.E. 1948. A revision of the British species of *Arphirus* Tottenham (subgenus of *Quedius* Stephens) (Col., Staphylinidae). *Entomologist's Monthly Magazine* 84: 241–58.
Still useful, but the treatment of the 'boops-group' is now out of date.

Tottenham, C.E. 1954. Staphylinidae (Piestinae to Euaesthetinae). *Handbooks for the Identification of British Insects* 4(8a): 79pp.
Covers Piestinae, Micropeplinae, Pseudopsinae, Phloeocharinae, Proteininae, Omaliinae, Oxytelinae, Oxyporinae, Steninae and Euaesthetinae. Inevitably out of date in part, but still useful for many genera. However, the keys to *Omalium*, *Carpelinus* and *Bledius* should be avoided if alternatives (e.g. Lohse, 1964, etc.) are available.

Ullrich, W.G. 1975. *Monographie der Gattung Tachinus Gravenhorst (Coleoptera:*

Staphylinidae), mit Bermerkungen zur Phylogenie und Verbreitung der Arten. Privately published, Kiel.
In German.

Welch, R.C. 1969. Two species of *Gyrophaena* (Col., Staphylinidae) new to Britain. *Entomologist's Monthly Magazine* **104**: 180–2.
Good figures of male genitalia and secondary sexual features.

Welch, R.C. 1995. *Cypha tarsalis. Entomologist's Record and Journal of Variation* **107**: 185–7.

Welch, R.C. 1997. The British species of the genus *Aleochara* Gravenhorst (Staphylinidae). *Coleopterist* **6**(1): 1–45.
An up-to-date treatment, with good keys and excellent figures.

White, I. M. 1977. The larvae of some British species of *Gyrophaena* Mannerheim (Coleoptera: Staphylinidae) with notes on the taxonomy and biology of the genus. *Zoological Journal of the Linnean Society of London* **60**: 297–317.
Key to adults of 19 British species as appendix.

Williams, S.A. 1968. *Gyrophaena pseudonana* Strand (Col., Staphylinidae) in Cambridgeshire: an addition to the British list. *Entomologist's Monthly Magazine* **103**: 250.
Distinguishing characters.

Williams, S.A. 1968. Notes on the British species of *Ochthephilum* Mulsant & Rey (Col., Staphylinidae). *Entomologist's Monthly Magazine* **104**: 261–2.
Distinguishing characters of the two British species of *Ochthephilum.*

Williams, S.A. 1969. The British species of the genus *Amischa* Thomson (Col., Staphylinidae), including *A. soror* Kraatz, an addition to the list. *Entomologist's Monthly Magazine* **105**: 38–42.
A very useful paper, but note that the nomenclature of some species is now out of date, and that *A. simillima* is not now regarded as a distinct species (see Muona, 1990).

Williams, S.A. 1970. Notes on the genus *Oligota* Mannerheim (Col., Staphylinidae) and key to the British species. *Entomologist's Monthly Magazine* **106**: 54–62.

Williams, S.A. 1979. *Ocyusa nitidiventris* (Fagel) (Col., Staphylinidae) new to Britain. *Proceedings and Transactions of the British Entomological and Natural History Society* **12**: 46–8.
Key to three species of *Ocyusa*, with figures of male genitalia and spermathecae of *O. nitidiventris.*

Williams, S.A. 1980. *Gnypeta ripicola* (Kiesenw.) (Col., Staphylinidae) new to Britain. *Entomologist's Monthly Magazine* **116**: 37–9.
Key to the five British species of *Gnypeta.*

Wunderle, P. 1990. Revision der mitteleuropäischen Arten der Gattung *Ischnoglossa* Kraatz, 1856 (Coleoptera, Staphylinidae, Aleocharinae). *Entomologische Blätter für Biologie und Systematik der Käfer* **86**: 51–68.
Does not include one British species (see Wunderle, 1992): in German.

Wunderle, P. 1992. Eine neue Art der Gattung *Ischnoglossa* Kraatz 1856 aus Turkei (Coleoptera, Staphylinidae, Aleocharinae). *Entomologische Blätter für Biologie und Systematik der Käfer* **88**: 49–52.
Illustrated description of species later reported from Britain; in German.

Zanetti, A. 1987. Coleoptera Staphylinidae Omaliinae. *Fauna d'Italia* 25: 472pp.
    **Extensively illustrated account of the approximately 200 Italian species. In Italian, but
    with all keys in both English and Italian.**

Family Pselaphidae

Besuchet, C. 1974. Pselaphidae. *Die Käfer Mitteleuropas.* Vol. 5: 305–62.
    The most convenient identification work for *Euplectus* and *Plectophloeus*.
Bowestead, H. & Eccles, T.M. 1987. *Euplectus bonvouloiri rosae* Raffray (Col.,
    Pselaphidae) new to Britain. *Entomologist's Monthly Magazine* 123: 109–11.
    Table to distinguish from *E. karsteni*; illustrated with photographs.
Jeannel, R. 1950. Coléoptères Pselaphides. *Faune de France* 53: 1–421.
Johnson, C. & Eccles, T.M. 1983. *Plectophloeus erichsoni occidentalis* Besuchet
    (Coleoptera: Pselaphidae) new to Britain. *Entomologist's Gazette* 34: 267–9.
    Good habitus and male genitalia figures.
Pearce, E. J. 1957. Pselaphidae. *Handbooks for the Identification of British Insects*
    4(9): 32pp.
    **Now out of date in taxonomy and nomenclature, but still a good identification work for
    all except *Euplectus* and *Plectophloeus* species.**

Superfamily Scarabaeoidea

Four families of this well-demarcated superfamily are represented in the
British Isles. The long-lived larvae of Lucanidae (stag beetles) feed on
decayed wood; the mostly nocturnal adults appear to feed little, but may
consume nectar or other sweet fluids. Adults and larvae of Trogidae feed on
dry animal remains, including those found in the nests of predaceous birds.
Except for *Odontaeus* (which is apparently associated with hypogeal fungi,
like most exotic members of the family), British Geotrupidae (dor beetles)
provision larval burrows with dung or other decaying organic matter such as
fungi or dead leaves. Members of the family Scarabaeidae exhibit a more
varied feeding biology. The most typical scarab beetles (Scarabaeinae) are
dung feeders, as are most members of the best-represented subfamily in
Britain, the Aphodiinae. Larvae of Melolonthinae and Rutelinae (chafers) are
soil-dwellers, where they feed on plant roots and other organic matter, sometimes reaching pest proportions, damaging pastures and even golf-course
fairways. The adults of these subfamilies feed on leaves and, at times of
abundance, some species such as the cockchafer (*Melolontha*) are capable of
completely defoliating trees and shrubs. Larvae of Cetoniinae (rose chafers
etc.) feed in soil or decaying wood; the mostly diurnal adults may feed on
nectar or other sweet fluids.

Jessop's (1986) handbook provides adequate coverage of all species, except for one additional species of *Psammodius* only recently recognised as British.

Baraud, J. 1977. Coléoptères Scarabaeoidea. Faune de l'Europe occidentale, Belgique. France. Grande-Bretagne. Italie. Peninsule Iberique. *Nouvelle Revue d'Entomologie* 7(3) (suppl.): 1–352.

Britton, E.B. 1956. Scarabaeoidea (Lucanidae, Trogidae, Geotrupidae, Scarabaeidae). *Handbooks for the Identification of British Insects* 5(2): 29pp.

Emden, F.I. van 1941. Larvae of British beetles. II. A key to the British Lamellicornia larvae. *Entomologist's Monthly Magazine* 77: 117–27; 181–92.
Covers 21 genera and many of the British species of Scarabaeoidea.

Jessop, L. 1986. Dung beetles and chafers Coleoptera: Scarabaeoidea. (New edn.) *Handbooks for the Identification of British Insects* 5(11): 53pp.
Generic keys to Scarabaeoidea including larvae.

Klausnitzer, B. 1995. *Die Hirschkäfer: Lucanidae.* Die Neue-Brehm Bücherei, Magdeburg. 109pp.
In German.

Machatschke, J.W. 1969. Lamellicornia. *Die Käfer Mitteleuropas.* Vol. 8: 265–371.
In German.

Paulian, R. & Baraud, J. 1982. *Encyclopédie Entomologique. Faune des Coléoptères de France II. Lucanidae et Scarabaeoidea.* Lechevalier, Paris.
In French.

Superfamily Scirtoidea (=Eucinetoidea) (see also water beetles)

This group of primitive Polyphaga comprises three families (Scirtidae, Eucinetidae and Clambidae) of rather disparate form and habits. Adult Scirtidae, loosely articulated and often rather short-lived, are most readily found on vegetation near their larval habitats. Their highly characteristic, somewhat isopod-like larvae, with multiannulate antennae and filter-feeding mouthparts, are exclusively aquatic. Many Eucinetidae, including the only British representative of the family, are associated with slime-mould fruiting bodies in wooded areas. The Clambidae are small, somewhat globular (especially when curled into a ball in their usual resting position) and are found in various types of litter and decaying vegetable debris. Some are at least partly mycophagous as adult and larva, and others may be associated with slime-moulds.

The Scirtidae (= Helodidae), except for the genera *Cyphon* and *Elodes*, are adequately covered by the standard works. For Clambidae, Johnson's (1966) Handbook covers all but one (very recently introduced) species. Kevan's (1962) key to *Cyphon* species includes all but one British species (see below). For *Elodes* reference should be made to recent Continental works.

Endrödi-Younga, S. 1971. Clambidae. *Die Käfer Mitteleuropas.* Vol. 3: 266–70.

Gardner, A.E. 1969. *Eucinetus meridionalis* Lap. (Col., Eucinetidae), a family and species new to Britain. *Entomologist's Gazette* **20**: 59–63.
Key to the three European species of the genus, with figures of *E. meridionalis*.

Hannappel, U. & Paulus, H.F. 1994. Familie Scirtidae. *Die Käfer Mitteleuropas.* Larvae. Vol. 2: 74–87.

Johnson, C. 1966. Clambidae. *Handbooks for the Identification of British Insects* **4**(6a): 13pp.

Johnson, C. 1997. *Clambus simsoni* Blackburn (Col., Clambidae) new to Britain, with notes on its wider distribution. *Entomologist's Monthly Magazine* **133**: 161–4.
Distinguishing characters; male genitalia figured.

Kevan, D.K. 1962. The British species of the genus *Cyphon* Paykull (Col., Helodidae), including three new to the British list. *Entomologist's Monthly Magazine* **98**: 114–21.

Klausnitzer, B. 1971. Beiträge zur Insekten-Fauna der DDR: Coleoptera – Helodidae. *Beiträge zür Entomologie* **21**: 477–94.
Also covers UK species.

Klausnitzer, B. 1974. Zur Kenntnis der Gattung *Helodes* Latr. (Col., Helodidae). *Entomologische Nachrichten, Dresden* **18**: 73–8.
Key to European species of *Elodes*.

Klausnitzer, B. 1975. Zur Kenntnis der Larven der mitteleuropäischen Helodidae. *Deutsche Entomologische Zeitschrift* **22**: 61–5.

Lohse, G.A. 1979. Fossipedes [families Helodidae and Eucinetidae]. *Die Käfer Mitteleuropas.* Vol. 6: 25–264.

Nyholm, T. 1972. Die nordeuropäischen Arten der Gattung *Cyphon* Paykull (Coleoptera). Taxonomie, Biologie, Okologie und Verbreitung. *Entomologica Scandinavica* (suppl.) **3**: 1–100.

Skidmore, P. 1985. *Cyphon kongsbergensis* Munster (Col., Scirtidae) in Scotland. *Entomologist's Monthly Magazine* **121**: 249–52.
Distinguishing characters; well illustrated.

Superfamily Byrrhoidea (including Dryopoidea) (see also water beetles)

Both adult and larval Byrrhidae (pill beetles) are slow-moving ground surface or litter dwellers of phytophagous habits, feeding mostly on mosses and liverworts, but some species also feed on roots of herbaceous plants. In addition to their highly convex body form and thick cuticle, adults receive protection from predation by their ability to retract all appendages into grooves on the ventral surface. In the remaining families of Byrrhoidea, often classed separately as the Dryopoidea, aquatic or riparian habits are the rule; various special respiratory features are present in many of their larvae, and sometimes also their adults. In the Elmidae (riffle beetles) both larvae and adults are essentially aquatic in habits, feeding on algae or vegetable debris. In the Psephenidae, the short-lived adults are terrestrial, but the algophagous larvae (water pennies) are highly modified for an aquatic existence. In the Dryopidae, on the other hand, the adults are largely aquatic, whereas the larvae are basically terrestrial. Both adults and larvae of Limnichidae and Heteroceridae are riparian rather than truly aquatic, feeding on algae and decaying vegetable matter. Heteroceridae, especially the adults, are modified for tunnel-dwelling and burrow in soft mud beside fresh or brackish water.

The standard works are adequate for identifying most Byrrhidae, except for three species relatively recently recognised as British (see below). All of the British species of Elmidae (= Elminthidae) are covered by Friday's (1988) key (see above section on water beetles), and all but one (*Oulimnius major*) by Holland's (1972) excellent Handbook. The Heteroceridae are adequately covered by Clarke's (1973) Handbook. The species of Dryopidae remain difficult to identify, and reference to Jäch (1992) or other Continental works (see below) illustrating the genitalia of all British species is essential.

Berthélemy, C. 1979. Elmidae de la region paléarctique occidentale: systématique et repartition (Coleoptera Dryopidae). *Annales de Limnologie* 15: 1–102.
In French.

Bollow, H. 1938–41. Monographie der palaearktischen Dryopidae, mit Berücksichtigung der eventuell transgredierenden Arten (Col.) *Mitteilungen der Münchener Entomologischen Gesellschaft* 28: 147–87, 319–71; 29: 109–45; 30: 24–71; 31: 1–88.
Includes Psephenidae and Elminthidae (= Elmidae); in German.

Bonadona, P. 1975. Les *Byrrhus (sensu lato)* de France (Col., Byrrhidae). *Entomologiste* 31: 193–209.
In French.

Clarke, R.O.S. 1973. Heteroceridae. *Handbooks for the Identification of British Insects* 5 (2c): 15pp.

Drechsel, U. 1979. Heteroceridae. *Die Käfer Mitteleuropas.* Vol. 8: 296–304.
In German.

Emden, F.I. van 1958. Über die Larvenmerkmale einige deutscher Byrrhidengattungen. *Mitteilungen der Deutschen Entomologischen Gesellschaft* 17: 39–40.
Key to four genera; no figures.

Holland, D. G. 1972. A key to the larvae, pupae and adults of the British species of Elminthidae. *Scientific Publications of the Freshwater Biological Association* 26: 58pp.

Jäch, M. 1992. Dryopidae. *Die Käfer Mitteleuropas.* Vol. 13: 67–9.
In German.

Johnson, C. 1965. The British species of the genus *Byrrhus* L., including *B. arietinus* Steffahny (Col., Byrrhidae) new to the British list. *Entomologist's Monthly Magazine* 101: 111–15.
    Good key to the British species, with useful figures.

Johnson, C. 1966. Taxonomic notes on British Coleoptera. No. 4 *Simplocaria maculosa* Erichson (Byrrhidae). *The Entomologist* 99: 155–6.
    Key to the British species of *Simplocaria*, with useful figures.

Johnson, C. 1978. Notes on Byrrhidae (Col.); with special reference to, and a species new to, the British fauna. *Entomologist's Record and Journal of Variation* 90: 141–7.
    Includes a useful key to the British species of *Curimopsis*.

Olmi, M. 1972. The palaearctic species of the genus *Dryops* Olivier (Coleoptera: Dryopidae). *Bolletino del Museo di Zoologia dell'Università di Torino* 5: 69–132.

Olmi, M. 1976. Coleoptera Dryopidae, Elminthidae. *Fauna d'Italia* 12: 280pp.
    Monographic treatment, including a section on larvae, with many excellent illustrations and colour photographs from life. In Italian.

Parry, J. 1980. A major new find by John Parry. *Balfour-Browne Club Newsletter* 18: 1–2.
    Brief note with figures of *Oulimnius major*.

Paulus, H.F. 1973. Revision der Familie Byrrhidae I. Zur Systematik und Faunistik der west-paleärktischen Vertreter der Gattung *Curimopsis* Ganglebauer 1902 (Col., Byrrhidae, Sincalyptinae). *Senckenbergiana Biologica* 54: 353–67.
    In German.

Paulus, H.F. 1979. Byrrhidae. *Die Käfer Mitteleuropas.* Vol. 6: 328–50.
    In German.

Pierre, F. 1945. La larve d'*Heterocerus aragonicus* Kiesw. et son milieu biologique (Col., Heteroceridae). *Revue Française d'Entomologie* 12: 166–74.
    Key to larvae of *Heterocerus*, with 31 figures; in French.

Steffan, A.W. 1979. Dryopidae. *Die Käfer Mitteleuropas.* Vol. 8: 265–94.
    Includes the family Elmidae.

Superfamily Buprestoidea: jewel beetles

Even compared with nearby countries of continental Europe, Buprestidae (the only family of Buprestoidea) are very poorly represented in the British Isles. The often attractively coloured adults are generally active by day and may sometimes be seen on flowers. The larvae are essentially phytophagous, feeding internally in plants, often in woody tissues. The British species are leaf-miners (*Aphanisticus* in rushes and sedges, *Trachys* in various trees, shrubs and herbs), or wood-feeders (*Agrilus*, *Anthaxia* and *Melanophila*).

All but one of the 13 species of Buprestidae known to occur in Britain are included in Levey's (1977) handbook.

Bílý, S. 1982. The Buprestidae (Coleoptera) of Fennoscandia and Denmark. *Fauna Entomologica Scandinavica* 10: 109pp.

Harde, K.W. *et al.* 1979. Buprestidae (Prachtkäfer). *Die Käfer Mitteleuropas.* Vol. 6: 204–48.
In German.

James, T.J. 1994. *Agrilus sulcicollis* Lacordaire (Buprestidae), a jewel beetle new to Britain. *Coleopterist* 3: 33–5.
Key and figures to distinguish from related British species of *Agrilus.*

Levey, B. 1977. Buprestidae. *Handbooks for the Identification of British Insects* 5(5a): 8pp.

Lompe, A. 1979. Tribus Agrilini. *Die Käfer Mitteleuropas.* Vol. 6: 230–43.
In German.

Palm, T. 1962. Zur Kenntnis der früheren Entwicklungsstadien schwedischer Käfer. 2. Buprestiden-Larven, die in Bäumen leben. *Opuscula Entomologica* 27: 65–78.
In German.

Schaefer, L. 1949. Les buprestides de France. Tableaux analytiques des Coléoptères de la faune franco-rhénane (France, Rhénanie, Belgique, Holland, Valais, Corse). *Miscellanea Entomologica*) (Supplément), Le Moult, Paris. 511pp.
In French.

Superfamily Elateroidea (including Cantharoidea): click beetles, soldier beetles and glowworms

Adults of the three families represented in the British Isles belonging to the Elateroidea in the former more exclusive sense are known as 'click beetles' owing to their jumping ability. This utilises a clicking mechanism that involves a prosternal peg and mesosternal cavity. Rather little is known of the biology of the small and relatively short-bodied species of Throscidae, although the larvae of one species (*Trixagus dermestoides*) feed on ectomy-corrhizal fungi on the roots of various trees. The few British species of Eucnemidae are all wood-associated; their larvae are liquid-feeders, but their precise food is unknown, and may include nutrient-rich fluid, such as slime-mould plasmodia, within decaying wood. Elateridae are rather varied in habitat and food, although all larvae (wireworms) are liquid feeders prac-tising extraoral digestion. Some wood-associated species are predators or at least facultative predators. Larvae of many genera are soil-dwelling and feed on roots; some of these (e.g. *Agriotes* species) may be crop or pasture pests. Adults of the families sometimes referred to a distinct superfamily Cantharoidea (Cantharidae, Drilidae, Lycidae and Lampyridae) are generally loosely articulated, soft-bodied and rather short-lived. Many of them produce defensive chemicals and this may be advertised by aposematic colouring. Cantharid (soldier beetle) larvae are of velvety appearance and are general predators in the soil, litter or dead wood; adults may feed on pollen or nectar as well as on other insects. Adult Lampyridae (glow worms) and Drilidae do not feed; their larvae are predators of snails. The larvae of Lycidae inhabit well-decayed wood; they are liquid feeders, but their precise food is uncertain.

Most British species may be identified satisfactorily with the 'standard' works, as only a few additional species have been recognised as British in recent years (see below). For Eucnemidae the revised key of Allen (1969) is essential. The species of *Ampedus* (Elateridae) with red elytra remain difficult to identify accurately (see Allen, 1969, Leseigneur, 1972, etc.). The recent atlas of Mendel & Clarke (1996) provides a revised checklist of British Elateroidea (in the narrow sense), with references to relevant recent published works. The standard works will suffice for identification of most species of Cantharidae and Lycidae, but see Continental and other works listed below and also Blair (1948) (see 'General Introduction' above). Useful information on identification sources and other literature on Cantharidae and related families may be obtained from the National Recording Scheme.

Allen, A.A. 1969. Notes on some British serricorn Coleoptera, with adjustments to the list. I. Sternoxia. *Entomologist's Monthly Magazine* **104**: 208–16.
Keys to British Eucnemidae.

Allen, A.A. 1990. Notes on, and a key to, the often-confused British species of *Ampedus* Germ. (Col.: Elateridae), with corrections of some erroneous records. *Entomologist's Record and Journal of Variation* **102**: 121–7.
Key to species of *Ampedus*; no figures.

Ashe, G.H. 1946. A new British *Cantharis* confused with *C. pallida* Goeze (= *bicolor* Br. Cat.) (Col., Cantharidae). *Entomologist's Monthly Magazine* **82**: 138–9.
Distinguishing characters of *C. cryptica*, the only species of its genus not included in the standard works, from *C. pallida*.

Cooter, J. 1983. *Zorochros flavipes* (Aubé) (Col., Elateridae) new to Britain. *Entomologist's Monthly Magazine* **119**: 233–6.
Well-illustrated key to the two British species of *Zorochros*.

Dahlgren, G. 1968. Beiträge zur Kenntnis der Gattung *Rhagonycha* (Col., Cantharidae). *Entomologische Blätter für Biologie und Systematik der Käfer* **64**: 93–124.
In German.

Emden, F.I. van 1943. Larvae of the British beetles. IV. Various small families. Eucnemidae. *Entomologist's Monthly Magazine* **79**: 218–19.

Emden, F.I. van 1945. Larvae of British beetles. V. Elateridae. *Entomologist's Monthly Magazine* **81**: 13–38.
Key to all but 10 of the British species, with 54 figures.

Emden, F.I. van 1956. Morphology and identification of the British larvae of the genus *Elater* (Col., Elateridae). *Entomologist's Monthly Magazine* **92**: 167–88.
Most British species of this genus [now *Ampedus*] covered, but several not under their current names.

Fitton, M.G. 1975. The larvae of the British genera of Cantharidae (Coleoptera). *Journal of Entomology* (B) **44**: 243–54.
Illustrated key to six genera, with 34 figures.

Franz, H. 1967. Beiträge zur Systematik der europäischen und nordwestafrikanischen *Agriotes* s. str. (Elateridae). *Entomologische Blätter für Biologie und Systematik der Käfer* **63**: 65–86.
In German.

Franz, H. 1967. Zur Kenntnis der mitteleuropäischen *Hypnoidus*-Arten aus dem Subgenus *Zorochrus* Thoms. *Entomologische Blätter für Biologie und Systematik der Käfer* **63**: 32–7.
In German.

Geisthardt, M. *et al.* 1979. Malacodermata (families Lycidae, Cantharidae, Lampyridae, and Drilidae). *Die Käfer Mitteleuropas.* Vol. 8: 9–53.

Korschefsky, R. 1941. Bestimmungstabelle der bekanntesten deutschen Elateridenlarven (Coleoptera, Elateridae). *Arbeiten über Morphologische und Taxonomische Entomologie* **8**: 217–330.
In German.

Korschefsky, R. 1951. Bestimmungstabelle der bekanntesten deutschen Lyciden-, Lampyriden- und Drilidenlarven (Coleoptera). *Beiträge zur Entomologie* **1**: 60–4.

Leiler, T.E. von 1976. Zur Kenntnis der Entwicklungsstadien und der Lebensweise nord und mitteleuropäischer Eucnemiden (Coleoptera). *Entomologische Blätter für Biologie und Systematik der Käfer* **72**: 10–50.
In German.

Leseigneur, L. 1972. Coléoptères Elateridae de la faune de France continentale et de Corse. *Bulletin Mensuel de la Société Linnéenne de Lyon* **41** (suppl.): 1–379.
In French.

Leseigneur, L. 1978. Les *Hypocoelus* (Eucnemidae) de la faune de France. Systématique et distribution. *Entomologiste* **34**: 105–23.
In French.

Lohse, G. A. 1979. Sternoxia [families Elateridae, Cerophytidae, Eucnemidae, and Throscidae]. *Die Käfer Mitteleuropas.* Vol. 6: 101–203.
In German.

Mendel, H. 1990. The status of *Ampedus pomonae* (Stephens), *A. praeustus* (F.) and *A. quercicola* (du Buysson) (Coleoptera: Elateridae) in the British Isles. *Entomologist's Gazette* **41**: 23–32.
Distinguishing characters, with one figure, of these much confused species.

Mendel, H. & Clarke, R.E. 1996. *Provisional atlas of the click beetles (Coleoptera: Elateroidea) of Britain and Ireland.* Ipswich Borough Council Museums, Ipswich.

Muona, J. 1993. Review of the phylogeny, classification and biology of the family Eucnemidae (Coleoptera). *Entomologica Scandinavica* (suppl.) **44**: 1–133.

Owen, J.A., Allen, A.A., Carter, I.S. & Hayek, C.M.F. von 1985. *Panspoeus guttatus* Sharp (Col., Elateridae) new to Britain. *Entomologist's Monthly Magazine* **121**: 91–5.
Distinguishing features illustrated by photographs.

Palm, T. 1960. Zur Kenntnis der früheren Entwicklungsstadien schwedischer Käfer. I. Bisher bekannte Eucnemiden-Larven. *Opuscula Entomologica* 25: 157–69.
In German.

Palm, T. 1972. Die skandinavischen Elateriden-Larven (Coleoptera). *Entomologica Scandinavica* (suppl.) 2: 1–63.
In German.

Platia, G. 1994. Coleoptera Elateridae. *Fauna d'Italia* 33: 429pp.
Well-illustrated key to the 243 Italian species; in Italian.

Superfamily Dermestoidea (including Derodontoidea): hide beetles, etc.

The larvae of Dermestidae all feed on dry material of animal origin, including dry carcasses, hair, feathers and insect remains. A number are found, under natural conditions, in the nests of mammals, birds or social insects; a few are specialist feeders on insect remains found in spiders' webs on old tree trunks. The adults of some species, notably *Anthrenus*, feed on nectar or pollen in flowers. A number of dermestids are significant pests of stored products such as hides, furs, skins and woollen fabrics. The only British species of Derodontidae (*Laricobius erichsoni*) is associated with coniferous trees where both adults and larvae prey on woolly aphids.

All resident British species and a good selection of those that regularly crop up as imports in cargoes are covered by an excellent handbook (Peacock, 1993).

Hammond, P.M. & Barham, C.S. 1982. *Laricobius erichsoni* Rosenhauer (Coleoptera: Derodontidae), a species and superfamily new to Britain. *Entomologist's Gazette* 33: 35–40.
Distinguishing characters; bionomic data; good habitus figure.

Lohse, G.A. 1979. Dermestidae. *Die Käfer Mitteleuropas.* Vol. 8: 304–27.

Mroczkowski, M. 1975. Dermestidae of Poland. *Fauna Polonica* 4: 1–163.
The most comprehensive recent treatment of European Dermestidae.

Peacock, E.R. 1976. *Dermestes peruvianus* Cast., *D. haemorrhoidalis* Küst. and other *Dermestes* spp. (Col., Dermestidae). *Entomologist's Monthly Magazine* 111: 1–14.

Peacock, E.R. 1979. *Attagenus smirnovi* Zhantiev (Coleoptera: Dermestidae) a species new to Britain, with keys to the adults and larvae of British *Attagenus*. *Entomologist's Gazette* 30: 131–6.

Peacock, E.R. 1993. Adults and larvae of hide, larder and carpet beetles and their relatives (Coleoptera: Dermestidae) and of derodontid beetles (Coleoptera: Derodontidae). *Handbooks for the Identification of British Insects.* 5(3): 144pp.
Subfamily, generic, specific and larval keys.

Superfamily Bostrichoidea (Teredilia): furniture beetles, spider beetles, etc.

Bostrichoid beetles are especially characterised by their ability to subsist on various very dry foods without needing to imbibe water. Most species are associated with the dry wood or bark of dead trees. Several species of Bostrichidae (including the Lyctinae or powder post beetles) and of Anobiidae are timber pests; the most notable of these is the furniture beetle or woodworm, *Anobium punctatum*. A few species of Anobiinae have become cosmopolitan stored-product pests; well-known examples are the bread beetle (*Stegobium paniceum*) and *Lasioderma serricorne*.

The subfamily Ptininae, whose members feed on various very dry materials, also includes stored-product pests such as *Ptinus tectus*. Most native species of the group are associated with nests and/or old tree cavities. Adult dorcatomine Anobiidae are generally of somewhat short and even globular form. Many of them (*Dorcatoma, Anitys*) feed on woody or fibrous fungal fruiting bodies causing heart-rot in large deciduous trees; *Caneocara* species are specialist feeders on puffball fungus fruiting bodies.

Most native British species may be identified by reference to the 'standard' works, supplemented if necessary by Volume 8 of *Die Käfer Mitteleuropas*. For the most difficult genera, e.g. *Ernobius* and *Ptinus*, reference should be made to the additional works listed below. A number of bostrichid beetle species that are regularly imported in cargoes are not included in the 'standard' works.

Allen, A.A. 1977. Notes on some British Serricorn Coleoptera, with adjustments to the list. 3. – A new British *Anobium*, with notes on three others in the family. *Entomologist's Monthly Magazine* **112** (1976): 151–4.

Baranowsky, R. 1985. Central and Northern European *Dorcatoma* (Coleoptera: Anobiidae), with a key and description of a new species. *Entomologica Scandinavica* **16**: 203–7.
    Well-illustrated key to nine species, including all those found in Britain.

Bellés, X. 1990. Coleoptera Ptinidae Gibbiinae. *Fauna Iberica*: 1–42.
    Well-illustrated key to the seven Iberian species; excellent habitus figures. In Spanish.

Böving, A.G. 1954. Mature larvae of the beetle-family Anobiidae. *Biologiske Meddelelser* **22** (2): 1–298. 50 pls.
    Key to genera and species of the world.

Emden, F.I. van 1943. Larvae of British beetles. IV. Various small families. Lyctidae, Bostrychidae. *Entomologist's Monthly Magazine* **79**: 265–9.

Espanol, F. 1992. Coleoptera Anobiidae. *Fauna Iberica* **2**: 1–195.
    Well illustrated key to the 92 Iberian species; excellent habitus figures. In Spanish.

Freude, H., Harde, K.W. & Lohse, G.A. (Eds) 1969. Lyctidae, Bostrychidae, Anobiidae, Ptinidae. *Die Käfer Mitteleuropas.* Vol. 8: 7–74.
In German.

Hall, D.W. & Howe, R.W. 1953. A revised key to the larvae of the Ptinidae associated with stored products. *Bulletin of Entomological Research* **44**: 85–96.

Hinton, H.E. 1941. The Ptinidae of economic importance. *Bulletin of Entomological Research* **31**: 331–81.

Hodge, P.J. & Parry, J.A. 1981. *Ptinus dubius* Sturm (Col., Ptinidae) new to Britain. *Entomologist's Monthly Magazine* **117**: 225–6.
Key to distinguish from *P. fur;* no figures.

Johnson, C. 1966. The Fennoscandian, Danish and British species of the genus *Ernobius* Thomson (Col., Anobiidae). *Opuscula Entomologica* **31**: 81–92.
Good key, with many useful figures.

Johnson, C. 1975. A review of the Palaearctic species of the genus *Ernobius* Thomson (Coleoptera: Anobiidae). *Entomologische Blätter für Biologie und Systematik der Käfer* **71** (2): 65–93.
Good key and illustrations of male genitalia.

Manton, S.M. 1945. The larvae of the Ptinidae associated with stored products. *Bulletin of Entomological Research* **35** (4): 341–65.
Key to genera and species.

Mendel, H. 1982. *Hemicoelus nitidus* (Hbst.) (Col., Anobiidae) new to Britain. *Entomologist's Monthly Magazine* **118**: 253–4.
Characters for distinguishing from *H. fulvicornis*, with figures.

Parkin, E.A. 1933. The larvae of some wood-boring Anobiidae (Coleoptera). *Bulletin of Entomological Research* **24**: 33–68.

Superfamilies Cleroidea and Lymexyloidea

Although Phloiophilidae (the single species *Phloiophilus edwardsi* feeds on *Phlebia* fungi on oak) and some Trogossitidae are mycophagous (in association with wood or arboricolous fungi), most cleroids are predaceous at all stages, or predaceous as larvae and pollen-feeding as adults (e.g. *Dasytes* and other Melyridae). Most species are associated with wood, where they may be very specific predators of various wood-boring insect larvae. The larvae of some others (e.g. a number of Melyridae) inhabit plant stems; some Cleridae (e.g. *Necrobia*) are associated with carrion, and the introduced trogossitid *Tenebroides mauritanicus*, although largely predaceous, is a cosmopolitan stored-product pest.

Adults of most British species of these two superfamilies may be reliably identified with the aid of the standard British works, although reference to the various relevant keys in Volume 8 of *Die Käfer Mitteleuropas* is to be recommended, as several additional species of Cleroidea have been found to occur in the British Isles in recent years (see Hodge & Jones, 1995, in general references, above). Coloured illustrations and comment on two of these (*Ostoma ferrugineum* and *Axinotarsus marginalis*) are provided by Hammond *et al.* (1989) (see general references, above). The introductory notes prepared for the Cleroidea, Lymexyloidea and Heteromera Recording Scheme (R.S. Key, 1986) also contain useful information on identification.

Allen, A.A. 1971. British Coleoptera: corrections and supplementary notes, including the addition of *Axinotarsus marginalis* Lap. (Melyridae) to our list. *Entomologist's Record and Journal of Variation* **83**: 46–51.
  Key to separate *A. marginalis* and *A. pulicarius*; no figures.
Crowson, R.A. 1964. A review of the classification of Cleroidea (Coleoptera), with descriptions of two new genera of Peltidae and several new larval types. *Transactions of the Royal Entomological Society of London* **116**: 275–327.
  Key to subfamilies, genera and species of world Melyridae.
Crowson, R.A. 1970. Further observations on Cleroidea (Coleoptera). *Proceedings of the Royal Entomological Society of London* (B) **39**: 1–20.
  Key to subfamilies of Trogossitidae of the world.
Emden, F.I. van 1943. Larvae of British beetles. IV. Various small families. Trogositidae, Cleridae. *Entomologist's Monthly Magazine* **79**: 214–18.
Emden, F.I. van 1943. Larvae of British beetles. IV. Various small families. Lymexylidae. *Entomologist's Monthly Magazine* **79**: 259–61.
Evers, A.M.J. 1971. Über die paläarktischen Arten der Gattungen *Sphinginus* Rey und *Fortunatius* nov. gen. [Col., Melyridae]. *Entomologische Blätter für Biologie und Systematik der Käfer* **67**: 21–41.
Evers, A.M.J. *et al.* 1979. Malacodermata [families Malachiidae, Melyridae and Cleridae]. *Die Käfer Mitteleuropas*. Vol. 8: 53–98.
  Includes the family Phloiophilidae; in German.
Evers, A.M.J. 1985. Aufteilung der paläarktischen Arten des Gattungskomplexes *Malachius* F. *Entomologische Blätter für Biologie und Systematik der Käfer* **81**: 1–40.
  Key to males of all Palaearctic species of *Malachius* and closely related genera. No figures; in German.
Key, R.S. 1983. *Troglops cephalotes* (Olivier) (Col., Melyridae) from Buckinghamshire, possibly a new British species. *Entomologist's Monthly Magazine* **123**: 179–80.
  Distinguishing characters; whole-body figure.
Lohse, G.A. 1979. Lymexylonidae. *Die Käfer Mitteleuropas*. Vol. 8: 100–1.
  In German.
Majer, K. 1982. Species of the genus *Aplocnemus* of Central Europe. *Deutsche Entomologische Zeitschrift* **29**: 421–45.
  Key includes all British species.
Vogt, H. 1979. Ostomidae [=Trogossitidae]. *Die Käfer Mitteleuropas*. Vol. 7: 14–18.
  In German.

Wheeler, Q.D. 1986. Revision of the genera of Lymexylidae (Coleoptera: Cucujiformia). *Bulletin of the American Museum of Natural History* **183**(2): 113–210.

Winkler, J.R. 1961. Die Buntkäfer (Cleridae). *Neue Brehm Bücherei* **281**: 1–108. In German.

Superfamily Cucujoidea (= Clavicornia)

Like the Tenebrionoidea, many cucujoids are essentially mycophagous as both adult and larva, but the group embraces a great diversity of feeding habits. Some Latridiidae are largely pollenivorous, and some Phalacridae and Crytophagidae as well as many Nitidulidae (Meligethinae, etc.) also feed on pollen or the sexual parts of flowering plants. Most Coccinellidae (ladybirds) are relatively specialised predators of aphids, mealybugs, scale insects or mites. Cucujidae and some Monotomidae are also predators. Sphindidae are all specialised feeders on slime-mould fruiting bodies and this habit is shared by some Latridiidae. The styliform mouthparts of some Cerylonidae appear to be adapted for sucking juices from fungal hyphae.

The Monotomidae (= Rhizophagidae) and Phalacridae are covered well by handbooks, as is the most difficult section of the Nitidulidae: pollen beetles of the subfamily Meligethinae. The ladybirds (Coccinellidae) are also well covered by a *Naturalist's Handbook* (supplemented for *Scymnus* and relatives by Pope's (1973) revision). However, for Corylophidae, the standard works will not suffice, and reference should be made to others listed below. Most species of the remaining smaller families of Cucujoidea (Sphindidae, Cucujidae, Laemophloeidae, Silvanidae, Biphyllidae, Byturidae, Erotylidae, Cerylonidae and Endomychidae) are identifiable using the standard works, supplemented by a few papers (see below).

This leaves three 'difficult' groups: Cryptophagidae, Latridiidae and Nitidulidae in part (notably Epuraeinae), for which no satisfactory comprehensive key works in English are available. For *Cryptophagus* (Cryptophagidae) the revision of Coombes & Woodroffe (1955) is helpful but Continental keys, e.g. that of Lohse (1967), may be easier to use. Recent revisional studies of Cryptophagidae Atomariinae by C. Johnson have put understanding of the composition of the British fauna of this difficult group on a sure footing, but no comprehensive key work to the group is yet available. For Nitidulidae Epuraeinae the most useful single reference is that of Spornraft (1967). Most British Latridiidae may be satisfactorily identified using von Peez's (1967) key, but see various other useful recent works in English listed below.

Allen, A.A. 1967. A clarification of the status of *Cartodere separanda* Reitt. (Col., Lathridiidae) and *C. schueppeli* Reitt. new to Britain. *Entomologist's Monthly Magazine* **102**: 192–8.
Tables and figures to separate *C. separanda* from *C. elongata* and *C. filiformis* from *C. schueppeli*.

Allen, A.A. 1970. Revisional notes on the British species of *Orthoperus* Steph. (Col., Corylophidae). *Entomologist's Record and Journal of Variation* **82**: 112–20.

Audisio, P. 1993. Coleoptera Nitidulidae – Kateretidae. *Fauna d'Italia* **32**: xvi + 971pp.
A monographic treatment of the Western Palaearctic species, extensively illustrated. In Italian, with keys in both Italian and English.

Coombs, C.W. & Woodroffe, G.E. 1955. A revision of the British species of the genus *Cryptophagus* (Herbst) (Coleoptera, Cryptophagidae). *Transactions of the Royal Entomological Society of London* **106**: 237–82.

Dajoz, R. 1977. Coléoptères: Colydiidae et Anommatidae paléarctiques. *Faune de l'Europe et du Bassin Méditerranéen* **8**: 280pp.

Eccles, T.M. & Bowestead, S. 1987. *Anommatus diecki* Reitter (Coleoptera: Cerylonidae) new to Britain. *Entomologist's Gazette* **38**: 225–227.
Key to separate the two British species of *Anommatus*; excellent habitus figure.

Emden, F.I. van 1928. Die Larve von *Phalacrus grossus* Er. und Bemerkungen zur Larvensystem der Clavicornia. *Entomologische Blätter für Biologie und Systematik der Käfer* **24**(1): 8–20.
Key to genera and some species of Phalacridae in Europe.

Emden, F.I. van 1949. Larvae of British beetles. VII. (Coccinellidae). *Entomologist's Monthly Magazine* **85**: 265–83.
Covers most British species, and includes an 'easy key for field use'; with 61 figures.

Gourreau, J.M. 1974. Systématique de la tribu des Scymnini (Coccinellidae) [de France]. *Annales de Zoologie – Ecologie Animale* (hors Séries): 1–221.

Hammond, P.M. 1971. *Rypobius ruficollis* Jacqu. (Col., Corylophidae), a genus and species new to Britain. *Entomologist's Gazette* **22**: 241–3.
Distinguishing characters illustrated.

Hinton, H.E. 1941. The Lathridiidae of economic importance. *Bulletin of Entomological Research* **32**: 191–247.

Johnson, C. 1966a. *Caenoscelis subdeplanata* Bris. (Col., Cryptophagidae): a beetle new to Britain. *The Entomologist* **99**: 129–31.
Illustrated key to the British species of the genus.

Johnson, C. 1966b. *Atomaria clavigera* Ganglb. (Col., Cryptophagidae) new to Britain. *The Entomologist* **99**: 230–231.
Figures and distinguishing characters from near relatives.

Johnson, C. 1967a. Additions and corrections to the British list of *Atomaria s. str.* (Col., Cryptophagidae), including a species new to science. *The Entomologist* **100**: 39–47.
Useful figures of male genitalia; does not include all species of *Atomaria* s. str. now known to occur in the UK.

Johnson, C. 1967b. The identity of *Brachypterolus linariae* (Stephens) (Col., Nitidulidae), with notes on its occurrence in Britain. *The Entomologist* 100: 142–4.
Key and male genitalia figures to distinguish from nearest relative.

Johnson, C. 1972. *Epuraea adumbrata* Mannerheim and *E. biguttata* (Thunberg) (Col., Nitidulidae) new to Britain. *The Entomologist* 105: 126–9.
Useful figures and key couplets to distinguish additional British species from their close relatives.

Johnson, C. 1973. The *Atomaria gibbula* group of species. (Coleoptera, Cryptophagidae). *Reichenbachia* 14: 125–41.
Contains key to Palaearctic species of this group, with figures of male genitalia.

Johnson, C. 1976. Nine species of Coleoptera new to Britain. *Entomologist's Monthly Magazine* 111: 177–84.
Distinguishing characters of the two British species of *Pocadius* (Nitidulidae) and of four additional British species of *Atomaria* (Cryptophagidae).

Johnson, C. 1986. Notes on some Palaearctic *Melanophthalma* Motschulsky (Coleoptera: Latridiidae), with special reference to *transversalis* auctt. *Entomologist's Gazette* 37: 117–25.
Distinguishing characters illustrated.

Johnson, C. 1988. Notes on some British *Cryptophagus* Herbst (Coleoptera: Cryptophagidae), including *confusus* Bruce new to Britain. *Entomologist's Gazette* 39: 329–35.
Distinguishing characters from *C. labilis*; male genitalic features of *dentatus*-group species illustrated.

Johnson, C. 1992. Additions and corrections to the British list of Coleoptera. *Entomologist's Record and Journal of Variation* 104: 305–10.
Additional species of *Caenoscelis* and *Cryptophagus* characterised.

Kirk-Spriggs, A.H. 1996. Pollen Beetles, Coleoptera: Kateretidae and Nitidulidae: Meligethinae. *Handbooks for the Identification of British Insects* 5(6a): 157pp.
A very full and extensively illustrated treatment; an essential work for identifying British *Meligethes*.

Klausnitzer, B. 1970. Zur Larvalsystematik der mitteleuropäischen Coccinellidae (Coleoptera). *Entomologische Abhandlungen und Berichte aus dem Staatlichen Museum für Tierkunde in Dresden* 38: 55–110.
Keys to and descriptions of 58 species; in German.

Klausnitzer, B. 1973. Bestimmungstabelle für mitteleuropäische Coccinelliden-Larven nach leicht sichtbaren Merkmalen. *Beiträge zur Entomologie* 23: 93–8.
Keys to 41 species by 'hand lens' characters; in German.

Levey, B. 1997. *Stephostethus alternans* (Mannerheim) (Latridiidae), a species new to Britain. *Coleopterist* 6: 49–51.
Good figures comparing the British species of *Stephostethus*.

Lohse, G.A. 1967. Cryptophagidae. *Die Käfer Mitteleuropas*. Vol. 7: 110–57.
Particularly useful for *Cryptophagus*; now somewhat out-of-date for Atomariinae; in German.

Lyszkowsky, R.M., Owen, J.A. & Taylor, S. 1992. *Corticaria abietorum* Motschulsky (Col., Lathridiidae) new to Britain. *Entomologist's Record and Journal of Variation* 104: 67–9.

Distinguishing characters, with outline whole-body figures of *C. linearis* and *C. abietorum.*

Majerus, M.E.N. & Kearns, P. 1989. *Ladybirds.* Naturalist's Handbooks No. 10. Richmond Publishing Co., Slough.

Majerus, M.E.N. 1994. *Ladybirds.* Collins New Naturalist, London.

Peacock, E.R. 1977. Rhizophagidae. *Handbooks for the Identification of British Insects* 5(5a): 19pp.

Peez, A. von 1967. Lathridiidae. *Die Käfer Mitteleuropas.* Vol. 7: 168–90.
Covers most British species satisfactorily; also includes Holoparamecinae, now regarded as Endomychidae.

Pope, R.D. 1973. The species of *Scymnus (s. str.), Scymnus (Pullus)* and *Nephus* (Col., Coccinellidae) occurring in the British Isles. *Entomologist's Monthly Magazine* 109: 3–39.
An essential work for identifying many of the small British ladybirds; well illustrated, and with maps and bionomic details.

Pope, R.D. 1953. Coccinellidae and Sphindidae. *Handbooks for the Identification of British Insects* 5(7): 12pp.
A rather spare treatment, now out-of-date for many of the smaller species, and largely replaced by later works.

Roberts, A.W.R. 1958. On the taxonomy of Erotylidae (Coleoptera), with special reference to the morphological characters of the larvae. *Transactions of the Royal Entomological Society of London* 88(3): 89–117.
Key to subfamilies and genera of the world.

Rücker, H.W. 1980. Bestimmungstabelle der Merophysiidae des Mittelmeeräumes und der angrenzenden Gebeite (Coleoptera: Merophysiidae). *Entomologische Blätter für Biologie und Systematik der Käfer* 75: 141–54.

Savöiskaya, G.I. & Klausnitzer, B. 1973. Morphology and taxonomy of the larvae with keys for their identification. *In:* Hodek, J. (Ed.) *Biology of Coccinellidae,* pp. 36–55. Prague.

Spornraft, K. 1967. Nitidulidae. *Die Käfer Mitteleuropas.* Vol. 7: 20–79.
Especially useful for Epuraeinae; in German.

Thompson, R.T. 1958. Phalacridae. *Handbooks for the Identification of British Insects* 5(5b): 17pp.

Tozer, E. R. 1973. On the British species of *Lathridius* Herbst (Col., Lathridiidae). *Entomologist's Monthly Magazine* 108: 193–199.

Vogt, H. *et al.* 1979. Clavicornia. *Die Käfer Mitteleuropas. Vol.* 7: 19–309.

Superfamily Tenebrionoidea (= Heteromera)

Tenebrionoid beetles embrace a wide range of feeding biologies, and occupy a wide range of habitat types. However, a large proportion are essentially mycophagous as both adult and larva, and are found in association with rotten wood or fungus fruiting bodies. Many of these (e.g. Melandryidae, Mordellidae, Scraptiidae) are relatively short-lived as adults. A number of others (e.g. Anthicidae, some Tenebrionidae) are essentially scavengers.

British Ciidae feed exclusively on the context tissues of durable fruiting bodies of bracket fungi. Larvae of some mordellid genera (e.g. *Mordellistena*) mine herbaceous plant stems and galls. The only British rhipiphorid *(Metoecus paradoxus)* is a parasitoid of social wasps, and British oil beetles (Meloidae) are also mostly parasitoids, of bees. A few species (e.g. of Colydiidae) are predators. The Tenebrionidae includes several serious pests (mealworms and flour beetles) of stored products.

Many of the families now included here are covered by keys in the Royal Entomological Society *Handbook* series (Buck, 1954; Brendell, 1975). These are adequate for Tenebrionidae, Tetratomidae, Melandryidae, Rhipiphoridae, Lagriidae, Meloidae, Mycteridae, Pythidae, Pyrochroidae, Salpingidae, Anthicidae and Aderidae, but do not cover Mycetophagidae, Colydiidae and Ciidae. In addition, the keys are now well out of date for Scraptiidae, Oedemeridae and (especially) Mordellidae. For these latter families reference may be made to various recent 'standard' Continental works, supplemented by papers listed below. For most Mycetophagidae and Colydiidae the standard British works will suffice for identification (but see additional references below). For Ciidae, reference to Lohse's (1967) key and/or Kevan's (1967) paper are essential, but may not solve all identification problems.

Allen, A.A. 1964. The genus *Synchita* Hellw. (Col., Colydiidae) in Britain; with an addition to the fauna and a new synonymy. *Entomologist's Monthly Magazine* **100**: 36–42.
  Distinguishing characters of the two British species; good figures.
Allen, A.A. 1975. Two species of *Anaspis* (Col., Mordellidae) new to Britain, with a consideration of the status of *A. hudsoni* Donis., etc. *Entomologist's Record and Journal of Variation* **87**: 269–274.
Allen, A.A. 1986. On the British species of *Mordellistena* Costa (Col.: Mordellidae) resembling *parvula* Gyll. *Entomologist's Record and Journal of Variation* **98**: 47–50.
  Key to the species of the *parvula*-group; no figures.
Allen, A.A. 1990. Notes on the species-pair *Cis festivus* Panz. and *C. vestitus* Mell. (Col.: Cisidae). *Entomologist's Record and Journal of Variation* **102**: 177–9.
  Distinguishing characters; no figures.
Allen, A.A. 1995. An apparently new species of *Mordellistena* (Col.: Mordellidae) in Britain. *Entomologist's Record and Journal of Variation* **107**: 25–7.
  Figures genitalia of new species of *Mordellistena*.
Allen, A.A. 1995. On the British *Mordellistena humeralis* L. (Col.: Mordellidae) and its allies. *Entomologist's Record and Journal of Variation* **107**: 181–4.
Aubrook, E.W. 1970. *Cis dentatus* Mell. (Col. Cisidae): an addition to the British list. *The Entomologist* **103**: 250–1.
  Distinguishing characters; whole-body figure.

Batten, R. 1986. A review of the British Mordellidae (Coleoptera). *Entomologist's Gazette* 37: 225–35.

Brief keys to all species, with 22 figures.

Bologna, M.A. 1991. Coleoptera Meloidae. *Fauna d'Italia* 28: 541pp.

Well illustrated; in Italian.

Brendell, M.J.D. 1975. Tenebrionidae. *Handbooks for the Identification of British Insects* 5(10): 22pp.

Bucciarelli, I. 1980. Coleoptera Anthicidae. *Fauna d'Italia* 16: 240pp.

Keys, extensively illustrated by habitus and male genitalia figures, for the nearly 100 species found in Italy; in Italian.

Buck, F.D. 1954. Lagriidae, Alleculidae, Tetratomidae, Melandryidae, Salpingidae, Pythidae, Mycteridae, Oedemeridae, Mordellidae, Scraptiidae, Pyrochroidae, Rhipiphoridae, Anthicidae, Aderidae and Meloidae. *Handbooks for the Identification of British Insects* 5(9): 30pp.

Cooter, J. 1991. *Mordellistena pygmaeola* Ermisch, 1956 (Coleoptera: Mordellidae) new to Britain. *Entomologist's Gazette* 42: 97–98.

Characters to distinguish from *M. parvula*; with male genitalia figures.

Crowson, R.A. 1963. Observations on British Tetratomidae (Col.), with a key to the larvae. *Entomologist's Monthly Magazine* 99: 82–86.

Illustrated key to larvae of three species of *Tetratoma*.

Duffy, E.A.J. 1946. Notes on the British species of *Pyrochroa* (Col., Pyrochroidae) with a key to their first-stage larvae. *Entomologist's Monthly Magazine* 82: 92–93.

One figure.

Emden, F.I. van 1943. Larvae of British beetles. IV. Various small families. Meloidae, Rhipiphoridae, Lagriidae, Synchroidae, Pyrochroidae. *Entomologist's Monthly Magazine* 79: 219–23; 259–65.

Emden, F.I. van 1947. Larvae of British beetles. VI. Tenebrionidae. *Entomologist's Monthly Magazine* 83: 154–71 (see also 84: 10).

Covers some 20 genera and the individual species of some of these; with 46 figures.

Harrison, T.D. 1996. *Eulagius filicornis* (Reitter) (Mycetophagidae) apparently established in Britain. *Coleopterist* 4: 65–67.

Includes good habitus figure.

Kaszab, Z. 1969. Heteromera. *Die Käfer Mitteleuropas* Vol. 8: 75–138, 160–264.

Kevan, D.K. 1967. On the apparent conspecificity of *Cis pygmaeus* (Marsh) and *C. rhododacylus* (Marsh) and on other closely allied species (Col., Ciidae). *Entomologist's Monthly Magazine* 102: 138–44.

Liebenow, K. 1979. Beiträge zur Insektenfauna der DDR: Coleoptera-Oedemeridae. *Beiträge zur Entomologie* 29: 249–66.

Lohse, G.A. 1969. Cisiden Studien IV. *Rhopalodontus perforatus* und seine Verwandten. *Entomologische Blätter für Biologie und Systematik der Käfer* 65: 48–52.

Mendel, H. 1990. The identification of British *Ischnomera* Stephens (Coleoptera: Oedemeridae). *Entomologist's Gazette* 41: 209–11.

Key to the four British species of *Ischnomera*.

Mendel, H. & Owen, J.A. 1987. *Cicones undata* Guér. (Coleoptera, Colydiidae) new to Britain. *Entomologist's Record and Journal of Variation* 99: 93–5.
Key to separate the two British species of *Cicones*, with figures.

Rozen, J.G. 1960. Phylogenetic-systematic study of larval Oedemeridae (Coleoptera). *Miscellaneous Publications of the Entomological Society of America* 1 (2): 35–68.
Key to genera and species of the world.

Skidmore, P. 1973. *Chrysanthia nigricornis* Westh. (Col. Oedemeridae), in Scotland, a genus and species new to the British list. *The Entomologist* 106: 234–7.
Distinguishing characters; habitus figure.

Vasquez, X.A. 1993. Coleoptera Oedemeridae, Pyrochroidae, Pythidae, Mycteridae. *Fauna Iberica* 5: 1–181.
Well illustrated keys to the 49 Iberian species of these families; excellent habitus figures; in Spanish.

Viedma, M.G. de 1966. Contribución al conocimiento de las larvas de Melandryidae de Europa (Coleoptera). *Eos, Madrid* 41: 483–506 (see also pp. 507–13).
In Spanish.

Welch, R.C. 1964. *Pycnomerus fuliginosus* Er. (Col., Colydiidae) new to Britain. *Entomologist's Monthly Magazine* 100: 57–60.
Distinguishing characters; habitus figure.

Superfamily Chrysomeloidea

Adult Cerambycidae may feed on flowers, foliage or bark and are attracted to sugary fluids, on which they may also feed. At the more important feeding stage, the often long-lived larvae are all essentially phytophagous. Most British species feed internally on bark, phloem, sapwood or heartwood of various trees and shrubs. Some of them are at least minor forestry pests and one, the house longhorn beetle (*Hylotrupes bajulus*), may be a serious (although very localised) pest of worked timbers in buildings. The larvae of a few British species (e.g. of *Phytoecia* and *Agapanthia*) feed internally in the stems of herbaceous plants. All Chrysomelidae (including Bruchidae) and Megalopodidae are essentially phytophagous with adults feeding externally and larvae either externally or internally on a wide variety of higher-plant tissues (and those of bryophytes). With the major exception of larval Cryptocephalinae, which specialise in dead leaves, most Chrysomelidae feed on living tissues. Some species achieve agricultural or horticultural pest status, and others have been employed in attempts at biological control of various weed plants. As well as being, in some instances, pests of field crops, seed-feeding larvae of Bruchinae may also be pests of stored pulses.

Family Cerambycidae (longhorn or timber beetles)

Most adult British longhorn beetles may be readily identified by using the 'standard' works. However, even Duffy's (1952) handbook is now out-of-date, and reference to one of the more recent Continental works, such as the cerambycid part of the *Fauna Entomologica Scandinavica* series, is recommended if in doubt. Hodge & Jones (1995) (see general references, above) provide a useful list of references to descriptions of exotic species that are found, not infrequently, as imports.

Bense, U. 1995. *Longhorn beetles: illustrated key to the Cerambycidae and Vesperidae of Europe*. Josef Margraf, Weikersheim.
   Many illustrations and distribution maps. Keys in German and in English.

Bílý, S. & Mehl, O. 1989. Longhorn Beetles (Coleoptera, Cerambycidae) of Fennoscandia and Denmark. *Fauna Entomologica Scandinavica* 22: 203pp.

Demelt, C. von 1966. Bockkäfer oder Cerambycidae. I. Biologie mitteleuropäischer Bockkäfer (Col., Cerambycidae) unter besonderer Berücksichtigung der Larven. *Tierwelt Deutschlands* 52(2): vii + 115pp.

Duffy, E.A.J. 1952. Cerambycidae. *Handbooks for the Identification of British Insects* 5(12): 18pp.
   Keys rather sparsely illustrated, difficult to use, and now somewhat out-of-date.

Duffy, E.A.J. 1953. *A monograph of the immature stages of British and imported timber beetles (Cerambycidae)*. British Museum (Natural History), London.
   Rather sparsely illustrated but still a standard work.

Emden, F.I. van 1939–40. Larvae of British beetles. 1. A key to the genera and most of the species of British cerambycid larvae. *Entomologist's Monthly Magazine* 75: 257–73; 76: 7–13.
   Key to five subfamilies, 36 genera, and to the species of some genera.

Harde, K.W. 1966. Cerambycidae. *Die Käfer Mitteleuropas*. Vol. 9: 7–94.
   In German.

Harrison, T.D. 1992. *Tetrops starkii* Chevrolat (Col., Cerambycidae) new to Britain. *Entomologist's Monthly Magazine* 128: 181–3.
   Illustrated tabular key to the two British species of *Tetrops*.

Klausnitzer, B. & Sander, F. 1978. *Die Bockkäfer Mitteleuropas, Cerambycidae*. A. Ziemsen Verlag, Wittenberg Lutherstadt.
   Key to subfamilies, genera and species in Europe. In German.

Picard, F. 1929. Coléoptères Cerambycidae. *Faune de France* 20: vii + 166pp.
   In French.

Svácha, P. & Danilevsky, M.L. 1987. Cerambycoid larvae of Europe and Soviet Union (Coleoptera, Cerambycoidea). Part I. *Acta Universitatis Carolinae, Biologica* 30 [1986]: 1–176.
   Key to families, genera and species of Prioninae and Aseminae.

Svácha, P. & Danilevsky, M.L. 1988. Cerambycoid larvae of Europe and Soviet Union (Coleoptera, Cerambycoidea). Part II. *Acta Universitatis Carolinae, Biologica* 31: 121–284.
Key to genera and species of Cerambycinae.

Svácha, P. & Danilevsky, M.L. 1989. Cerambycoid larvae of Europe and Soviet Union (Coleoptera, Cerambycoidea. Part III. *Acta Universitatis Carolinae, Biologica* 32: 1–205.
Key to genera and species of Lepturinae.

Families Chrysomelidae (including Bruchidae) and Megalopodidae: leaf beetles, flea beetles, tortoise beetles, seed beetles

No comprehensive account in English is available for the identification of British leaf beetles, although most species may be identified by reference to recent Continental works (e.g. Mohr, 1966). However, for some subfamilies, tribes or genera good up-to-date keys to both adults and larvae in English are already available (see below). In addition, others covering Cryptocephalinae, Chrysomelinae (in part) and *Psylliodes* (by M.L. Cox), and on species of *Chaetocnema* and *Longitarsus* (by R.G. Booth) are in preparation or in press. Identification of species in the large and difficult genus *Longitarsus* remains something of a specialist matter despite recent revisional studies, as all currently available keys have some shortcomings. Further information on identification sources and help with identifications of Chrysomelidae may be obtained through the Leaf and Seed Beetle National Recording Scheme; see Recording Schemes (above) and Cox (1992).

Aldridge, R.J.W. & Pope, R.D. 1986. The British species of *Bruchidius* Schilsky (Coleoptera: Bruchidae). *Entomologist's Gazette* 37: 181–93.
Key to the three British species; extensively illustrated, including SEM pictures.

Allen, A.A. 1967. Two new species of *Longitarsus* Latr. (Col., Chrysomelidae) in Britain. *Entomologist's Monthly Magazine* 103: 75–82.

Allen, A.A. 1976. Notes on some British Chrysomelidae (Col.) including amendments and additions to the list. *Entomologist's Record and Journal of Variation* 88: 294–9.
Taxonomic notes, including distinguishing characters of *Psylliodes napi* and *P. weberi*.

Booth, R.G. 1994. *Longitarsus longiseta* Weise rediscovered and *Longitarsus obliteratoides* Gruev (Chrysomelidae) new to Britain. *Coleopterist* 3(1): 4–5.

Brandl, P. 1981. Bruchidae (Samenkäfer). *Die Käfer Mitteleuropas* Vol. 10: 7–21.
In German.

Cox, M.L. 1991. The larvae of the British *Phaedon* (Coleoptera: Chrysomelidae, Chrysomelinae). *Entomologist's Gazette* 42: 267–80.
Includes a key to adults of the four British species.

Cox, M.L. 1992. Progress report on the Bruchidae/Chrysomelidae recording scheme. *Coleopterist* 1(1): 18–24.
Includes confirmation of one species of *Longitarsus* as British, with distribution map.

Cox, M.L. 1995. Identification of the *Oulema 'melanopus'* species group (Chrysomelidae). *Coleopterist* 4 (2): 271–276.
   Good figures of male genitalia and spermatheca of the two British species.

Cox, M.L. 1995. *Psylliodes cucullata* (Illiger, 1807) (Coleoptera: Chrysomelidae, Alticinae), a species new to Britain. *Entomologist's Gazette* 46: 271–6.
   Distinguishing characters; extensively illustrated.

Döberl, M. 1994. Chrysomelidae: Alticinae. *Die Käfer Mitteleuropas.* Vol. 3. Supplementband und Katalogteil, pp. 92–142.
   Many of Mohr's (1966) keys revised; with many genitalia figures; in German.

Hodge, P.J. 1997. *Bruchidius varius* (Oliver) (Chrysomelidae) new to the British Isles. *Coleopterist* 5(3): 65–8.
   Good illustrations.

Hoffmann, A. 1945. Coléoptères Bruchides et Anthribides. *Faune de France* 44: 1–184.
   In French.

Kangas, E. & Rutanen, I. 1993. Identification of females of the Finnish species of *Altica* Müller (Coleoptera, Chrysomelidae). *Entomologica Fennica* 4: 115–29.
   Includes only three of the seven British species

Kendall, P. 1982. *Bromius obscurus* (L.) in Britain (Col., Chrysomelidae*)*. *Entomologist's Monthly Magazine* 117: 233–4.
   Distinguishing characters; habitus figures.

Kevan, D.K. 1963. The British species of the genus *Haltica* Geoffroy (Col., Chrysomelidae). *Entomologist's Monthly Magazine* 98: 189–96.
   Well-illustrated key.

Kevan, D.K. 1967. The British species of the genus *Longitarsus* Latreille (Col., Chrysomelidae). *Entomologist's Monthly Magazine* 103: 83–110.
   The most up-to-date key in English to the British species of this difficult flea beetle genus, but some of the figures are inaccurate and the key is unreliable in places.

Kippenberg, H. 1994. Chrysomelidae. *Die Käfer Mitteleuropas.* Vol. 14: 17–92.
   Many of Mohr's (1966) keys revised; with many genitalia figures; in German.

Laboissière, V. 1934. Galerucinae de la faune française. *Annales de la Société Entomologique de France* 103: 1–108.
   Includes all British species, with habitus figures; in French.

Lambelet, J. 1975. Les *Phyllodecta* de la faune française (Col., Chrysomelidae). *Entomologiste* 31: 154–8.
   Key to three out of four British species, with male genitalia figures; in French.

Menzies, I.S. & Cox, M.L. 1996. Notes on the natural history, distribution and identification of British reed beetles. *British Journal of Entomology and Natural History* 9: 137–62.
   Excellent and extensively illustrated key to all 21 British species of Donaciinae; colour photos of 16 species in natural settings.

Mohr, K.H. 1966. Chrysomelidae. *Die Käfer Mitteleuropas* Vol. 9: 95–280.
   Keys include most British species (but see Döberl, 1994 and Kippenberg, 1994); in German.

Morris, M.G. 1970. *Phyllodecta polaris* Schneider (Col., Chrysomelidae) new to the British Isles from Wester Ross and Inverness-shire, Scotland. *Entomologist's Monthly Magazine* 106: 48–53.

Illustrated key to the four British species of *Phyllodecta*.

Shute, S.L. 1976. *Longitarsus jacobaeae* Waterhouse: identity and distribution (Col., Chrysomelidae). *Entomologist's Monthly Magazine* 111: 33–9.

Distinguishing features of two species, with 16 useful figures.

Shute, S.L. 1976. A note on the specific status of *Psylliodes luridipennis* Kutschera (Col., Chrysomelidae). *Entomologist's Monthly Magazine* 111: 123–7.

Distinguishing features of four species of *Psylliodes*, illustrated with 24 figures.

Strand, A. 1962. Hannens genitalorgan hos de nordiske *Longitarsus-arter* (Col., Chrysomelidae). *Norsk Entomologisk Tidsskrift* 12: 25–6.

Photographs of aedeagi of most British species; in Norwegian.

Families Chrysomelidae and Megalopodidae: larvae

Böving, A.G. 1927. On the classification of the Mylabridae-larvae (Coleoptera: Mylabridae). *Proceedings of the Entomological Society of Washington* 29 (6): 133–42.

Key to genera and species of Bruchidae; cosmopolitan.

Cox, M.L. 1981. Notes on the biology of *Orsodacne* Latreille with a subfamily key to the larvae of the British Chrysomelidae (Coleoptera). *Entomologist's Gazette* 32: 123–35.

Revised subfamily key, and illustrated key to the British species of *Orsodacne*.

Cox, M.L. 1982. Larvae of the British genera of chrysomeline beetles (Coleoptera, Chrysomelidae). *Systematic Entomology* 7: 297–310.

Key to 11 genera.

Cox, M.L. 1991. The larvae of the British *Phaedon* (Coleoptera: Chrysomelidae, Chrysomelinae). *Entomologist's Gazette* 42: 267–80.

Key to the larvae of the four British species; profusely illustrated.

Cox, M.L. 1996. The pupae of Chrysomeloidea. *In* P.H. Jolivet and M.L. Cox (Eds.) *Chrysomelidae Biology, vol. 1: The Classification, Phylogeny and Genetics*, SPB Academic Publishing, Amsterdam, The Netherlands: pp. 119–265.

Contains descriptions of pupae of 80 British species, with most of these illustrated.

Cox, M.L. 1997. The larva of the flea beetle, *Mniophila muscorum* (Koch, 1803) (Coleoptera: Chrysomelidae, Alticinae), not a leaf-miner. *Entomologist's Gazette* 48: 275–83.

Illustrated description of first instar larva.

Emden, F. I. van 1962. Key to species of British Cassidinae larvae (Col., Chrysomelidae). *Entomologist's Monthly Magazine* 98: 33–6.

All but two British species included; 11 figures.

Hennig, W. 1938. Übersicht über die Larven der wichtigsten deutschen Chrysomelinen. *Arbeiten über Physiologische und Angewandte Entomologie* 5(2): 85–136.

Key to genera and species of European Chrysomelinae; includes all British genera and most species, with some figures of chaetotaxy; in German.

Henriksen, K.L. 1927. Larver. Chrysomelidae. *Danmarks Fauna* 31: 290–376.
   In Danish.

Marshall, J.E. 1979. The larvae of the British species of *Chrysolina* (Chrysomelidae). *Systematic Entomology* 4: 409–17.
   Key to 15 species; good illustrations.

Marshall, J.E. 1981. A key to some larvae of the British Galerucinae and Halticinae (Coleoptera: Chrysomelidae). *Entomologist's Gazette* 31: 275–83.
   Covers all genera and some species of externally feeding Galerucinae and Alticinae.

Paterson, N.F. 1931. The bionomics and comparative morphology of the early stages of certain Chrysomelidae (Coleoptera, Phytophaga). *Proceedings of the Zoological Society of London* 1931: 879–949.
   Contains subfamily and tribal keys to larvae and pupae of British species, with illustrated descriptions of some 36 species.

Superfamily Curculionoidea: weevils

Adult weevils are characterised by the elongation of the anterior part of the head to form a rostrum, with accompanying mouthpart modifications. They are unusual among the beetles in using the mouthparts and rostrum in oviposition, as these (rather than ovipositors) are used to make incisions in plant tissues into which the eggs of many species are laid. Phytophagous habits, in the broad sense, are almost universal in the group, notable exceptions being the mycophagous or predaceous Anthribidae, and those Scolytinae (bark beetles) and Platypodinae that feed on 'ambrosia' fungi. Almost every major group of vascular plants (including ferns) and almost every type of plant tissue is fed upon by one or other species of weevil. Their larvae, almost all of them completely lacking legs, mostly feed internally in plant tissues including roots, stems, buds, leaves, flowers, fruit and seeds, some species provoking the development of galls. Some soil-dwelling species feed externally on plant roots and a number of species, especially of Cossoninae and Scolytinae, specialise in woody tissues, some feeding on dead and decaying wood. Some wood-feeding Curculionoidea are forestry pests, and a range of other species achieve pest status in field crops, orchards and food stores, either as a direct result of their depredations or as vectors of plant diseases. Like leaf-beetles, some weevils have been employed successfully to control weeds, especially those invading lakes and other water bodies.

The higher classification of the weevils is in a state of flux and unlikely to be settled for some time to come (see Morris, 1995; Thompson, 1992, etc.). The arrangement adopted here is a relatively conservative one and is essentially that of Lawrence & Newton (1995) (see Coleoptera introduction), with just five families recognised: Nemonychidae, Anthribidae, Attelabidae, Brentidae (including Apioninae and Nanophyinae) and Curculionidae (including Scolytidae, Platypodidae and Raymondionymidae).

Two excellent volumes dealing with weevils in the Royal Entomological Society's *Handbooks* series by M.G. Morris (1990, 1997) have already appeared, and a final volume by the same author is in preparation. For the groups covered by the first *Handbook* reference to other works is kept to a minimum. Relatively few references also are given for the broad-nosed weevils covered in the second book (Morris, 1997). Duffy's *Handbook* on Scolytidae and Platypodidae is now out-of-date, but there are several Continental and other works that cover all British species of these groups (now regarded as subfamilies of Curculionidae) that may be consulted (see below).

Allen, A.A. 1947. *Dorytomus filirostris* Gyll. (Col., Curculionidae), a weevil new to Britain. *Entomologist's Monthly Magazine* **83**: 52.
  Distinguishing characters; no figures.
Allen, A.A. 1950. *Coeliodes nigritarsis* Hartmann (Col., Curculionidae) in Scotland: an addition to the British fauna. *Entomologist's Monthly Magazine* **86**: 88–9.
  Key to distinguish from related species; no figures.
Allen, A.A. 1968. *Dorytomus affinis* Payk. (Col., Curculionidae) in Kent and notes on its British allies. *Entomologist's Monthly Magazine* **103**: 264–7.
  Key to distinguish *D. dejeani* and *D. taeniatus.*
Allen, A.A. 1970. *Ernoporus caucasicus* Lind. and *Leperisinus orni* Fuchs (Col., Scolytidae) in Britain. *Entomologist's Monthly Magazine* **105**: 245–9.
  Keys to the British species of *Ernoporus* and *Leperisinus*; no figures.
Allen, A.A. 1974. *Rhinoncus albicinctus* Gyll. (Col., Curculionidae). *Entomologist's Monthly Magazine* **109**: 188–90.
  Distinguishing characters; habitus figure.
Allen, A.A. 1981. *Tychius crassirostris* Kirsch, a weevil new to Britain; with some remarks on the problem of the British '*T. haematopus*'. *Entomologist's Record and Journal of Variation* **93**: 161–4.
  Table to distinguish three British *Tychius* species; no figures.
Allen, A.A. 1992. On *Bagous arduus* Sharp and *B. rudis* Sharp (Col.: Curculionidae). *Entomologist's Record and Journal of Variation* **104**: 199–201.
  The identity of Sharp's species discussed; male genitalia figured.
Balachowsky, A. 1949. Scolytides. *Faune de France* **50**: 1–320.
  Covers virtually all British and many other species; good habitus figures and lists of known hosts. Nomenclature now out of date; in French.
Bruce, N. 1968. The nordic species of the beetle genus *Bagous* (Coleoptera Curculionidae) with a key. *Entomologisk Tidskrift.* **89**: 229–41.
Burke, H.R. & Clark, W.E. (Eds.) *Curculio.*
  A twice yearly newsletter dealing with Curculionoidea on a world basis, providing infor-

mation on current research, literature, etc. Available from Dept of Entomology, Texas A & M University, College Station, Texas 77843, USA.

Caldara, R. 1990. Revisione tassonomica delle specie paleartiche del genere *Tychius* Germar (Coleoptera Curculionidae). *Memorie della Società Italiana di Scienze Naturali e del Museo Civico di Storia Naturale de Milano* **25**: 53–218.
Monographic treatment of Palaearctic species, including all those found in Britain; in Italian.

Clemons, L. 1983. *Gronops inaequalis* Boheman (Col., Curculionidae). *Entomologist's Record and Journal of Variation* **95**: 213–215.
Distinguishing characters from *G. lunatus*; habitus figure.

Dieckmann, L. 1972. Beiträge zur Insektenfauna der DDR: Coleoptera-Curculionidae: Ceutorhynchinae. *Beiträge zur Entomologie* **22**: 3–128.
Keys to the German species, illustrated by 141 figures; in German.

Dieckmann, L. 1973. Die westpaläarktischen *Thamiocolus*-Arten (Coleoptera, Curculionidae). *Beiträge zur Entomologie* **23**: 245–73.
Key to 17 species, illustrated with 51 figures; in German.

Dieckmann, L. 1980. Beiträge zur Insektenfauna der DDR: (Coleoptera, Curculionidae: Brachycerinae, Otiorhynchinae, Brachyderinae). *Beiträge zur Entomologie* **30**: 145–310.
Keys to about 200 species, illustrated by 172 figures; in German.

Dieckmann, L. 1983. Beiträge zur Insektenfauna der DDR: Coleoptera – Curculionidae (Tanymecinae, Leptopiinae, Cleoninae, Tanyrhynchinae, Cossoninae, Raymondionyminae, Bagoinae, Tanysphyrinae). *Beiträge zur Entomologie* **33**: 257–381.
In German.

Dieckmann, L. 1986. Beiträge zur Insektenfauna der DDR: Coleoptera Curculionidae (Erirhinae). *Beiträge zur Entomologie* **36**: 119–81.
Keys to genera and species known from Germany, illustrated with 52 figures; in German.

Dieckmann, L. 1988. Beiträge zur Insektenfauna der DDR: Coleoptera Curculionidae (Curculioninae: Ellescini, Acalyptini, Tychiini, Anthonomini, Curculionini). *Beiträge zur Entomologie* **38**: 365–468.
Keys to the German species, illustrated by 124 figures; in German.

Duffy, E.A.J. 1953. Scolytidae and Platypodidae. *Handbooks for the Identification of British Insects* **5**(15): 20pp.
Now rather out-of-date.

Emden, F.I. van 1938. On the taxonomy of Rhynchophora larvae (Coleoptera). *Transactions of the Royal Entomological Society of London* **87**: 1–37.

Emden, F.I. van 1952. On the taxonomy of Rhynchophora larvae: Adelognatha and Alophinae (Insecta, Coleoptera). *Proceedings of the Zoological Society of London* **122**: 651–759.

Folwaczny, B. 1973. Bestimmungstabelle der paläarktischen Cossoninae (Coleoptera, Curculionidae). *Entomologische Blätter für Biologie und Systematik der Käfer* **69**: 65–180.
Useful habitus figures; includes Dryophthorinae; in German.

Fowler, V.W. 1964. The identification of *Otiorrhynchus* larvae from blackcurrant roots with descriptions of the larvae of *O. clavipes* (Bonsd.) and *O. singularis* (L.) (Col., Curculionidae). *Entomologist's Monthly Magazine* 99: 210–12.

Fowles, A.P. 1994. A provisional key to weevils of the genus *Hypera* (Germar) (Curculionidae) recorded from Great Britain. *Coleopterist* 3: 15–20.

Fowles, A.P. & Morris, M.G. 1994. *Apion (Helianthapion) aciculare* Germar (Col., Apionidae), a weevil new to Britain. *Entomologist's Monthly Magazine* 130: 177–81.
    Amendment to Morris's (1990) key, with good habitus figure.

Frieser, R. von 1981. Anthribidae (Breitmaulrüssler). *Die Käfer Mitteleuropas.* Vol. 10: 22–4.
    In German.

Hoffmann, A. 1945. Coléoptères Bruchides et Anthribides. *Faune de France* 44: 1–184.
    In French.

Hoffmann, A. 1950, 1954, 1958. Coléoptères Curculionides. *Faune de France* 52: 1–486; 59: 487–1208; 62: 1209–839.
    A well-illustrated monograph with many useful habitus figures and much biological information; extensively updated by Tempère & Pericart (1989); in French.

Jermiin, L.S. & Mahler, V. 1993. Revised descriptions of the morphology of *Trachyphloeus bifoveolatus, T. angustisetulus, T. spinimanus* and *T. digitalis* (Coleoptera: Curculionidae). *Entomologist's Gazette* 44: 139–53.
    Key to separate these 'difficult' species; profusely illustrated.

Johnson, C. 1982. *Phytobius olssoni* Israelson (Coleoptera, Curculionidae) new to Britain. *Entomologist's Gazette* 33: 221–2.
    Key to distinguish from *C. quadrituberculatus*; male genitalia figured.

Kalina, V. 1970. A contribution to the knowledge of the larvae of European bark beetles (Coleoptera, Scolytidae). *Acta Entomologica Bohemoslovaca* 67: 116–32.
    Illustrated key to species of *Phloeophthorus, Cryphalus* and *Pityogenes*.

Kenward, H.K. 1990. A belated record of *Procas granulicollis* Walton (Col., Curculionidae) from Galloway, with a discussion of the British *Procas* spp. *Entomologist's Monthly Magazine* 126: 21–5.
    Distinguishing characters of the two British *Procas* species.

Kevan, D.K. 1960. The British species of the genus *Sitona* Germar (Col., Curculionidae). *Entomologist's Monthly Magazine* 95: 251–61 (see also 100: 91–3).
    Well-illustrated keys.

Lee, C.-Y. & Morimoto, K. 1988. Larvae of the weevil family Curculionidae of Japan Part 1. Key to genera and the short-nosed group (Insecta: Coleoptera). *Journal of the Faculty of Agriculture, Kyushu University* 33(1–2): 109–30.
    Keys to subfamilies, genera, tribes and species.

Lohse, G.A. *et al.* 1981. Curculionidae (Rüsselkafer). *Die Käfer Mitteleuropas.* Vol. 10: 102–279.
    Keys to subfamilies, genera, and species of ten subfamilies (Rhinomacerinae to Leptopinae); in German.

et al. 1983. Familienreihe: Rhynchophora (Schluss). *Die Käfer Mitteleuropas.* Vol. 11: 1–342.
In German.

Massee, A.M. & Gardner, A.E. 1963. *Ips cembrae* Heer (Col., Scolytidae) in Britain. *Entomologist's Monthly Magazine* **98**: 225–6.
Description and key to distinguish from other British species of *Ips.*

Michalski, J. 1973. *Revision of the palaearctic species of the genus Scolytus Geoffroy (Coleoptera, Scolytidae).* Polish Academy of Sciences, Warsaw and Cracow.

Morris, M.G. 1966. *Ceuthorhynchus unguicularis* C. G. Thomson (Col., Curculionidae) new to the British Isles, from the Suffolk Breckland and the Burren, Co. Clare. *Entomologist's Monthly Magazine* **101**: 279–286.
Key to the 12 British species of *Ceutorhynchus* s. str.

Morris, M.G. 1977. The British species of *Anthonomus* Germar (Col., Curculionidae). *Entomologist's Monthly Magazine* **112**: 19–40.
Well illustrated keys to the 12 British species of *Anthonomus*, with extensive notes.

Morris, M.G. 1990. Orthocerous Weevils. Coleoptera; Curculionidae (Nemonychidae, Anthribidae, Urodontidae, Attelabidae and Apionidae). *Handbooks for the Identification of British Insects.* 5(16): 108pp.
A well illustrated modern handbook covering all but one (see Fowles & Morris, 1994) of the species of these families known to occur in Britain; an essential key work.

Morris, M.G. 1991. A taxonomic check list of the British Ceutorhynchinae, with notes, particularly on host plant relationships (Coleoptera: Curculionidae). *Entomologist's Gazette* **42**: 255–65.
Revised check list, with 93 species referred to 29 genera; a useful starting point for identifying species in this 'difficult' group.

Morris, M.G. 1991. *Weevils.* Naturalists Handbooks No. 16. Richmond Publishing, Slough.

Morris, M.G. 1993. A review of the British species of Rhynchaeninae (Col., Curculionidae*). Entomologist's Monthly Magazine* **129**: 177–97.
Key to 19 species; well illustrated.

Morris, M.G. 1995. Recent advances in the higher systematics of Curculionoidea, as they affect the British fauna. *Coleopterist* **4**(1): 21–30.

Morris, M.G. 1997. Broad-nosed Weevils. Coleoptera: Curculionidae (Entiminae). *Handbooks for the Identification of British Insects* 5(17a): 106pp.
A well illustrated modern handbook; an essential key work.

Nash, D.R. 1983. The genus *Hylastes* Erichson (Col., Scolytidae). *Coleopterist's Newsletter* **14**: 5–6.
Practical key to British species of *Hylastes.*

Pesarini, C. 1980. Le specie paleartiche occidentali della tribù Phyllobiini (Coleoptera Curculionidae). *Bolletino di Zoologia Agraria e Bachicoltura* (Ser. II) **15**: 49–230.
Illustrated key to species of *Phyllobius*; in Italian.

Pfeffer, A. 1995. *Zentral- und westpaläarktische Borken- und Kernkäfer (Coleoptera: Scolytidae, Platypodidae).* Pro Entomologia, Basel.
The most comprehensive of recent works covering British bark-beetles; in German.

Roudier, A. 1980. Les *Sitona* Germar 1817 du groupe de *Sitona humeralis* Stephens 1831. *Bulletin de la Société Entomologique de France* **85**: 207–217.

Key illustrated with male genitalia figures; in French.

Schedl, K. 1981. Scolytidae (Borken- und Ambrosiakäfer). Platypodidae (Kernkäfer). *Die Käfer Mitteleuropas* Vol. 10: 34–99, 100–101.

In German.

Scherf, H. 1964. Die Entwicklungsstadien der mitteleuropäischen Curculioniden (Morphologie, Bionomie, Ökologie). *Abhandlungen von der Senckenbergischen Naturforschenden Gesellschaft* **506**: 1–335.

Key to species based on bionomic characters, including larval leaf-mines, galls, leaf-rolls and other workings.

Tempère, G. & Pericart, J. 1989. Coléoptères Curculionidae (4me partie). *Faune de France* **74**: 1–534.

Includes updated keys to many of the more difficult genera, and a revised checklist of French Curculionidae *sensu lato*; in French.

Thompson, R.T. 1989. A preliminary study of the weevil genus *Euophryum* Broun (Coleoptera: Curculionidae: Cossoninae). *New Zealand Journal of Zoology* **16**: 65–79.

Distinguishing characters of the two species of *Euophryum* established in Britain, and of these from *Pentarthrum*; excellent illustrations.

Thompson, R.T. 1992. Observations on the morphology and classification of weevils (Coleoptera: Curculionoidea) with a key to major groups. *Journal of Natural History* **26**: 835–91.

See comments under 'Curculionoidea' heading.

Thompson, R.T. 1995. Raymondionymidae (Col., Curculionoidea) confirmed as British. *Entomologist's Monthly Magazine* **131**: 61–64.

Includes photograph of *Raymondionymus* (included here in Curculionidae).

Welch, R.C. 1990. *Macrorhyncolus littoralis* (Broun) (Col., Curculionidae), a littoral weevil new to the Palaearctic region from two sites in Kent. *Entomologist's Monthly Magazine* **126**: 97–101.

Distinguishing characters of this cossonine weevil illustrated.

Winter, T.G. 1990. *Crypturgus subscribosus* Eggers (Col., Scolytidae) a bark beetle new to Britain. *Entomologist's Monthly Magazine* **126**: 209–211.

Distinguishing characters; no figures.

Strepsiptera: the stylops

(*ca.* 20 species in 3 families)

PETER M. HAMMOND AND STUART J. HINE

Although included in a number of works on Coleoptera as a superfamily (Stylopoidea) of that Order, the Strepsiptera remain a group of uncertain position within the Endopterygota. However, recent work points to a closer relationship of the Strepsiptera to the Diptera than to either the Coleoptera or Hymenoptera.

As obligate endoparasites of other insects (the hosts of known British species are all aculeate Hymenoptera or Homoptera), Strepsiptera are rarely met with as such. However, hosts that have been 'stylopised' are readily recognisable and are seen more frequently. Tiny first-instar larvae are sometimes encountered, occasionally in numbers, on flower-heads, and the free-living, winged and short-lived adult males, most frequently those of *Elenchus tenuicornis*, are occasionally captured in flight.

The three families represented in Britain are: Stylopidae, Halictophagidae, Elenchidae.

References

Crowson, R.A. 1954. The classification of the families of British Coleoptera. Superfamily 19: Stylopidae (Order Strepsiptera). *Entomologist's Monthly Magazine* **90**: 57–63.

Kathirithamby, J. 1989. Review of the Order Strepsiptera. *Systematic Entomology* **14**: 41–92.

The most up-to-date introduction to and general account of the Order; includes keys to families, subfamilies and most genera.

Kinzelbach, R.K. 1969. Stylopidae, Fächerflügler (= Ordnung: Strepsiptera). *Die Käfer Mitteleuropas.* Vol. 8: 139–59.

Covers the Central European species of the five families represented in Europe, with useful habitus figures of adult males; in German.

Kinzelbach, R.K. 1978. Strepsiptera. *Tierwelt Deutschlands* **65**: 116pp.

A fuller account than the author's previous works on the topic; in German.

Perkins, R.C.L. 1918. Synopsis of British Strepsiptera of the genera *Stylops* and *Halictoxenus. Entomologist's Monthly Magazine* **54**: 67–76.

Covers 14 species, with illustrated key to males of eight species (see also pp. 107–8, dealing with one additional species).

Mecoptera: the scorpionflies

(4 species in 2 families)

PETER C. BARNARD

The Mecoptera form a distinct group of insects, easily recognised by their long, downward-pointing 'beak' which has the mouthparts at the tip. The panorpids take their common name of scorpionflies from the swollen genital capsules of the male, which are held recurved over the body like a scorpion's tail; they are a common sight in rough grassland, woodland margins, etc. in spring and summer. The striking wing-markings are not as reliable for identification as once thought, and all modern keys emphasise the need to examine the genitalia. The sole British member of the Boreidae is known as the snow-flea, as it is mature and active through the winter months and is relatively easy to see on the surface of lying snow: it is undoubtedly more common than currently reported, but is overlooked because of its small size and inconspicuous habit, living among mosses on moors and heathland.

The two families in Britain are Boreidae and Panorpidae.

Fraser, F.C. 1959. Mecoptera, Megaloptera and Neuroptera. *Handbooks for the Identification of British Insects* **1** (12, 13): 40pp.
Fraser's key to males still works if used with care, but Plant (1997) must be used to identify females reliably.

Plant, C.W. 1994. *Provisional atlas of the lacewings and allied insects (Neuroptera, Megaloptera, Raphidioptera and Mecoptera) of Britain and Ireland.* Biological Records Centre, Huntingdon.
Useful summary of known distributions, particularly emphasising the widespread distribution of *Boreus hyemalis*.

Plant, C.W. 1997. A key to the adults of British lacewings and their allies (Neuroptera, Megaloptera, Raphidioptera and Mecoptera). *Field Studies* **9**: 179–269.
Better than Fraser (1959) because it includes a good key to females.

Trichoptera: the caddisflies

(198 species in 18 families)

PETER C. BARNARD

The caddisflies are one of the larger groups of freshwater insects when compared with the Odonata, Ephemeroptera or Plecoptera, for example. Of the 198 species currently recorded from the British Isles, three (*Apatania auricula, Limnephilus fuscinervis* and *Tinodes maculicornis*) are found only in Ireland, and four (*Hydropsyche bulgaromanorum, H. exocellata, Orthotrichia tragetti* and *Oxyethira distinctella*) are almost certainly extinct. Caddis larvae are found in all kinds of fresh water, although each species may have narrow habitat requirements, and they can therefore be useful indicators of water quality. Most species live in characteristic transportable cases, and the shape and construction can be a useful guide to identification at least to generic level. However, 46 of the British species do not build such cases, but instead construct silk nets to trap food, or live in silk galleries attached to rocks, or else are free-living predators. One remarkable case-building species, *Enoicyla pusilla*, is terrestrial and lives in damp leaf-litter in the Wyre Forest area of Hereford & Worcester. It is not easy to identify some of the case-building Limnephilidae to species level, and a good stereo microscope is essential for specific determination of most larvae. However, there are good keys available: Wallace *et al.* (1990) for the case-bearing larvae, and Edington & Hildrew (1995) for the caseless species.

Adult caddisflies are rather inconspicuous moth-like insects with wing-lengths from 3 mm to 28 mm. Most are nocturnal or crepuscular, and many species are attracted to moth-collectors' light-traps. The day-flying species, particularly the 'long-horned' Leptoceridae with conspicuously long antennae, are frequently seen swarming around suitable freshwater sites. Many species are important as natural food for fish such as trout, and fly-fishermen use many artificial caddisflies or 'sedges'. These species therefore have common names, as listed by Price (1989) and Goddard (1988, 1991). Apart from these commoner, more conspicuous species, identification can be very difficult, even with the aid of a good stereo microscope, partly because the existing literature is not easy to use and partly because many species can be separated only by minute differences in the genitalia. The family-level classification is based mainly on the number of spurs on the legs and on the wing venation, and even these can be difficult for the novice to see clearly. The most recent key to adult caddis of the British Isles is by Macan (1973) although this is now out of print: it also contains some errors and inconsistencies, which were pointed out in the latest checklist (Barnard, 1985) and in Wallace's (1991) review of the status of the British species. Some workers still find the much older work by Mosely (1939) useful even though the taxonomy is now very out of date. The atlas of European species by Malicky (1983) contains genitalia figures of all the British species (excluding the females of some important groups such as *Hydropsyche*) but is mainly of use to the specialist. Other useful papers are listed below, with notes on their particular value.

The eighteen families in Britain are: Rhyacophilidae, Glossosomatidae, Hydroptilidae, Philopotamidae, Psychomyiidae, Ecnomidae, Polycentropodidae, Hydropsychidae, Phryganeidae, Brachycentridae, Lepidostomatidae, Limnephilidae, Goeridae, Beraeidae, Sericostomatidae, Odontoceridae, Molannidae, Leptoceridae.

References

Barnard, P.C. 1985. An annotated check-list of the Trichoptera of Britain and Ireland. *Entomologist's Gazette* 36: 31–45.
  The only recent change to this list is that *Apatania nielseni* is no longer regarded as a species separate from *A. muliebris*, thus reducing the number of British species to 198: see the next reference.
Barnard, P.C. & O'Connor, J.P. 1987. The populations of *Apatania muliebris* in the British Isles (Trichoptera: Limnephilidae). *Entomologist's Gazette* 38: 263–8.
  See note above.
Edington, J.M. & Hildrew, A.G. 1995. Caseless Caddis Larvae of the British Isles. *Scientific Publications of the Freshwater Biological Association* 53: 134pp.
  The only modern work on the caseless larvae, with good keys and comprehensive information on life histories and biology. Essential for anyone working on caddis larvae.
Fisher, D. 1977. Identification of adult females of *Tinodes* in Britain (Trichoptera: Psychomyiidae). *Systematic Entomology* 2: 105–10.
  The only key to females of this group, not covered by Macan, although Fisher's drawings are reproduced in Malicky's atlas.

Goddard, J. 1988. *John Goddard's waterside guide.* Unwin Hyman Ltd, London.
A handy pocket-sized guide to the commoner freshwater insects, with notes on their biology, many colour photos, and illustrations of the corresponding artificial flies.

Goddard, J. 1991. *Trout flies of Britain and Europe.* A & C Black, London.
Concise accounts of the biology and identification of the commoner insects of interest to fishermen, with 500 colour photographs of the insects and artificial flies.

Hildrew, A.G. & Morgan, J.C. 1974. The taxonomy of British Hydropsychidae (Trichoptera). *Journal of Entomology* (B) 43: 217–29.
Good for identifying *Hydropsyche* males, which are not properly covered by Macan's (1973) book, but see also Tobias (1972) for both sexes.

Macan, T.T. 1973. A key to the adults of the British Trichoptera. *Scientific Publications of the Freshwater Biological Association* 28: 151pp.
Although the most up-to-date work on adult caddis, it is best used with the other works listed here.

Malicky, H. 1983. Atlas of European Trichoptera. *Series Entomologica* 24: 298pp.
This expensive volume is of value to the specialist because for many of the species it provides better illustrations of male genitalia than Macan's book.

Marshall, J.E. 1978. Trichoptera: Hydroptilidae. *Handbooks for the Identification of British Insects* 1(14a): 31pp.
More reliable than Macan's book for the identification of these 'micro-caddis', and was brought up to date by the following paper.

Marshall, J.E. 1979. The female of *Hydroptila tigurina* Ris (Trichoptera: Hydroptilidae). *Entomologist's Gazette* 30: 213–14.
Fills the only gap in Marshall's handbook to this family.

Mosely, M.E. 1939. *The British caddis flies (Trichoptera).* Routledge, London.
Although now very out of date, this pioneering book on the adult caddisflies is still of some value for identifying some groups of species.

O'Connor, J.P. 1982. *Ithytrichia clavata* (Trichoptera: Hydroptilidae), a caddisfly new to Ireland. *Irish Naturalists' Journal* 20: 548–9.
Contains a small correction to Marshall's (1978) key.

O'Connor, J.P. 1987. A review of the Irish Trichoptera. *In* Bournaud, M. & Tachet, H. (Eds) Proceedings of the 5th International Symposium on Trichoptera. *Series Entomologica* 39: 73–7.
A review of the status and distribution of the Irish species.

O'Connor, J.P. & Barnard, P.C. 1981. *Limnephilus tauricus* Schmid (Trichoptera: Limnephilidae) new to Great Britain, with a key to the *L. hirsutus* (Pictet) group in the British Isles. *Entomologist's Gazette* 32: 115–19.
Supplements the limnephilid key in Macan's (1973) book.

Price, Taff 1989. *The angler's sedge: tying and fishing the caddis.* Blandford Press, London.
Like Goddard's books this only covers the species of interest to anglers, with emphasis on the artificial flies, but is unusual in concentrating on the caddisflies, or sedges.

Tobias, W. 1972. Zür Kenntnis europäischer Hydropsychidae (Trichoptera). *Senckenbergiana Biologica* 53: 59–89, 245–68, 391–401.

This three-part paper is useful for identifying both sexes of Hydropsychidae, as these are not adequately covered in Macan's (1973) book.

Wallace, I.D. 1991. A review of the Trichoptera of Great Britain. *Research and Survey in Nature Conservation* 32: 59pp. Nature Conservancy Council, Peterborough.

As well as reviewing the status and distribution of all the British species, this work also contains a valuable bibliography to the key literature on this group. Dr Ian Wallace runs the Trichoptera Recording Scheme in Britain.

Wallace, I.D., Wallace, B. & Philipson, G.N. 1990. A key to the case-bearing caddis larvae of Britain and Ireland. *Scientific Publications of the Freshwater Biological Association* 51: 237pp.

The only modern work on the case-building larvae, with good keys and comprehensive information on life histories and biology. Like Edington & Hildrew's book, essential for anyone working on caddis larvae.

Lepidoptera: the moths and butterflies

(*ca.* 2500 species in 58 families)

MARK S. PARSONS, GADEN S. ROBINSON,
MARTIN R. HONEY AND DAVID J. CARTER

The Lepidoptera are one of the largest and most important groups of phytophagous insects in Britain. The adults and larvae are generally conspicuous and familiar and there has been a great deal of both taxonomic and biological work on the group in the past two hundred years. The British fauna is well known but one or two species new to Britain are still discovered each year.

Lepidoptera are to be found in all terrestrial habitat types, and a few species have aquatic early stages. Larvae feed either concealed (leaf-mining, feeding internally in stems, fruits, seeds or even tree-trunks, living in a webbed fold of leaf or in spun-together flowers, or living in a silk web) or exposed on a very wide range of green plants. Larvae of a few species (notably in the families Incurvariidae, Tineidae and Oecophoridae) feed on fungi or plant or animal detritus. Larvae of some species (especially those of Psychidae and Coleophoridae) build a portable case and feed from this, incrementing the case as the larva grows.

Higher classification of British Lepidoptera

Suborder Zeugloptera
Superfamily Micropterigoidea: Micropterigidae

Suborder Glossata
Superfamily Eriocranioidea: Eriocraniidae
Superfamily Hepialoidea: Hepialidae
Superfamily Nepticuloidea: Nepticulidae, Opostegidae
Superfamily Tischerioidea: Tischeriidae

Superfamily Incurvarioidea: Heliozelidae, Adelidae, Incurvariidae, Prodoxidae
Superfamily Tineoidea: Psychidae, Tineidae
Superfamily Gracillarioidea: Gracillariidae, Roeslerstammiidae, Bucculatricidae, Douglasiidae
Superfamily Yponomeutoidea: Yponomeutidae, Ypsolophidae, Lyonetiidae, Glyphipterigidae, Heliodinidae
Superfamily Gelechioidea: Oecophoridae, Ethmiidae, Elachistidae, Coleophoridae, Momphidae, Scythrididae, Blastobasidae, Cosmopterigidae, Gelechiidae
Superfamily Cossoidea: Cossidae
Superfamily Tortricoidea: Tortricidae
Superfamily Sesioidea: Sesiidae, Choreutidae
Superfamily Zygaenoidea: Zygaenidae, Limacodidae
Superfamily Schreckensteinioidea: Schreckensteiniidae
Superfamily Epermenioidea: Epermeniidae
Superfamily Alucitoidea: Alucitidae
Superfamily Pyraloidea: Pyralidae
Superfamily Pterophoroidea: Pterophoridae
Superfamily Hesperioidea: Hesperiidae
Superfamily Papilionoidea: Papilionidae, Pieridae, Nymphalidae, Lycaenidae
Superfamily Drepanoidea: Drepanidae
Superfamily Geometroidea: Geometridae
Superfamily Bombycoidea: Lasiocampidae, Saturniidae, Endromidae, Sphingidae
Superfamily Noctuoidea: Notodontidae, Lymantriidae, Arctiidae, Ctenuchidae, Herminiidae, Noctuidae

The higher classification of some parts of the Lepidoptera is still the subject of debate (the system given above is based on Emmet & Heath, 1991; see 'Immature Stages' section, below), and several other group-names are commonly used in the literature. For example, the Ditrysia contains all the superfamilies from the Tineoidea onwards in the above list, and the butterflies are often given the group name Rhopalocera, which comprises the Hesperioidea and Papilionoidea. Traditionally the Lepidoptera have been divided into three artificial groups for the purposes of practical study: the butterflies; the larger moths, or Macrolepidoptera (Noctuoidea, Geometroidea, Bombycoidea, Hepialoidea, Cossoidea, Sesioidea and Zygaenoidea); and the smaller moths, or Microlepidoptera (all other moth superfamilies). The first two groups have received considerably more attention than the last but there has been a renaissance of interest in Microlepidoptera in the past 15 years. The bibliography below is arranged by these divisions and subdivided further into works that deal with identification and those that deal with distribution and other topics. The main divisions are: General Lepidoptera, Immature stages, Microlepidoptera, Macrolepidoptera, and Butterflies, with the Microlepidoptera and Macrolepidoptera also subdivided into superfamilies.

Most butterflies and larger moths can be identified by comparison with illustrations and no equipment more sophisticated than a hand lens is needed. However, the separation of certain groups of species necessitates dissection and comparison of the genitalia, and a stereo microscope is needed for this. Study of many families of Microlepidoptera is difficult without a microscope, and dissection is often required for identification, particularly in the absence of life-history data.

The current definitive work for the identification of British Lepidoptera is the series of volumes of *Moths and butterflies of Great Britain and Ireland* ('MBGBI') of which 7 of 12 projected volumes have been published (Emmet, 1996; Emmet & Heath, 1985, 1989, 1991; Heath, 1976; Heath & Emmet, 1979, 1983). Major groups not yet covered by this series are most Gelechioidea (Coleophoridae and Elachistidae were dealt with by Emmet, 1996), Pyraloidea, Pterophoroidea and Geometroidea. However, there are other good modern works that also provide the means to identify butterflies (e.g. Thomas, 1986; Thomas & Lewington, 1991; Whalley, 1996), the larger moths (Skinner, 1984) and, among the Microlepidoptera, the Pyraloidea (Goater, 1986) and the Tortricoidea (Bradley, Tremewan & Smith, 1973, 1979). Thus the only major gaps are the Gelechioidea (but this superfamily will shortly be dealt with in volume 4 of MBGBI, in preparation) and the Pterophoroidea. Many of the papers listed in the bibliography for Microlepidoptera provide piecemeal coverage of the Gelechioidea; the Pterophoroidea may be identified from Beirne (1952) and Gielis (1996).

Other references listed in the 'identification' sections are complementary to the large basic works cited above in that they provide alternative illustrations, additional figures of genitalia, or add newly discovered resident British species not included in the key works.

The sections dealing with distribution and other topics encompass major regional or faunistic lists published subsequent to Chalmers-Hunt's bibliographical catalogue of local lists (1989).

There is no full synonymic checklist of the British Lepidoptera; synonymy in Kloet & Hincks (1972) is restricted to those synonyms relevant in the British context. Bradley (1998) is an updated and annotated version of the Bradley & Fletcher (1979) list.

Lepidoptera: general references (including checklists)

Bradley, J. D. 1998. *Checklist of Lepidoptera recorded from the British Isles.* Bradley & Bradley, Hampshire.
  Provides an annotated and numbered list, with a few synonyms.
Bradley, J.D. & Fletcher, D.S. 1979. *A recorder's log book or label list of British butterflies and moths.* Curwen Books, Plaistow.
  Provides a numbered checklist and label list to the British Lepidoptera. Not synonymic.

Bradley, J.D., Fletcher, D.S. & Hall-Smith, D.H. 1983. *A recorder's log book or label list of British butterflies and moths. Index together with addenda and corrigenda.* Leicestershire Museums Service, Leicester.
Provides an index and addenda and corrigenda to Bradley & Fletcher (1979). Not synonymic.

Bradley, J.D. & Fletcher, D.S. 1986. *An indexed list of British butterflies and moths.* Kedleston Press, Orpington.
Supersedes the two works above. Contains recent UK-relevant synonyms only. Marred by four pages of addenda/corrigenda and an omitted column (species nos. 1753–1766).

Dickson, R. 1976. *A lepidopterist's handbook.* Amateur Entomologist's Society, Middlesex.
An introduction to the study of Lepidoptera with chapters covering fieldwork, breeding, identification and recording.

Emmet, A.M. 1987. Addenda and corrigenda to the British list of Lepidoptera. *Entomologist's Gazette* 38: 31–52.
Provides a corrigenda to Bradley & Fletcher (1986) and Kloet & Hincks (1972).

Ford, E.B. 1972. *Moths* (3rd edn). Collins New Naturalist, London.
A general introduction to the natural history of moths, including structure and physiology, dispersal, geographical races and evolution. See also Young (1997).

Karsholt, O. & Razowski, J. (eds). 1996. *The Lepidoptera of Europe: a distributional checklist.* Apollo Books, Stenstrup.
A checklist to the Lepidoptera of Europe giving distribution on a country basis. Some limited synonymic notes included.

Kloet, G.S. & Hincks, W.D. 1972. A check-list of British Insects. 2. Lepidoptera. (2nd edn). *Handbooks for the Identification of British Insects* 11(2): 153pp.
Outdated, but this is the only British checklist to contain synonyms, although only those relevant in a British context.

Meyrick, E. 1928. *A revised handbook of British Lepidoptera.* Watkins & Doncaster, London.
A standard work, which has now been surpassed in many areas, but still useful as an aid to the identification of those microlepidoptera not covered by more recent publications, e.g. much of the Gelechiidae.

Novák, I. 1980. *A field guide in colour to butterflies and moths.* Octopus Books, London.
Broad non-specialist introduction to the whole of the Lepidoptera. Not comprehensive; coverage is central European.

Pierce, F.N. & Beirne, B.P. 1941. *The genitalia of the British Rhopalocera and larger moths.* F.N. Pierce, Oundle.
Contains illustrations, crude by modern standards, of the male and female genitalia of the butterflies and some families of moths, including the Notodontidae and Arctiidae.

Scoble, M.J. 1995. *The Lepidoptera: form, function and diversity.* Oxford University Press, Oxford.
A general treatise of the Lepidoptera covering morphology, environmental importance and classification.

Young, M. 1997. *The natural history of moths.* Poyser, London.
A good general introduction to the natural history of moths, including chapters on origin and distribution, dispersal and migration, life cycles, studying moths and conservation. See also Ford (1972).

Distribution

This section includes general distribution references, using Chalmers-Hunt (1989) as a baseline. Other county lists covering just microlepidoptera, macrolepidoptera or butterflies are given under the relevant section.

Arnold, V.W., Baker, C.R.B., Manning, D.V. & Woiwod, I.P. 1997. *The butterflies and moths of Bedfordshire.* The Bedfordshire Natural History Society. Bedford.
Summarises the history and distribution of over 1300 species found in Bedfordshire.

Baker, B.R. 1994. *The butterflies and moths of Berkshire.* Hedera Press, Uffington.
Discusses the distribution by presenting the records of over 1650 species recorded within the vice-county.

Chalmers-Hunt, J.M. 1989. *Local lists of Lepidoptera or a bibliographical catalogue of local lists and regional accounts of the butterflies and moths of the British Isles.* Hedera Press, Uffington.
Covers approximately 3000 titles, citing the known county and regional lists and local accounts of Lepidoptera in the British Isles.

Goater, B. 1992. *The butterflies and moths of Hampshire and the Isle of Wight: additions and corrections.* Joint Nature Conservation Committee, Peterborough. (UK Nature Conservation Committee, No. 7).
Provides an update to Goater (1974), *The butterflies and moths of Hampshire and the Isle of Wight.* Classey, Faringdon.

Goodey, B. & Firmin, J. 1992. *Lepidoptera of north east Essex.* Colchester Natural History Society, Colchester.
Records the species listed on a 10 km square basis and summarises distribution and status in this part of the county.

Hancock, E.G. 1990. Lepidoptera in vice-county 74 (Wigtown), June 1989. *Entomologist's Record and Journal of Variation* **102**: 107–9.
A useful list from an under-recorded vice-county.

Heckford, R.J. & Langmaid, J.R. 1991. Lepidoptera in Ireland, 1989. *Entomologist's Gazette* **42**: 15–29.
A useful list of species found in three Irish vice-counties.

Horton, G.A.N. 1994. *Monmouthshire Lepidoptera: the butterflies and moths of Gwent.* Comma International Biological Systems.
Provides the first comprehensive list of the Lepidoptera of Monmouthshire covering over 1150 species.

Johnson, R. 1996. *The butterflies and moths of Lincolnshire. The micro-moths and species review to 1996.* Lincolnshire Naturalist's Union, Lincoln.
A supplement to Duddington & Johnson (1983), *The butterflies and larger moths of*

Lincolnshire and South Humberside (Lincolnshire Naturalists' Trust, Lincoln) providing
coverage of the microlepidoptera of Lincolnshire.

Langmaid, J.R. 1989. Lepidoptera in Ireland, August 1986. *Entomologist's Gazette* 40:
307–13.
A useful list of species found in nine Irish vice-counties.

Palmer, R.M. & Young, M.R. 1991. Lepidoptera of Aberdeenshire and
Kincardineshire. 6th appendix. *Entomologist's Record and Journal of Variation*
103: 125–7.
Updating previous lists and recording 22 species new to the area.

Palmer, R.M. & Young, M.R. 1994. Lepidoptera of Aberdeenshire and
Kincardineshire. 7th appendix. *Entomologist's Record and Journal of Variation*
106: 85–9.
Updating previous lists and adding 11 species new to the area.

Peet, T.N.D. 1988. An introduction to Guernsey Lepidoptera. *Entomologist's Record
and Journal of Variation* 100: 21–4.
Discusses species found on Guernsey but not known at that time as breeding species on
the British mainland, and provides references to figures to the species. Also discusses a
selection of the more interesting microlepidoptera found on the island.

Pennington, M. 1997. Insects in Shetland. Lepidoptera. *Shetland Entomological
Group. Newsletter* 13: 6–11.
An annotated list of the species found on the Shetland Islands.

Plant, C.W. 1995. A review of the butterflies and larger moths of the London area for
1992–1994. *London Naturalist* 74: 145–57.
Provides an update to Plant (1993) and lists the important additions and corrections to
the London fauna. A few butterfly records of note are also included.

Plant, C.W. 1997. A review of the butterflies and moths (Lepidoptera) of the London
area for 1995 and 1996. *London Naturalist* 76: 157–74.
Provides further updates to Plant (1993) with some butterfly records.

Price, J.M. 1993. Lepidoptera of the Midland (Birmingham) Plateau. A concise
history 1890 – 1990. *Proceedings of the Birmingham Natural History Society*
26(3/4): 121– 277.
Discusses the distribution of just over 1300 species recorded from this area.

Riley, A.M. 1991. *A natural history of the butterflies and moths of Shropshire*. Swan
Hill Press, Shrewsbury.
Lists over 1250 species that have been recorded in Shropshire, giving the status of the
macrolepidoptera and listing records for the microlepidoptera.

Riley, A.M. & Palmer, R.M. 1994. Recent significant additions and corrigenda to the
list of Lepidoptera recorded in Shropshire. *Entomologist's Gazette* 45: 167–82.
Provides an update to Riley (1991).

Robbins, J. 1990. *The moths and butterflies of Exmoor National Park*. Exmoor Natural
History Society, Minehead.
A booklet listing over 1000 species recorded from the National Park.

Robbins, J. 1992. *Provisional atlas of the Lepidoptera of Warwickshire*. Part 3: *the
smaller moths and the more primitive larger moths*. Warwickshire Biological
Records Centre, Warwick.
Discusses status of these groups of moths and maps their distribution in Warwickshire.

Smith, F.N.H. 1997. *The moths and butterflies of Cornwall and the Isles of Scilly.* Gem
Publishing, Wallingford.
A summary of over 1500 species recorded from Cornwall and the Isles of Scilly, with
introductory chapters covering migration, distribution, etc.

Spalding, A. 1992. *Cornwall's butterfly and moth heritage.* Twelveheads Press, Truro.
A brief guide to the factors influencing the distribution of butterflies and moths in
Cornwall, discussing the habitat types and typical species found.

Sutton, S.L. & Beaumont, H.E. 1989. *Butterflies and moths of Yorkshire. Distribution
and conservation.* Yorkshire Naturalists' Union, Doncaster.
A summary of just under 2000 species recorded from Yorkshire, with records or/and summaries of status given for each species.

Young, M.R., Harper, M.W. & Christie, I. 1990. Lepidoptera on Colonsay and Oronsay,
Inner Hebrides. *Entomologist's Record and Journal of Variation* 102: 281–4.
A useful list of species found on the islands, including a number of additions to the fauna
of this group of islands.

Immature stages

Some of the references in this section simply give details of life history,
others figure the early stages, and in some cases also give details of life histories.

Buckler, W. 1886–1901. *The larvae of the British butterflies and moths.* Ray Society,
London. (Volumes 1–9): **1**, *The butterflies* (1886); **2**, *The Sphinges or hawk-moths
and part of the Bombyces* (1887);. **3**, *The concluding portion of the Bombyces*
(1889); **4**, *The first portion of the Noctuae* (1891); **5**, *The second portion of the
Noctuae* (1893); **6**, *The third and concluding portion of the Noctuae* (1895); **7**, *The
first portion of the Geometrae* (1897); **8**, *The concluding portion of the Geometrae*
(1899); **9**, *The Deltoides, Pyrales, Crambites, Tortrices, Tineae and Pterophori,
concluding the work* (1901).
The definitive work on the larvae of the British Lepidoptera, with exquisite colour plates.
This work has never been superseded, but care should be taken in updating the nomenclature. The date given on the spine of each volume precedes the year of publication in
every case and the volumes should be dated from the title-pages.

Carter, D.J. & Hargreaves, B. 1986. *A field guide to caterpillars of butterflies and moths
in Britain and Europe.* Collins, London.
Gives illustrations of the larvae and adults of a significant proportion of the British
Lepidoptera fauna, the illustrations being arranged by foodplant.

Emmet, A.M. (Ed.) 1988. *A field guide to the smaller British Lepidoptera.* 2nd edition.
British Entomological and Natural History Society, London.
Although not an identification guide as such, this book lists hostplants, details of larval
feeding and pupation site together with months in which ova, larvae, pupae and adults
occur, and can therefore be useful as an aid to identification.

Emmet, A.M. & Heath, J. (Eds) 1991. *The moths and butterflies of Great Britain and
Ireland.* 7(2). Lasiocampidae – Thyatiridae. Harley Books, Colchester.
Contains the definitive tables of life-history data.

Haggett, G.M. 1981. *Larvae of the British Lepidoptera not figured by Buckler.* British
Entomological & Natural History Society, London.
Provides illustrations and descriptions of the various instars of 78 taxa not figured in
Buckler (1886–1901).

Haggett, G.M. 1992. The early stages of *Hypenodes humidalis* Doubleday (*turfosalis*
Wocke) (Lepidoptera: Noctuidae). *Entomologist's Gazette* **43**: 95–98.
Includes black and white figures of the larva and an illustration of the pupa and
cocoon.

Porter, J. 1997. *The colour identification guide to caterpillars of the British Isles.*
Viking, Harmondsworth.
An important work figuring the larvae of the butterflies and macrolepidoptera, covering
over 850 species. Text gives information on foodplants and habitats. A companion
volume to Skinner (1984).

Scorer, A.G. 1913. *The entomologist's log-book.* George Routledge & Sons, London.
A dictionary, listed by English and Latin names, of the life histories and food plants of
the British macrolepidoptera. Useful, but now largely replaced by Emmet & Heath
(1991).

Stehr, F.W. 1987. *Immature insects.* Vol. 1. Kendall/Hunt, Dubuque.
This American work provides detailed keys to families and may be useful for identifying
microlepidoptera larvae in conjunction with Emmet (1988).

Stokoe, W.J. & Stovin, G.H.T. 1948. *The caterpillars of British moths.* 2 vols. Warne,
London.
Although long out of print, with indifferent colour plates, provides figures of eggs, larvae
and pupae of most British Macrolepidoptera.

Thomas, J.A. & Lewington, R. 1991. *The butterflies of Britain and Ireland.* Dorling
Kindersley, London.
A useful reference work covering all butterflies found in the British Isles. Illustrates the
butterflies from egg to adult and discusses distribution, habits and history.

Microlepidoptera: identification

See also the Immature Stages section above. Several works listed in this
section are used as a baseline, e.g. Agassiz (1992), and earlier works are not
usually listed. Several continental references are given which, although not
constituting a complete listing, are considered most useful in aiding
identification. Other continental publications, e.g. the *Microlepidoptera
Palaearctica* series, can also be useful for the identification of a range of
species, although they are not included here.

Following general references, and those covering a range of microlepidop-
teran families, the remaining references are listed under the appropriate
superfamily.

Agassiz, D.J.L. 1980. Presidential address 2. Some easily confused British microlepidoptera. *Proceedings of the British Entomological and Natural History Society*
13(3/4): 77–87.

Includes illustrations that aid the identification of some difficult species groups, e.g. *Pseudatemelia* spp. Information on *Oecogonia* species has been superseded: see Agassiz (1992) for reference.

Agassiz, D.J.L. 1992. Additions to the British microlepidoptera. *British Journal of Entomology and Natural History* 5: 1–13.

An extremely useful bibliography to additions to the British microlepidoptera fauna since Ellerton (1970).

Ellerton, J. 1970. Presidential address. Microlepidoptera added to the British list since L.T. Ford's review. *Proceedings and Transactions of the British Entomological and Natural History Society* 3(2): 33–41.

An extremely useful bibliography to additions to the British microlepidoptera fauna since Ford (1949).

Emmet, A.M. (Ed.) 1988. *A field guide to the smaller British Lepidoptera.* (2nd edn). British Entomological and Natural History Society, London.

Although not an identification guide as such, this book lists host plants, details of larval feeding and pupation site together with months in which ova, larvae, pupae and adults occur, and can therefore be useful as an aid to identification.

Emmet, A.M. (Ed.) 1996. *The moths and butterflies of Great Britain and Ireland.* Vol. 3, Yponomeutidae – Elachistidae. Harley Books, Colchester.

Comprehensive, detailed treatment with a wealth of biological and distributional information; illustrations of adults considerably improved in comparison with previous two microlepidoptera volumes, figures of genitalia included for Coleophoridae together with superb figures of larval cases; also includes chapter on invasions.

Ford, L.T. 1949. Presidential address. *Proceedings of the South London Entomological and Natural History Society* 1947–48: 48–58.

An extremely useful bibliography of additions to the British microlepidoptera fauna since Meyrick (1928).

Ford, L.T. 1954. The Glyphipterigidae and allied families. *Proceedings and Transactions of the South London Entomological and Natural History Society* 1952–53: 90–9.

Comprehensive treatment of the British species but nomenclature outdated.

Heath, J. (Ed.) 1976. *The moths and butterflies of Great Britain and Ireland.* Vol. 1, Micropterigidae – Heliozelidae. Curwen Press/Blackwell Scientific Publications, Oxford.

Comprehensive, detailed treatment with a wealth of biological and distributional information; but illustrations of adults poor, and lacking figures of genitalia. For Nepticulidae, use in conjunction with Johansson *et al.* (1990). Also includes introductory sections on morphology, habitats and conservation.

Heath, J. & Emmet, A.M. (Eds) 1985. *The moths and butterflies of Great Britain and Ireland.* Vol. 2, Cossidae – Heliodinidae. Harley Books, Colchester.

Comments as for Heath (1976). Also includes introductory chapter on aposematic Lepidoptera.

Meyrick, E. 1928. *A revised handbook of British Lepidoptera.* Watkins & Doncaster, London.

A standard work, which has now been surpassed in many areas, but still useful as an aid to the identification of those microlepidoptera not covered by more recent publications, e.g. much of the Gelechiidae.

Novák, I. 1980. *A field guide in colour to butterflies and moths.* Octopus Books, London.
Broad non-specialist introduction to the whole of the Lepidoptera. Not comprehensive; coverage is central European.

Parsons, M.S. 1996. *A review of the scarce and threatened ethmiine, stathmopodine and gelechiid moths of Great Britain.* 130pp. Joint Nature Conservation Committee, Peterborough. (UK Nature Conservation No. 16.)
Provides detail on distribution, biology, management and conservation on the rarer native species found in Great Britain. Identification references and an extensive bibliography are given.

Pierce, F.N. & Metcalfe, J.W. 1935. *The genitalia of the tineid families of the Lepidoptera of the British Islands.* F.N. Pierce, Oundle.
Provides line drawings, crude by modern standards, of the male and female genitalia of many families of the microlepidoptera, including the Gelechiidae and Oecophoridae.

Pitkin, B.R. 1986. Techniques for the preparation of complex male genitalia in microlepidoptera. *Entomologist's Gazette* 37: 173–9.
Gives details and illustrations of a technique that can be a valuable aid to identification.

Sokoloff, P. 1980. *Practical hints for collecting and studying microlepidoptera.* Amateur Entomologist's Society, Middlesex.
A pamphlet that introduces the study of microlepidoptera, covering topics such as collecting adults and immature stages, breeding and identification.

Superfamily Nepticuloidea

Bland, K.P. 1997. *Stigmella pretiosa* (Heinemann, 1862) (Lepidoptera: Nepticulidae) mining *Geum rivale* L. in southern Scotland – a species new to the British Isles. *Entomologist's Gazette* 48: 81–3.
Includes an illustration of the male genitalia and the larval mine.

Dickerson, B. 1995. *Ectoedemia amani* Svensson, 1966 (Lep.: Nepticulidae) new to Britain. *Entomologist's Record and Journal of Variation* 107: 163–4.
Not illustrated.

Johansson, R., Nielsen, E.S., van Nieukerken, E.J. & Gustafsson, B. 1990. The Nepticulidae and Opostegidae (Lepidoptera) of North West Europe. 2 vols. *Fauna Entomologica Scandinavica* 23(1): 1–413; 23(2): 415–739.
Superb, comprehensive treatment with exemplary illustrations of adults and genitalia together with biological information.

Superfamily Tineoidea

Hättenschwiler, P. 1997. Die Sackträger der Schweiz (Lepidoptera, Psychidae). *Schmetterlinge und ihre lebensräume Arten-Gefährdung-Schutz.* Band 2.
Contains many of the British species of Psychidae with colour photographs of live adults and immature stages; some diagnostic features given. In German.

Superfamily Yponomeutoidea

Agassiz, D.J.L. 1987. The British Argyresthiinae and Yponomeutinae. *Proceedings and Transactions of the British Entomological and Natural History Society* **20**(1): 1–26.
Comprehensive treatment superseded for the most part by MBGBI Vol. 3 but with useful colour photographs of adults.

Ford, L.T. 1951. The Plutellidae. *Proceedings and Transactions of the South London Entomological and Natural History Society* **1949–50**: 85–93.
In its time a comprehensive treatment, now superseded by MBGBI Vol. 3.

Superfamily Gelechioidea

Bengtsson, B.Å. 1984. The Scythrididae (Lepidoptera) of northern Europe. *Fauna Entomologica Scandinavica* **13**: 1–137.
Comprehensive, with good colour figures; genitalia figures mediocre.

Bengtsson, B.Å. 1997. Scythrididae. *In* Huemer, P., Karsholt, O., Lyneborg, L. (Eds) *Microlepidoptera of Europe* **2**: 1–301.
Covers 237 species of Scythrididae from Europe and North Africa. Includes colour illustrations and genitalia line drawings. Supersedes Bengtsson (1984).

Buszko, J. & Bengtsson, B.Å. 1992. First records of some Lepidoptera in Poland. *Polskie Pismo Entomologiczne* **61**: 47–56.
Provides useful line drawings of the wing pattern and male and female genitalia of both *Neofriseria peliella* and *N. singula*.

Hanneman, H.-J. 1995. Kleinschmetterlinge oder Microlepidoptera. IV. Flachleibmotten (Depressariidae). *Die Tierwelt Deutschlands* **69**: 1–192.
Descriptions in German; illustrations of adults (colour) and genitalia; maps.

Hanneman, H.-J. 1997. Kleinschmetterlinge oder Microlepidoptera. V. Oecophoridae, Chimabachidae, Carcinidae, Ethmiidae, Stathmopodidae. *Die Tierwelt Deutschlands* **70**: 1–163.
Descriptions in German; illustrations of adults (colour) and genitalia; maps.

Harper, M.W. 1993. *Pleurota aristella* (Linnaeus) (Lepidoptera: Oecophoridae) resident in Jersey. *Entomologist's Gazette* **44**: 11–13.
Includes black and white figure of *P. aristella* and *P. bicostella*.

Harper, M.W. 1994. *Mompha bradleyi* Riedl (Lepidoptera: Momphidae) new to Britain, with some observations on its life history. *Entomologist's Gazette* **45**: 151–6.
Contains a black and white figure of the adult and the larval gall and gives illustrations of the male and female genitalia of *M. bradleyi* and *M. divisella*.

Heckford, R.J. 1994. *Batrachedra parvulipunctella* Chrétien (Lepidoptera: Momphidae), a surprising addition to the British list. *Entomologist's Gazette* **45**: 261–5.
Illustrates the male genitalia of *B. parvulipunctella* and *B. pinicolella*.

Hoare, R.J.B. 1995. 1994 annual exhibition. Imperial College London SW7 – 22 October 1994. *British Journal of Entomology and Natural History* **8**: 190.
The plate associated with this exhibit report figures *Caryocolum blandelloides*.

Huemer, P. 1988. A taxonomic revision of *Caryocolum* (Lepidoptera: Gelechiidae). *Bulletin of the British Museum (Natural History)* (Entomology) **57**(3): 439–571.
All 63 species of *Caryocolum* are covered. This includes black and white figures of the adults and the male and female genitalia.

Huemer, P. 1993. The British species of *Caryocolum* Gregor & Povolný. *British Journal of Entomology and Natural History* **6**(4): 145–57, pl. V.
Covers all British species; photographic colour plate of adults and figures of genitalia of all species.

Jacobs, S.N.A. 1950. The British Oecophoridae Part I, and allied genera. *Proceedings and Transactions of the South London Entomological and Natural History Society* **1948–49**: 123–41.

Jacobs, S.N.A. 1951. The British Oecophoridae Part II. *Proceedings and Transactions of the South London Entomological and Natural History Society* **1949–50**: 187–203.

Jacobs, S.N.A. 1956. The British Oecophoridae Part III. *Proceedings and Transactions of the South London Entomological and Natural History Society* **1954–55**: 50–76.
These three papers provide a nearly comprehensive treatment but nomenclature is now somewhat outdated; colour plates (paintings) of adults; no genitalia figures. Use in conjunction with Hannemann (1995).

Koster, S & Biesenbaum, W. 1994. Momphidae. *Die Lepidopterenfauna der Rheinlande und Westfalens* **3**: 1–103.
Provides colour illustrations of adults, with illustrations of male and female genitalia and larval feeding damage of all British species except *M. subdivisella*.

Palm, E. 1989. *Nordeuropas prydvinger (Lepidoptera: Oecophoridae)*. Fauna Bøger, Copenhagen. (Danmarks Dyreliv Bind 4.)
Gives colour photographs of adults of the Oecophoridae, including the Ethmiidae, along with some genitalia photographs and a few line drawings which aid separating difficult species. Text in Danish with a limited English summary.

Sattler, K. 1997. The history of *Ethmia pyrausta* (Pallas, 1771) (Lepidoptera: Ethmiidae) in Britain and its biology. *Entomologist's Gazette* **48**: 89–92.
Gives a black and white figure of the adult.

Sokoloff, P.A. 1985. An introduction to the Gelechiidae. *Proceedings and Transactions of the British Entomological and Natural History Society* **18**(2/4): 99–106.
Covers the genera *Teleiodes* and *Teleiopsis*, with colour figures of adults; no genitalia figures.

Sokoloff, P.A. & Bradford, E.S. 1990. The British species of *Metzneria, Paltodora, Isophrictis, Apodia, Eulamprotes* and *Argolamprotes* (Lepidoptera: Gelechiidae). *British Journal of Entomology and Natural History* **3**(1): 23–8.
Descriptions and colour figures of adults; no genitalia figures.

Sokoloff, P.A. & Bradford, E.S. 1993. The British species of *Monochroa, Chrysoesthia, Ptocheuusa* and *Sitotroga* (Lepidoptera: Gelechiidae). *British Journal of Entomology and Natural History* **6**(2): 37–44.
Descriptions and colour figures of adults; no genitalia figures.

Sterling, P.H. 1997. *Cosmopterix scribaiella* Zeller (Lepidoptera: Cosmopterigidae) new to the British Isles. *Entomologist's Gazette* **48**: 205–7.
Gives a key to all the British species of the genus *Cosmopterix*.

Traugott-Olsen, E. & Nielsen, E.S. 1977. The Elachistidae (Lepidoptera) of
Fennoscandia and Denmark. *Fauna Entomologica Scandinavica* 6: 1–299.
Comprehensive treatment with excellent colour pictures of adults; more recent changes
of nomenclature are unlikely to stand the test of time.

Wakely, S. 1945. Notes on the genus *Mompha*. *Proceedings and Transactions of the
South London Entomological and Natural History Society* **1944–45**: 81–4, pl. V.
Descriptions and colour figures of adults; no genitalia figures.

Superfamily Tortricoidea

Bentinck, G.A.G. & Diakonoff, A. 1968. De Nederlandse bladrollers (Tortricidae).
Monografieën van de Nederlandsche Entomologische Vereeniging 3: 1–201, pls 1–99.
Has useful illustrations of male and female genitalia of the majority of Tortricidae
species occurring in the British Isles. Adults are also figured; text in Dutch.

Bradley, J.D., Tremewan, W.G. & Smith, A. 1973. *British Tortricoid moths. Cochylidae
and Tortricidae: Tortricinae.* Ray Society, London.

Bradley, J.D., Tremewan, W.G. & Smith, A. 1979. *British Tortricoid moths. Tortricidae:
Olethreutinae.* Ray Society, London.
These two well-illustrated volumes provide comprehensive coverage of the British tor-
tricids. There have been a few additions to the fauna since their publication, see Agassiz
(1992), Heckford (1993), Langmaid (1994).

Heckford, R.J. 1993. *Cydia amplana* (Hübner) (Lepidoptera: Tortricidae): the first
confirmed British record. *Entomologist's Gazette* 44: 107–10.
Includes a black and white figure of the adult.

Langmaid, J.R. 1994. *Cochylis molliculana* Zeller (Lepidoptera: Tortricidae) new to
the British fauna. *Entomologist's Gazette* 45: 255–8.
Includes black and white figures of the adult and male genitalia.

Larsen, K. & Vihelmsen, F. 1986. De Danske viklere (Tortricidae). *Lepidoptera* 5(1):
22–35.

Larsen, K. & Vihelmsen, F. 1986. De Danske viklere (Tortricidae) II. *Lepidoptera* 5(2):
61–71.

Larsen, K. & Vihelmsen, F. 1987. De Danske viklere (Tortricidae) III. *Lepidoptera* 5(3):
98–107.

Larsen, K. & Vihelmsen, F. 1987. De Danske viklere (Tortricidae) IV. *Lepidoptera* 5(4):
134–43.

Larsen, K. & Vihelmsen, F. 1988. De Danske viklere (Tortricidae) V. *Lepidoptera* 5(5):
177–87.

Larsen, K. & Vihelmsen, F. 1988. De Danske viklere (Tortricidae) VI. *Lepidoptera* 5(6):
226–31.

Larsen, K. & Vihelmsen, F. 1989. De Danske viklere (Tortricidae) VII. *Lepidoptera*
5(7/8): 270–80.

Larsen, K. & Vihelmsen, F. 1990. De Danske viklere (Tortricidae) VIII. *Lepidoptera*
5(9): 308–28.
Includes colour photographs of adults of the entire Danish Tortricidae fauna. Very useful
for identifying 'difficult' species not easily recognised from colour paintings in British
publications; genitalia figures poor; text in Danish.

Pierce, F.N. & Metcalfe, J.W. 1922. *The genitalia of the group Tortricidae of the Lepidoptera of the British Islands.* F.N. Pierce, Oundle.
Provides line drawings, crude by modern standards, of the male and female genitalia of the Tortricidae.

Superfamily Epermenioidea

Godfray, H.C.J. & Sterling, P.H. 1993. The British Epermeniidae. *British Journal of Entomology and Natural History* **6**: 141–3, pl. V.
Text superseded by MBGBI Vol. 3 but photographic colour plate complements later painted illustrations.

Superfamily Pyraloidea

Agassiz, D.J.L. 1996. *Eccopisa effractella* Zeller (Lepidoptera: Pyralidae) new to the British Isles. *Entomologist's Gazette* **47**: 181–3.
Includes an illustration of the underside of the wings and the male and female genitalia.

Beirne, B.P. 1952. *British pyralid and plume moths.* F. Warne, London.
Comprehensive when published but now out of print; nomenclature outdated and figures mediocre; superseded by Goater (1986).

Burrow, R. 1996. Some notable additions to the macrolepidoptera of Jersey. *Entomologist's Record and Journal of Variation* **108**: 133–6.
Includes photograph of one species (*Evergestis limbata*) not figured by Goater (1986).

Clancy, S.P.C. & Tunmore, M. 1997. Gallery. *Atropos* **2**: 43.
Includes figure of *Duponchelia fovealis*, a species not covered in Goater (1986).

Goater, B. 1986. *British pyralid moths – a guide to their identification.* Harley Books, Colchester.
Comprehensive account; descriptions and photographic colour illustrations; figures of genitalia of 'critical species'.

Palm, E. 1986. *Nordeuropas pyralider (Lepidoptera: Pyralidae).* Fauna Bøger, Copenhagen. (Danmarks Dyreliv Bind 3.)
Gives colour photographs of adults of the Pyralidae along with some genitalia illustrations and many line drawings of external characters that separate difficult species. Text in Danish with a limited English summary.

Parsons, M.S. 1993. *A review of the scarce and threatened pyralid moths of Great Britain.* 97pp. Joint Nature Conservation Committee, Peterborough. (UK Nature Conservation No. 11.)
Provides detail on distribution, biology, management and conservation of the rarer native species found in Great Britain. Identification references and an extensive bibliography are given.

Pierce, F.N. & Metcalfe, J.W. 1938. *The genitalia of the British pyrales with the deltoids and plumes.* F.N. Pierce, Oundle.
Provides line drawings, crude by modern standards, of the male and female genitalia.

Skinner, B. 1995. Pyralid moths in profile: Part 1 – *Sciota adelphella* (Fischer von Röslerstamm). *Entomologist's Record and Journal of Variation* **107**: 147–9.

Summarises the history of the species in Great Britain and gives a description of the life history of *S. adelphella* (not figured in Goater, 1986). Figures the larva and pupa and the adult moth, along with an adult of *S. hostilis* for comparison.

Skinner, B. 1995. Pyralid moths in profile: Part 2 – *Acrobasis tumidana* (Denis & Schiffermüller). *Entomologist's Record and Journal of Variation* 107: 241–3.
Summarises the history of the species in Great Britain, listing all the authenticated records.

Skinner, B. 1996. Pyralid moths in profile: Part 3 – *Udea fulvalis* (Hübner). *Entomologist's Record and Journal of Variation* 108: 108–9.
Summarises the known records of the species in Great Britain and gives a description of the life history. Figures the larva, cocoon and pupa.

Skinner, B. 1996. Pyralid moths in profile: Part 4 – *Salebriopsis albicilla* (Herrich-Schäffer). *Entomologist's Record and Journal of Variation* 108: 110–11.
Summarises the distribution and past history of the species in Great Britain and gives a description of the life history. Figures the larva, cocoon and pupa.

Slamka, F. 1995. *Die Zünslerartigen (Pyraloidea) Mitteleuropas*. Slamka, Bratislava.
Covers over 400 species of pyralids from central Europe. Includes genitalia figures and photographs of adults of many species found in Britain. Text is in German.

Superfamily Pterophoroidea

Beirne, B.P. 1952. *British pyralid and plume moths*. Wayside & Woodland Series, Warne, London.
Comprehensive when published but now out of print; nomenclature outdated and figures mediocre. Superseded by Gielis (1996).

Gielis, C. 1996. Pterophoridae. *In* Huemer, P., Karsholt, O., & Lyneborg, L. (Eds) *Microlepidoptera of Europe* 1: 1–222.
Covers all species found in Europe (excluding the former Soviet Union), Canary Islands and Madeira. Includes colour photographs of the adults and genitalia line drawings. Photographs of the early stages of a few species are also included.

Hart, C. 1996. The status of *Crombrugghia laetus* (Zeller, 1847) (Lep.: Pterophoridae) in Britain with a review of known records and notes on its separation from *C. distans* (Zeller, 1847). *Entomologist's Record and Journal of Variation* 108: 113–17.
Contains figures of the adults and the male and female genitalia of *C. laetus* and *C. distans*.

Pierce, F.N. & Metcalfe, J.W. 1938. *The genitalia of the British pyrales with the deltoids and plumes*. F.N. Pierce, Oundle.
Provides line drawings, crude by modern standards, of the male and female genitalia of the Pyralidae and Pterophoridae, as well as some species of the Hypeninae.

Microlepidoptera: distribution, conservation etc.

See also the general Lepidoptera section above, which includes county lists that also cover the microlepidoptera. Chalmers-Hunt (1989) is used as a baseline for this section.

Bond, K.G.M. 1995. Irish Microlepidoptera check-list. *Bulletin of the Irish Biogeographical Society* **18**(2): 176–262.
Lists just under 800 species and briefly discusses those considered to be of doubtful status in Ireland.

Bond, K.G.M. 1996. Previously unpublished records of Microlepidoptera to be added to the Irish list. *Irish Naturalists' Journal* **25**(6): 194–207.
Adds 39 species to Ireland and confirms a further 15 previously considered to be of doubtful status in Ireland.

Chalmers-Hunt, J.M. 1989. *Local lists of Lepidoptera or a bibliographical catalogue of local lists and regional accounts of the butterflies and moths of the British Isles.* Hedera Press, Uffington.
Covers approximately 3000 titles, citing the known county and regional lists and local accounts of Lepidoptera in the British Isles.

Dunn, T.C. & Parrack, J.D. 1992. *The moths and butterflies of Northumberland and Durham. Part two: Microlepidoptera.* The Northern Naturalists Union. (The Vasculum – Supplement No. 3.)
Discusses distribution and status over the three vice-counties.

Emmet, A.M. 1995. Records of Scottish microlepidoptera from south-western Scotland, July 1994. *Entomologist's Record and Journal of Variation* **107**: 5–11.
A useful list of species covering three vice-counties.

Emmet, A.M. & Langmaid, J.R. 1997. Microlepidoptera in Ireland, September 1995. *Entomologist's Gazette* **48**: 147–55.
A useful list of species from sixteen vice-counties, and adding four species to the Irish list.

Harrison, F. & Sterling, M.J. 1988. *Butterflies and moths of Derbyshire.* Part 3. pp. 205–345. Derbyshire Entomological Society, Derby.
Summarises records of over 640 species of microlepidoptera recorded in Derbyshire.

Lucas, S.R. 1994. A check list of the microlepidoptera of Carmarthenshire (VC44). *Entomologist's Record and Journal of Variation* **106**: 161–9.
Lists 423 species recorded from Carmarthenshire.

Parsons, M.S. 1993. *A review of the scarce and threatened pyralid moths of Great Britain.* 97pp. Joint Nature Conservation Committee, Peterborough. (UK Nature Conservation No. 11.)
Provides detail on distribution, biology, management and conservation of the rarer native species found in Great Britain. Identification references and an extensive bibliography are given.

Parsons, M.S. 1996. *A review of the scarce and threatened ethmiine, stathmopodine and gelechiid moths of Great Britain.* 130pp. Joint Nature Conservation Committee, Peterborough. (UK Nature Conservation No. 16.)
Provides detail on distribution, biology, management and conservation on the rarer native species found in Great Britain. Identification references and an extensive bibliography are given.

Skinner, B. 1995. Pyralid moths in profile: Part 1 – *Sciota adelphella* (Fischer von Röslerstamm). *Entomologist's Record and Journal of Variation* **107**: 147–9.

Summarises the history of the species in Great Britain and gives a description of the life history of *S. adelphella*. Figures the larva and pupa and the adult moth, along with an adult of *S. hostilis* for comparison.

Skinner, B. 1995. Pyralid moths in profile: Part 2 – *Acrobasis tumidana* (Denis & Schiffermüller). *Entomologist's Record and Journal of Variation* 107: 241–3.

Summarises the history of the species in Great Britain, listing all the authenticated records.

Skinner, B. 1996. Pyralid moths in profile: Part 3 – *Udea fulvalis* (Hübner). *Entomologist's Record and Journal of Variation* 108: 108–9.

Summarises the known records of the species in Great Britain and gives a description of the life history. Figures the larva, cocoon and pupa.

Skinner, B. 1996. Pyralid moths in profile: Part 4 – *Salebriopsis albicilla* (Herrich-Schäffer). *Entomologist's Record and Journal of Variation* 108: 110–11.

Summarises the distribution and past history of the species in Great Britain and gives a description of the life history. Figures the larva, cocoon and pupa.

Slade, D. 1997. Microlepidoptera recorded from South Wales in 1995, including four species new to Wales. *Entomologist's Record and Journal of Variation* 109: 31–9.

A useful list of species from several South Wales vice-counties.

Macrolepidoptera: identification

A range of continental literature references is given here as they sometimes include additional characters to aid identification, and can also form a useful comparison with British works. See also the immature stages section above.

Following the general references, and those covering a range of macrolepidopteran families, the remaining references are listed under the appropriate superfamily.

Anon. 1996. Plate 1. *Atropos* 1: 34 & pl. 1.

Figures several taxa that are not included in Skinner (1984), e.g. *Drepana curvatula, Thera cupressata* and *Odonthognophos dumetata* ssp. *hibernica*.

Brooks, M. 1991. *A complete guide to British moths*. Jonathan Cape, London.

Gives figures of adults of a substantial proportion of the British Macrolepidoptera and also provides figures of the early stages of a representative 80 species.

Burrow, R. 1996. Some notable additions to the macrolepidoptera of Jersey. *Entomologist's Record and Journal of Variation* 108: 133–6.

Includes photographs of several species not figured by Skinner (1984), including *Cryphia algae* and *Acontia lucida*.

Clancy, S.P.C. & Tunmore, M. 1997. Gallery. *Atropos* 2: 43.

Includes figures of species not covered in Skinner (1984), e.g. *Agrotis herzogi* and *Dendrolimus pini*.

Emmet, A.M. & Heath, J. (Eds) 1991. *The moths and butterflies of Great Britain and Ireland*. 7(2). Lasiocampidae – Thyatiridae. Harley Books, Colchester.

Comprehensive treatment with a wealth of biological and distributional information. Also includes a chapter on resting posture in Lepidoptera, and a life-history chart for all the British Lepidoptera.

Heath, J. 1969. Lepidoptera distribution maps scheme guide to the critical species. *Entomologist's Gazette* **20**: 89–95.

Provides line drawings of wing characters and genitalia, which aid identification of some groups of moths including *Pheosia gnoma* and *P. tremula*.

Heath, J. 1970. Lepidoptera distribution maps scheme guide to the critical species. Part III. *Entomologist's Gazette* **21**: 102–5.

Gives line drawings of the genitalia and wing patterns of a small number of difficult species. Includes *Watsonalla binaria* and *W. cultraria*.

Heath, J. 1972. Lepidoptera distribution maps scheme guide to the critical species. Part VI. *Entomologist's Gazette* **23**: 126–8.

Provides line drawings of the wing shape or wing pattern of a small number of difficult species, including *Epirrhoe alternata* and *E. rivata*.

Heath, J. & Emmet, A.M. (Eds) 1979. *The moths and butterflies of Great Britain and Ireland.* **9**. Sphingidae – Noctuidae (Noctuinae and Hadeninae). Harley Books, Colchester.

Comprehensive, detailed treatment with a wealth of biological and distributional information; but some illustrations of adults mediocre, and lacking figures of genitalia. Use in conjunction with Fibiger (1990, 1993), Skinner (1984) and Skou (1991). Also includes a chapter on eversible structures.

Heath, J. & Emmet, A.M. (Eds) 1983. *The moths and butterflies of Great Britain and Ireland.* **10**. Noctuidae (Cuculliinae to Hypeninae) and Agaristidae. Harley Books, Colchester.

Comprehensive, detailed treatment with a wealth of biological and distributional information; but some illustrations of adults mediocre, and lacking figures of genitalia. Use in conjunction with Ronkay & Ronkay (1994, 1995), Skinner (1984) and Skou (1991). Also includes a chapter on migrant Lepidoptera.

Heath, J. & Emmet, A.M. (Eds) 1985. *The moths and butterflies of Great Britain and Ireland.* **2**. Cossidae – Heliodinidae. Harley Books, Colchester.

Comprehensive, detailed treatment with a wealth of biological and distributional information; but illustrations of adults poor, and lacking figures of genitalia. Also includes introductory chapter on aposematic Lepidoptera.

Pierce, F.N. & Beirne, B.P. 1941. *The genitalia of the British Rhopalocera and larger moths.* F.N. Pierce, Oundle.

Contains illustrations, crude by modern standards, of the male and female genitalia of the butterflies and some families of moths, including the Notodontidae and Arctiidae.

Skinner, B. 1984. *Colour identification guide to moths of the British Isles.* Viking, Harmondsworth.

The current standard work for the identification of the macrolepidoptera of the British Isles, replacing South (1961). A series of fine plates depicting most species found in Britain, together with a brief summary of the ecology and distribution. Some line drawings are provided as an aid to the identification of some species. [Second revised edition, 1998]

South, R. 1961. *The moths of the British Isles.* 2 vols. (new edition.) Frederick Warne & Co., London.

Illustrates the adults in colour together with many black and white illustrations of the early stages. Text contains information on biology and distribution. Nomenclature very much out-of-date.

Tams, W.H.T. 1941. Some British moths reviewed. *Amateur Entomologist* 5: 1–20.
Illustrates the genitalia of a range of difficult species groups, e.g. *Luperina testacea, L. gueneei* and *L. dumerilii.*

Superfamily Sesioidea

Fibiger, M. & Kristensen, N.P. 1974. The Sesiidae (Lepidoptera) of Fennoscandia and Denmark. *Fauna Entomologica Scandinavica* 2: 1–91.
Covers the British species and includes genitalia illustrations; complementary to MBGBI 2.

Superfamily Geometroidea

British Entomological & Natural History Society. 1981. *An identification guide to the British pugs.* British Entomological & Natural History Society, London.
Figures the adults and includes illustrations of the male and female genitalia and the male abdominal plates.

Clancy, S.P.C. 1996. The Lydd beauty *Peribatodes manuelaria* (H.-S.) in Britain. *Atropos* 1: 41–2.
Includes a figure of the adult (now referred to as *P. ilicaria*) together with that of *P. rhomboidaria* and *P. secundaria* for comparison.

Pierce, F.N. 1914. *The genitalia of the group Geometridae of the Lepidoptera of the British Islands.* F.N. Pierce, Liverpool.
Contains line drawings, crude by modern standards, of the male and female genitalia of most of the British species of Geometridae.

Riley, A.M. 1985. *Eupithecia ultimaria* Boisduval (Lepidoptera: Geometridae): a pug new to the British list. *Entomologist's Gazette* 36: 259–61.
Gives line drawing of the adult and the male genitalia.

Riley, A.M. 1991. *Eupithecia ultimaria* Boisduval (Lepidoptera: Geometridae), a third record for the British Isles and the first mainland capture. *Entomologist's Gazette* 42: 289–90.
Illustrates the female genitalia.

Skou, P. 1986. The geometrid moths of north Europe. *Entomonograph* 6.
Covers the known Drepanidae, including Thyatirinae, and Geometridae of Denmark, Norway, Sweden and Finland. A useful comparison guide to works such as Skinner (1984) containing figures of the adults, many useful line drawings, some larval and habitat photographs and brief notes of habitat and biology.

Slade, B.E. & Agassiz, D.J.L. 1992. *Eupithecia sinuosaria* Eversmann (Lepidoptera: Geometridae) new to the British Isles. *Entomologist's Record and Journal of Variation* 104: 287–8.
Provides a black and white illustration of the adult.

Superfamily Noctuoidea

Barrett, R.J. 1986. *Antichloris eriphia* Fab. (Lepidoptera: Ctenuchidae), first record for Britain. *Entomologist's Record and Journal of Variation* **98**: 240.

Brown, D. 1996a. A brief history of Radford's flame shoulder *Ochropleura leucogaster* (Freyer) in the British Isles. *Atropos* **1**: 28–9.
  Includes a figure and an illustration of the adult together with that of *O. plecta* for comparison.

Brown, D. 1996b. A brief history of the Tunbridge Wells gem *Chrysodeixis acuta* (Walker) in Great Britain. *Atropos* **1**: 32–6.
  Includes a figure and an illustration of the adult together with that of *C. chalcites* for comparison.

Burrow, R. 1996. *Dryobota labecula* (Esper) the oak rustic (Lep.: Noctuidae): a new breeding species to the British list from the Channel Islands. *Entomologist's Record and Journal of Variation* **108**: 136–7.
  The species is figured in Burrow (1996): see general macrolepidoptera references above.

Cade, M. 1996. Identification of dark mottled willow *Spodoptera cilium* (Guen.). *Atropos* **1**: 38–9.
  Includes a figure of the adult with that of *S. exigua* for comparison.

Clancy, S.P.C. 1997. Separation of copper underwing *Amphipyra pyramidea* (L.) and Svensson's Copper underwing *A. berbera svenssoni* (Fletcher). *Atropos* **2**: 30–1.
  Includes line drawings of an upperside forewing character that is an identification aid.

Clancy, S.P.C. & Honey, M.R. 1997. The plumed fan-foot *Pechipogo plumigeralis* (Hübner) – a moth new to Britain and Ireland. *Atropos* **2**: 32–3.
  Includes figures of the adult.

Clancy, S.P.C. & Honey, M.R. 1997. The streaked plusia *Trichoplusia vittata* (Wallengren) – a moth new to Britain and Europe. *Atropos* **2**: 33–4.
  Include a figure of the adult.

Classey, E.W. 1954. Separation characters of some British Noctuid moths. *Proceedings and Transactions of the South London Entomological and Natural History Society* **1952–3**: 64–72.
  Illustrates the genitalia and adults of a range of difficult species of Noctuidae.

Craik, J.C.A. 1984. Differences between the females of *Amphipyra pyramidea* L. and *A. berbera* Rungs: a correction to MBGBI Volume 10. *Entomologist's Record and Journal of Variation* **96**: 160–1.
  Details an error given in Heath & Emmet (1983) relating to identification using the female genitalia.

Dyar, J.J. 1993. *Noctua janthina* ([Denis & Schiffermüller]) confused with *Noctua janthe* (Borkhausen, 1792). *Entomologist's Record and Journal of Variation* **105**: 171–3.
  Includes an illustration of the hindwings of *N. janthina* and *N. janthe* and notes that only *N. janthe* has, to date, been found in the British Isles, not *N. janthina*.

Fibiger, M. 1990. *Noctuidae Europaeae*. Vol.1, *Noctuinae I*. Entomological Press, Sorø.

The first part of the Noctuinae, providing a summary of the ecology and distribution of each species, together with fine figures of the adults. In English and French.

Fibiger, M. 1993. *Noctuidae Europaeae.* Vol. 2, *Noctuinae II.* Entomological Press, Sorø.

The second part of the Noctuinae, providing a summary of the ecology and distribution of each species, together with fine figures of the adults. In English and French.

Fibiger, M. 1997. *Noctuidae Europaea.* Vol. 3, *Noctuinae III.* Entomological Press, Sorø.

Very clear figures of the male and female genitalia of a range of Noctuinae.

Goater, B. 1994. The genus *Earias* Hübner, (1825) (Lepidoptera: Noctuidae) in Britain and Europe. *Entomologist's Record and Journal of Variation* **106**: 233–9.

Discusses the confusion over the *Earias* species recorded in the British Isles and illustrates the male and female genitalia of *E. clorana, E. vernaria, E. insulana* and *E. biplaga.*

Goater, B. & Skinner, B. 1995. A new subspecies of *Luperina nickerlii* Freyer, 1845 from south-east England, with notes on the other subspecies found in Britain, Ireland and mainland Europe. *Entomologist's Record and Journal of Variation* **107**: 127–31.

Describes subspecies *demuthi* and figures adults.

Haggett, G.M. & Smith, C.C. 1993. *Agrochola haematidea* Duponchel (Lepidoptera: Noctuidae, Cucullinae) new to Britain. *Entomologist's Gazette* **44**: 183–203.

Illustrates the male and female genitalia, the ova and the pupa. Gives colour illustrations of various larval instars and figures the adult and the larva.

Heath, J. 1971. Lepidoptera distribution maps scheme guide to the critical species. Part IV. *Entomologist's Gazette* **22**: 19–22.

Gives line drawings of the wing patterns and genitalia of a small number of difficult species, including *Hadena bicruris* and *H. rivularis.*

Heath, J. & Cooke, R. 1969. Lepidoptera distribution maps scheme guide to the critical species. Part II. The genera *Oligia* Hb. and *Mesoligia* Boursin (*Procus* and *Miana* auct.). *Entomologist's Gazette* **20**: 263–9.

Gives line drawings of the male genitalia of the species in these genera.

Heath, J. & Skelton, M.J. 1971. Lepidoptera distribution maps scheme guide to the critical species. Part V. *Entomologist's Gazette* **22**: 109–10.

Gives line drawings of the wing patterns of a small number of difficult species, including the British *Furcula* species.

Honey, M.R. & Sterling, M. 1994. *Paradasena virgulana* (Mabille) (Lepidoptera: Noctuidae), a species not previously found in the wild in Britain. *British Journal of Entomology and Natural History* **7**: 33–4.

Gives a black and white figure of the adult.

Jordan, M.J.R. 1986. The genitalia of the species pair *Mesapamea secalis* (L.) and *Mesapamea secalella* Remm, (Lep.: Noctuidae). *Entomologist's Record and Journal of Variation* **98**: 41–4.

Illustrates the male and female genitalia of both *M. secalis* and *M. didyma* (= *secalella*) and figures part of the male genitalia of both species.

Jordan, M.J.R. 1989. *Mesapamea remmi* Rezbanyai-Reser, 1985. (Lep.: Noctuidae) a species new to Britain. *Entomologist's Record and Journal of Variation* **101**: 161–5.

Figures the male aedeagus of *M. remmi*, *M. secalis* and *M. didyma* and part of the female genitalia of *M. remmi*.

Mikkola, K. & Goater, B. 1988. The taxonomic status of *Apamea exulis* (Duponchel) and *A. assimilis* (Doubleday) in relation to *A. maillardi* (Geyer) and *A. zeta* (Treitschke) (Lepidoptera: Noctuidae). *Entomologist's Gazette* 39: 249–57.

Summarises the taxa from this complex found in the British Isles. Illustrates the male and female genitalia of *A. zeta marmorata* and gives black and white figures of *A. zeta marmorata* and *A. zeta assimilis*.

Owen, D.F. 1993. Identification of *Amphipyra pyramidea* (L.) and *A. berbera* Fletcher (Lepidoptera: Noctuidae) *Entomologist's Record and Journal of Variation* 105: 133–4.

Gives a line drawing of the upperside of the forewing illustrating a character useful in the determination of the taxa.

Pierce, F.N. 1909. *The genitalia of the group Noctuidae of the Lepidoptera of the British Islands*. A.W. Duncan, Liverpool.

Contains line drawings, crude by modern standards, of the male genitalia of most of the British species of Noctuidae.

Pierce, F.N. 1942. *The genitalia of the group Noctuidae of the Lepidoptera of the British Islands*. F.N. Pierce, Oundle.

Contains line drawings, crude by modern standards, of the female genitalia of most of the British species of Noctuidae.

Riley, A.M. 1987. A further diagnostic feature for the separation of *Eilema lurideoa* Zincken and *E. complana* L. (Lep.: Arctiidae). *Entomologist's Record and Journal of Variation* 99: 28.

Gives illustrations of a thoracic character difference in each species.

Ronkay, G. & Ronkay, L. 1994. *Noctuidae Europaeae*. Vol. 6, *Cucullinae I*. Entomological Press, Sorø.

The first part of the Cucullinae, providing a summary of the ecology and distribution of each species, together with fine figures of the adults. In English and French.

Ronkay, G. & Ronkay, L. 1995. *Noctuidae Europaeae*. Vol. 7, *Cucullinae II*. Entomological Press, Sorø.

The second part of the Cucullinae, providing a summary of the ecology and distribution of each species, together with fine figures of the adults. In English and French.

Skou, P. 1991. *Nordens ugler*. Apollo Books, Stenstrup. (Danmarks Dyreliv Bind 5.)

Covers the known Noctuidae of Denmark, Norway, Sweden and Finland. A useful comparison guide to works such as Skinner (1984) containing figures of the adults, many useful line drawings, some larval and habitat photographs and brief notes of habitat and biology.

Tillotson, I.J.L. 1983. *Catocala nymphagoga* Esper and *Herminia zelleralis* Wocke: two species of Noctuidae new to Britain. *Entomologist's Record and Journal of Variation* 95: 133–4.

Provides references to figures of each species.

Waite, P. 1989. First British record of *Polymixis gemmea* (Treitschke, 1825) (Lepidoptera: Noctuidae). *British Journal of Entomology and Natural History* 2: 129–30.

Includes a black and white figure of the adult.

Walker, D. 1997. Pale-shouldered cloud *Chloantha hyperici* (Denis & Schiffermüller) at Dungeness, Kent – the first British record. *Atropos* 2: 55–6.
Includes a figure of the adult.

Winter, P.Q. 1988. An additional aid to the identification of *Amphipyra pyramidea* (L.) and *A. berbera svenssoni* (Fletcher) (Lepidoptera: Noctuidae). *British Journal of Entomology and Natural History* 1: 97–9.
Gives the differences found in the palps of both taxa together with illustrations.

Macrolepidoptera: distribution, etc.

See also general Lepidoptera distribution references above. Chalmers-Hunt (1989) is used as a baseline for this section.

Cadbury, J. 1993. The conservation importance of RSPB nature reserves for moths. *RSPB Conservation Review* 7: 85–91.
A paper discussing the importance of RSPB reserves for the larger moths. This paper includes sections on species diversity and internationally important species.

Chalmers-Hunt, J.M. 1989. *Local lists of Lepidoptera or a bibliographical catalogue of local lists and regional accounts of the butterflies and moths of the British Isles.* Hedera Press, Uffington.
Covers approximately 3000 titles, citing the known county and regional lists and local accounts of Lepidoptera in the British Isles.

Collins, G.A.C. 1997. *The larger moths of Surrey.* Surrey Wildlife Trust, Pirbright.
Discusses distribution and status of the larger moths of Surrey.

Harrison, F. & Sterling, M.J. 1986. *Butterflies and moths of Derbyshire. Part 2.* Pp. 69–204. Derbyshire Entomological Society, Derby.
Summarises the status and records of over 500 species recorded from Derbyshire.

Leverton, R. 1993. The Macrolepidoptera of Banffshire: a supplement. *Entomologist's Record and Journal of Variation* 105: 97–104.
A useful list of additions to the vice-counties fauna.

Plant, C.W. 1993. *Larger moths of the London area.* London Natural History Society, London.
Discusses the status and distribution of over 700 species found in the London area. There have been subsequent updates in the *London Naturalist.*

Rutherford, C.I. 1994. *Macro-moths in Cheshire 1961 to 1993.* The Lancashire and Cheshire Entomological Society.
Gives a brief summary of status and distribution of over 500 species recorded from Cheshire.

Svedson, P. (Ed.) & Fibiger, M. 1992. *The distribution of European Macrolepidoptera – Faunistica Lepidopterorum Europaeorum.* Vol. 1, *Noctuinae 1.* Copenhagen. European Faunistical Press.
Gives maps of the European distribution of part of the Noctuinae, including some species which occur in the British Isles.

Viles, I. 1995. Macrolepidoptera added to the county list since the publication of Butterflies and moths of Derbyshire. *Journal of the Derbyshire Entomological Society* 121: 18–20.
Gives details of sixteen species added to the county since Harrison & Sterling (1986).

Butterflies: identification

There is a wide range of publications covering the identification of butter-flies. The following list is a selection including good pocket guides and more comprehensive treatments.

Brooks, M. & Knight, C. 1982. *A complete guide to British butterflies.* Jonathan Cape, London.
Discusses the biology of each species and contains photographs of adults and the early instars.

Emmet, A.M. & Heath, J. (Eds) 1989. *The moths and butterflies of Great Britain and Ireland.* 7(1). Hesperiidae – Nymphalidae. Harley Books, Colchester.
Comprehensive, detailed treatment with a wealth of biological and distributional information. Also includes a chapter on insect re-establishment.

Heath, J. 1969. Lepidoptera distribution maps scheme guide to the critical species. *Entomologist's Gazette* 20: 89–95.
Gives a line drawings of the antennal characters for *Thymelicus lineola* and *T. sylvestris.*

Higgins, L.G. & Riley, N.D. 1970. *A field guide to the butterflies of Britain and Europe.* Collins, London.
A compact guide aimed at aiding the identification of butterflies from the western Palaearctic region.

Howarth, T.G. 1984. *Colour identification guide to butterflies of the British Isles.* Viking, Harmondsworth (revised edition).
Includes illustrations of typical adults along with many varieties, as well as early stages, and includes a life-history table.

Pierce, F.N. & Beirne, B.P. 1941. *The genitalia of the British Rhopalocera and larger moths.* F.N. Pierce, Oundle.
Contains illustrations, crude by modern standards, of the male and female genitalia of the butterflies.

Thomas, J.A. 1986. *Butterflies of the British Isles.* Hamlyn, London.
Pocket guide to the butterflies with illustrations of the adults with additional figures of living adults together with information on ecology and distribution.

Thomas, J.A. & Lewington, R. 1991. *The butterflies of Britain and Ireland.* Dorling Kindersley, London.
A useful reference work covering all butterflies found in the British Isles. Illustrates the butterflies from egg to adult and discusses distribution, habits and history.

Tolman, T. & Lewington, R. 1997. *Butterflies of Britain and Europe.* Collins Field Guide. Harper Collins, London.
Provides data on distribution, habitats and life histories of 440 species with illustrations of adults.

Whalley, P.E.S. 1996. *Mitchell Beazley pocket guide to butterflies.* (revised edn). Mitchell Beazley, London.
Covers the British and continental European butterflies with colour illustrations of adults.

Butterflies: distribution, etc.

See also the general Lepidoptera distribution section above, which includes county lists that also cover butterflies. Chalmers-Hunt (1989) is used as a baseline for this section. Some of the more general references provide a guide to broader topics associated with the study of butterflies and are often a useful source of further reading material.

Asher, J. 1994. *The butterflies of Berkshire, Buckinghamshire and Oxfordshire.* Pisces Publications, Newbury.
  Discusses the distribution and status of the butterfly fauna of the three counties.

Bristow, C.R., Mitchell, S.H. & Bolton, D.E. 1993. *Devon butterflies.* Devon Books, Tiverton.
  Discusses distribution and status of the Devon butterfly fauna.

Chalmers-Hunt, J.M. 1989. *Local lists of Lepidoptera or a bibliographical catalogue of local lists and regional accounts of the butterflies and moths of the British Isles.* Hedera Press, Uffington.
  Covers approximately 3000 titles, citing the known county and regional lists and local accounts of Lepidoptera in the British Isles.

Collins, G.A. 1995. *Butterflies of Surrey.* Surrey Wildlife Trust, Woking.
  Discusses the distribution and status of the Surrey butterfly fauna.

Corke, D. 1997. *The butterflies of Essex.* Lopinga Books, Wimbish.
  Covers distribution and history, with chapters on the causes of butterfly decline and an action plan for Essex butterfly conservation. Also includes a checklist of Essex Lepidoptera.

Dennis, R.L.H. 1977. *The British butterflies: their origin and establishment.* Classey, Faringdon.
  Includes chapters on faunal and floral analysis, including climatic changes and the activities of man, zoogeography, adaptations to the environment and geographical variation.

Dennis, R.L.H. (Ed.). 1992. *The ecology of butterflies in Britain.* Oxford University Press, Oxford.
  Includes chapters on adult behaviour, butterfly populations, monitoring butterfly movements and conservation.

Dennis, R.L.H. & Shreeve, T.G. 1996. *Butterflies on British and Irish offshore islands: ecology and landscape.* Gem Publishing Company, Wallingford.
  Lists the species found on each of 219 islands and investigates the ecological significance of the species records.

Feltwell, J. 1995. *The conservation of butterflies in Britain, past and present.* Wildlife Matters, Battle.
  Concentrating on conservation and ecological issues, this book provides an insight into the evolution of butterfly conservation in Britain and includes chapters on habitat management and butterflies and the law.

Ford, E.B. 1945. *Butterflies.* Collins New Naturalist, London.
  A general introduction to the butterflies covering topics such as structure and development, habits and protective devices, distribution, dispersal, and races and subspecies of British species.

Frost, M.P. & Madge, S.C. 1991. *Butterflies in south-east Cornwall.* The Caradon Field
& Natural History Club.
  A booklet discussing the distribution and status of the butterfly fauna of south-east
  Cornwall.

Fuller, M. 1995. *The butterflies of Wiltshire. Their history, status and distribution
1982–1994.* Pisces Publications, Newbury.
  Discusses the distribution and status of the butterfly fauna of Wiltshire.

Gay, J. & Gay, P. 1996. *Atlas of Sussex butterflies: with a commentary on their changing
conservation status.* Butterfly Conservation, Sussex Branch, Henfield.
  Discusses distribution and status of the butterfly fauna of Sussex based on records collected between 1989 and 1994.

Harding, P.T. & Green, S.V. 1991. *Recent surveys and research on butterflies in Britain
and Ireland: a species index and bibliography.* Biological Records Centre, Abbots
Ripton.
  A useful bibliography which provides references to publications and unpublished reports
  covering topics such as ecological studies and conservation.

Hickin, N. 1992. *The butterflies of Ireland. A field guide.* Roberts Rineheart Publishers,
Schull.
  Discusses the species found in Ireland, details their biology and gives distribution maps
  for the resident species.

Penhallurick, R.D. 1996. *The butterflies of Cornwall and the Isles of Scilly.* Dyllansow
Pengwella, Truro.
  Discusses the distribution and status of the butterfly fauna of Cornwall.

Philp, E.G. 1993. The butterflies of Kent: an atlas of their distribution. *Transactions of
the Kent Field Club* 12: i-iv, 1–58.
  Discusses the status of the Kent butterfly fauna and gives distribution maps.

Pollard, E. & Yates, T.J. 1993. *Monitoring butterflies for ecology and conservation.*
Chapman & Hall, London.
  Describes the results of the Butterfly Monitoring Scheme, including chapters on status,
  colonisation and extinction, migration and population ecology.

Pullin, A.S. 1995. *Ecology and conservation of Butterflies.* Chapman & Hall, London.
  Aims to give access to up-to-date information on a range of topics relating to butterfly
  conservation. Contains contributions from outside the British Isles.

Sawford, B. 1987. *The butterflies of Hertfordshire.* Castlemead Publications, Ware.
  Discusses distribution and status of the Hertfordshire butterfly fauna.

Smout, A.-M. & Kinnear, P.K. 1994. *Butterflies of Fife: a provisional atlas.* Fife Nature,
Glenrothes.
  Discusses distribution and status of the butterfly fauna in Fife.

Whalley, P.E.S. 1980. *Butterfly watching.* Severn House Publishers, London.
  Methods for observing, rearing, studying and photographing butterflies.

Whalley, P.E.S. 1996. *Butterflies of Gwynedd.* First Hydro, Llanberis.
  Discusses the distribution and status of the butterfly fauna of Gwynedd.

Diptera: the flies

(6643 species in 104 families)

NIGEL P. WYATT AND JOHN E. CHAINEY

The flies are the second largest order of insects in the British Isles with over 6600 species having been recorded. The order Diptera is nowadays usually divided into two suborders, Nematocera and Brachycera, although previously the most advanced families were grouped into a third sub-order, the Cyclorrhapha. Although particular species may be very specialised, flies as a group are perhaps the most biologically diverse of any insect order. Examples include predators of insects and other arthropods, snail-killers, fungivores, frugivores, dung-feeders, carrion feeders, leaf-miners, gall-feeders, ectoparasites of birds and mammals, endoparasites of other insects, arthropods and mammals. Adults in a few families are blood-feeders, biting humans and various other animals, and some of these are vectors of disease, more especially in the tropics. Not surprisingly, flies are found in almost any environment and are often abundant. However, it should also be noted that the biology of many species is unknown.

Flies are extremely varied in appearance, and include such familiar insects as crane-flies, mosquitoes, bluebottles, hoverflies and house-flies. Although a few groups (e.g. hoverflies) include large and colourful species that are popular with amateur entomologists, and are well covered in the literature, many are very small and difficult to identify. Some families are well defined (e.g. Tabanidae), but the family-level classification of other groups is often unresolved. A stereo microscope is essential for the study of most families, and some (e.g. many Ceratopogonidae, Sciaridae and Phoridae) should ideally be slide-mounted for examination with a high-powered compound microscope. Unfortunately, the British species of relatively few families are covered by modern monographs or handbooks. For many groups, including some large and economically important ones such as gall-midges (Cecidomyiidae), the relevant literature is old or piecemeal and widely scattered. The most comprehensive works for many families are still those (by authors such as Edwards and Collin) published several decades ago, which are becoming increasingly out of date or difficult to obtain. The result of these factors is that most Dipterists tend to specialise in just a few chosen families.

This bibliography is systematically arranged and largely based on the classification adopted by McAlpine *et al.* (1981), but also includes a few recent changes, for example the 'splitting' of Empididae.

Higher classification of the British Diptera

Suborder Nematocera
   Infraorder Tipulomorpha
Superfamily Tipuloidea: Tipulidae, Cylindrotomidae, Pediciidae, Limoniidae
   Infraorder Bibionomorpha
Superfamily Bibionoidea: Bibionidae
Superfamily Sciaroidea: Bolitophilidae, Diadociidae, Ditomyiidae, Keroplatidae, Mycetophilidae, Sciaridae, Cecidomyiidae
   Infraorder Psychodomorpha
Superfamily Psychodoidea: Psychodidae
Superfamily Trichoceroidea: Trichoceridae
Superfamily Anisopodoidea: Anisopodidae, Mycetobiidae
Superfamily Scatopsoidea: Scatopsidae
   Infraorder Ptychopteromorpha
Superfamily Ptychopteroidea: Ptychopteridae
   Infraorder Culicomorpha
Superfamily Culicoidea: Dixidae, Chaoboridae, Culicidae
Superfamily Chironomoidea: Thaumaleidae, Simuliidae, Ceratopogonidae, Chironomidae

Suborder Brachycera
   Infraorder Xylophagomorpha
Superfamily Xylophagoidea: Xylophagidae
   Infraorder Tabanomorpha
Superfamily Tabanoidea: Athericidae, Rhagionidae, Tabanidae
Superfamily Stratiomyoidea: Xylomyidae, Stratiomyidae
   Infraorder Asilomorpha
Superfamily Nemestrinoidea: Acroceridae
Superfamily Asiloidea: Bombyliidae, Therevidae, Scenopinidae, Asilidae
Superfamily Empidoidea: Atelestidae, Hybotidae, Empididae, Microphoridae, Dolichopodidae
   Infraorder Muscomorpha

Aschiza
Superfamily Platypezoidea: Opetiidae, Platypezidae, Phoridae
Superfamily Lonchopteroidea: Lonchopteridae
Superfamily Syrphoidea: Syrphidae, Pipunculidae
Superfamily Conopoidea: Conopidae
Schizophora: Calyptratae
Superfamily Hippoboscoidea: Hippoboscidae, Nycteribiidae
Superfamily Muscoidea: Scathophagidae, Anthomyiidae, Fanniidae, Muscidae
Superfamily Oestroidea: Calliphoridae, Gasterophilidae, Rhinophoridae,
    Sarcophagidae, Tachinidae, Oestridae
Schizophora: Acalyptratae
Superfamily Nerioidea: Pseudopomyzidae, Micropezidae
Superfamily Diopsoidea: Tanypezidae, Strongylophthalmyiidae, Megamerinidae,
    Psilidae
Superfamily Tephritoidea: Lonchaeidae, Pallopteridae, Piophilidae, Ulidiidae,
    Platystomatidae, Tephritidae
Superfamily Lauxanioidea: Lauxaniidae, Chamaemyiidae
Superfamily Sciomyzoidea: Coelopidae, Dryomyzidae, Helcomyzidae,
    Phaeomyzidae, Sciomyzidae, Sepsidae
Superfamily Opomyzoidea: Clusiidae, Acartophthalmidae, Odiniidae, Agromyzidae,
    Opomyzidae, Anthomyzidae, Aulacigastridae, Stenomicridae, Periscelidae,
    Asteiidae
Superfamily Carnioidea: Milichiidae, Carnidae, Braulidae, Tethinidae, Canacidae,
    Chloropidae
Superfamily Sphaeroceroidea: Heleomyzidae (= Helomyzidae), Chyromyidae,
    Sphaeroceridae
Superfamily Ephydroidea: Drosophilidae, Campichoetidae, Diastatidae, Camillidae,
    Ephydridae

General references

References covering a broad range of dipterous families.

Brindle, A. 1966. Taxonomic notes on the larvae of British Diptera. Revisional notes. *The Entomologist* **99**: 225–7.
    Additional notes to previously published works on Xylophagidae, *Pachygaster* (Stratiomyidae) and Psilidae, also provides a revised key to larvae of Ptychopteridae incorporating *Ptychoptera minuta*.
Colyer, C.N. & Hammond, C.O. 1968. *Flies of the British Isles.* Warne & Co., London.
    Now out of date, but nevertheless still the most comprehensive work on the British Diptera.
Edwards, F.W., Oldroyd, H. & Smart, J. 1939. *British blood-sucking flies.* British Museum (Natural History), London.
    Includes *Culicoides* (Ceratopogonidae), Culicidae, Simuliidae, Tabanidae, with keys and 45 colour plates of whole flies.

Kloet, G.S. & Hincks, W.D. 1976. A checklist of British insects. Second edition (completely revised) Part 5: Diptera and Siphonaptera. *Handbooks for the identification of British insects* 11(5): 139pp.

> The most up-to-date published checklist available, although now inevitably becoming somewhat out of date for some groups. A new checklist is in preparation.

Lindner, E. (Ed.) *Die Fliegen der palaearktischen Region*. Stuttgart.

> (In German.) A huge work in several parts including keys to Palaearctic species of all families of Diptera; many parts include the most comprehensive keys available covering the British fauna (e.g. Anthomyiidae and several acalyptrate groups; full references to these can be found below in the appropriate section of this bibliography), although some parts are now over 70 years old and consequently are now out of date.

McAlpine, J.F. *et al.* (Eds) 1981. *Manual of Nearctic Diptera* Vol. 1, Agriculture Canada Monograph No. 27: 1–674.

> Contains keys to family level including those that occur in Britain, enhanced by excellent illustrations.

Oldroyd, H. 1964. *The natural history of flies.* Weidenfeld & Nicholson, London.

> Not an identification guide, but an excellent introduction to the diverse biologies found in the Diptera.

Oldroyd, H. 1970. Diptera I. Introduction and key to families. (Third edition, rewritten and enlarged.) *Handbooks for the identification of British insects* 9(1): 104pp.

> Includes keys to families occurring in Britain, but does not include a few recent changes; for example, Acartophthalmidae are still included here within Clusiidae.

Smith, K.G.V. 1989. An introduction to the immature stages of British flies. Diptera larvae, with notes on eggs, puparia and pupae. *Handbooks for the identification of British insects* 10(14): 280pp.

> Keys to larvae, at least to family level and also to subfamily in some cases: this is the most comprehensive guide to the biology of British Diptera currently available.

Soos, A. & Papp, L. (Eds) 1984–93. *Catalogue of Palaearctic Diptera.* 13 vols. Elsevier.

> The most comprehensive and recent catalogue covering the Diptera of the palaearctic region.

Stubbs, A.E. & Chandler, P.J. 1978. A Dipterist's Handbook. *The Amateur Entomologist* 15: 1–255.

> Miscellaneous information relating to flies: collection and preservation techniques, various aspects of biology, ecology and behaviour.

Unwin, D.M. 1981. A key to the families of British Diptera. *Field Studies* 5: 513–53.

> An AIDGAP key to families that aims to avoid 'difficult' characters such as costal breaks.

Zumpt, F., 1965. *Myiasis in man and animals in the old world.* Butterworth & Co., London.

> Includes keys, with detailed descriptions of larvae, and information on biology and pathogenesis, of myiasis-causing Diptera, including all British species of Oestridae and Gasterophilidae.

Suborder Nematocera

Generally considered to be the most primitive Diptera. Members of this group are characterised by an elongate antenna with multi-segmented flagellum. Most have a significantly more complex wing venation than do other Diptera, and many are slender, long-legged insects. This group contains among others crane-flies, mosquitoes and various 'midges' and 'gnats'. The larvae are usually characterised by having a complete and non-retractile head capsule. Much of the identification literature needs revision, although new and improved keys have been published for some families.

General

Works covering more than one infraorder of Nematocera.

Freeman, P. & Lane, R.P. 1985. Bibionid and scatopsid flies. Diptera: Bibionidae and Scatopsidae. *Handbooks for the identification of British insects* **9**(7): 74pp.
Keys and check-lists to British species.

Coe, R.L., Freeman, P. & Mattingly, P.F. 1950. Diptera 2. Nematocera: families Tipulidae to Chironomidae. *Handbooks for the identification of British insects* **9** (2): 216pp.
Keys to British species covering Tipulidae, Trichoceridae, Anisopodidae, Psychodidae, Ptychopteridae, Dixidae, Chaoboridae, Culicidae and Chironomidae: many parts now superseded by more recent works, but still the most comprehensive keys for some groups, notably Tipulidae.

Infraorder Tipulomorpha

The larvae of this group often occur in soil where they sometimes attack the roots of plants, but some occur in moss or decaying wood, and a few are aquatic. Although many feed on plant matter, some are predators of other small soft-bodied invertebrates. The adults have reduced mouthparts but at least some may be able to feed on flowers where the nectar is exposed. The group is represented in Britain by four families, which until recently were all included in the Tipulidae (crane-flies).

Brindle, A. 1960. The larvae and pupae of the British Tipulinae. *Transactions of the Society for British Entomology* **14**(3): 63–114.
Includes keys to larvae and pupae of over 60 species.

Brindle, A. 1967. The larvae and pupae of the British Cylindrotominae and Limoniinae (Diptera, Tipulidae). *Transactions of the Society for British Entomology* **17**(7): 151–216.
Larvae and pupae keyed to genera, and some to species level.

Stubbs, A.E. 1972–74. Introduction to craneflies. Parts 1–7. *Bulletin of the Amateur Entomologists Society* **31**: 46–53, 83–93; **32**: 15–23, 58–63, 101–7; **33**: 19–23, 142–5.
Contains notes giving diagnostic characters, distribution and habitat preference of most British tipulids apart from most Limoniinae.

Stubbs, A.E. 1992. *Provisional atlas of the long-palped craneflies (Diptera: Tipulinae) of Britain and Ireland.* Biological Records Centre, Huntingdon.
  Distribution maps to British and Irish species.
Stubbs, A.E. 1996. Test key for *Tipula.* Cranefly Recording Scheme.
  Keys British species, couplets with line drawings to illustrate diagnostic characters.

Infraorder Bibionomorpha

The larvae of this group feed on roots, leaf litter, or wood debris (Bibionidae), mostly in fungi but sometimes in rotting wood, mosses, birds' nests, etc. (Mycetophilidae or 'fungus gnats', and their close relatives), on roots, in fungi, birds' nests, and various forms of moist decaying plant matter (Sciaridae), phytophagous and often gall-producing, or in decaying plant matter, and in some cases predaceous or parasitic on other small invertebrates (Cecidomyiidae or gall-midges). Most adults are flower feeders.

Barnes, H.F. 1946–56. *Gall midges of economic importance.* 7 vols. London.
  Gives accounts of the economically important species. See also Nijveldt (1969).
Brindle, A. 1962. Taxonomic notes on the larvae of British Diptera. 6. The family Bibionidae. *The Entomologist* 95: 22–6.
  Keys to ten species.
Chandler, P.J. 1977. Studies of some fungus gnats (Diptera: Mycetophilidae) including nine additions to the British list. *Systematic Entomology* 2: 67–93.
  Includes key to *Anatella* and description of a new British species of *Mycetophila.*
Chandler, P.J. 1978. Notes on the Holarctic species of *Pseudexechia* Tuomikoski (Diptera: Mycetophilidae), with the description of a new British species. *Entomologist's Record and Journal of Variation* 90: 44–51.
  Key to Holarctic species of this genus of Mycetophilidae.
Chandler, P.J. 1981. The European and North American species of *Epicypta* Winnertz (Diptera: Mycetophilidae). *Entomologica Scandinavica* 12: 199–212.
  Includes key to Holarctic species of this genus of Mycetophilidae.
Edwards, F.W. 1925. British fungus-gnats (Diptera, Mycetophilidae), with a revised generic classification of the family. *Transactions of the Entomological Society of London* 1924: 505–602.
  The keys to the subfamily Mycetophilinae (treated as a separate family by some authors) are still the most comprehensive available for British species in this group.
Edwards, F.W. 1938. On the British Lestremiinae, with notes on exotic specimens. (Diptera, Cecidomyiidae). 1–7. *Proceedings of the Royal Entomological Society of London* (B)7: 18–24, 25–32, 102–8, 173–82, 199–210, 229–43, 253–65.
  Contains accounts of the British species in this group.
Freeman, P. 1983. Sciarid flies. Diptera, Sciaridae. *Handbooks for the identification of British insects* 9(6): 68pp.
  Keys and checklist to British species.

Harris, K.M. 1966. Gall midge genera of economic importance Part 1: Introduction and subfamily Cecidomyiinae; supertribe Cecidomyiidi. *Transactions of the Royal Entomological Society of London* 118: 313–58.
Gives accounts of relevant genera, and diagnostic characters for economically important species.

Hutson, A.M., Ackland, D.M. & Kidd, L.N. 1980. Mycetophilidae (Bolitophilinae, Ditomyiinae, Diadocidiinae, Keroplatinae, Sciophilinae and Manotinae). Diptera, Nematocera. *Handbooks for the identification of British insects* 9(3): 111pp.
Keys and checklist to all British species of Mycetophilidae except for the subfamily Mycetophilinae, also includes list of recorded associations with fungi; includes species placed by some authors in families Bolitophilidae, Ditomyiidae, Diadocidiidae, Keroplatidae, Macroceridae and Manotidae)

Kieffer, J.J. 1913. Diptera Fam. Cecidomyidae. In Wytsman, P., *Genera Insectorum* 152: 1–346. (In French.)
Still the most comprehensive keys to genera of this group available, covering the world fauna, but inevitably is now out of date.

Mohn, E. 1955. Beitrage zur Systematik der Larven der Itonididae (=Cecidomyiidae, Diptera) 1.Teil: Porricondylinae und Itonidinae Mitteleuropas. (In German) *Zoologica* 38: 138–247.
Includes detailed descriptions of larvae of Porricondylinae and Cecidomyiinae.

Mohn, E. 1966–71. Cecidomyiidae (= Itonididae). Cecidomyiinae (part). *In* Lindner, E. (Ed.) *Die Fliegen der Palaearktischen Region* 2(2): 1–248. (In German.)
The start of a much-needed comprehensive work on this subfamily, but still far from complete.

Nijveldt, W. 1969. *Gall midges of economic importance*. VIII. *Gall midges – miscellaneous*. Crosby Lockwood & Son Ltd, London.
Accounts of economically important species: the final volume of this series, the rest written by Barnes (1946–56).

Panelius, S. 1965. A revision of the European gall midges of the subfamily Porricondylinae (Diptera: Itonidae) *Acta Zoologica Fennica* 113: 1–157.
Includes keys to European species.

Infraorder Psychodomorpha

The larvae feed mainly in moist decaying organic matter, especially of vegetable origin, but those of some Trichoceridae are occasionally found on carrion. Adults are mainly flower-feeders, although some (non-British) Psychodidae are blood-sucking. The group is represented in Britain by five families: Psychodidae, Trichoceridae, Anisopodidae, Mycetobiidae and Scatopsidae.

Brindle, A. 1962. Taxonomic notes on the larvae of British Diptera. 11. Trichoceridae and Anisopodidae. *The Entomologist* 95: 286–8.

Key to species covering five of the nine British trichocerids and three of the five anisopodids.

Withers, P. 1989. Moth flies. Diptera: Psychodidae. *Dipterist's Digest* 4: 1–83.
Keys to British species.

Infraorder Ptychopteromorpha

The larvae of the single family Ptychopteridae are semi-aquatic, in saturated mud at the bottom of shallow freshwater pools or at the edges of streams. They have an elongated respiratory siphon enabling them to breathe while submerged. The cranefly-like adults are most often seen resting on vegetation in damp habitats.

Brindle, A. 1962. Taxonomic notes on the larvae of British Diptera. 9. The family Ptychopteridae. *The Entomologist* 95: 212–16.
Keys to the larvae of five of the seven British species, adds a sixth in a subsequent paper of miscellaneous revisionary notes (1966, see above).

Stubbs, A.E. 1993. *Provisional atlas of the ptychopterid craneflies (Diptera: Ptychopteridae) of Britain and Ireland.* Biological Records Centre, Huntingdon.
Keys to British and Irish species, and distribution maps.

Infraorder Culicomorpha

The larvae are mostly aquatic, although those of some Ceratopogonidae and Chironomidae occur in damp terrestrial habitats. Adult mosquitoes (Culicidae), blackflies (Simuliidae) and Ceratopogonidae have piercing mouthparts, and females of many species suck blood to obtain protein for egg production, while males are mainly nectar-feeders. Mouthparts of other families are reduced, and many adults of the large family Chironomidae (non-biting 'midges') do not feed at all.

Boorman, J. & Rowland, C. 1988. A key to the British genera of Ceratopogonidae (Diptera). *Entomologist's Gazette* 39: 65–73.
Key and checklist to British genera.

Brindle, A. 1962. Taxonomic notes on the larvae of British Diptera. 8. The subfamily Chaoborinae (Culicidae). *The Entomologist* 95: 178–82.
Keys to the larvae of all six British species of Chaoboridae.

Campbell, J.A. & Pelham-Clinton, E.C. 1960. A taxonomic review of the British species of *Culicoides* Latreille (Diptera Ceratopogonidae). *Proceedings of the Royal Society of Edinburgh* B67(3): 181–302.
Keys to adults and species descriptions.

Cranston, P.S. 1982. A key to the larvae of the British Orthocladiinae (Chironomidae). *Scientific Publications of the Freshwater Biological Association* 45: 152pp.
Covers 92 species, approximately two-thirds of the British orthocladiine fauna.

Cranston, P.S., Ramsdale, C.D., Snow, K.R. & White, G.B. 1987 Adults, larvae and

pupae of British mosquitoes (Culicidae). *Scientific Publications of the Freshwater Biological Association* **48**: 152pp.
Keys to adult males based on external and genitalic characters, adult females, larvae and pupae, also with section providing more detailed accounts of each British species.

Crosskey, R.W. 1990. *The natural history of blackflies.* John Wiley & Sons, Chichester.
A synthesis of the current knowledge of blackflies, especially aspects of their biology, and medical and veterinary importance.

Crosskey, R.W. 1991. A new checklist of the blackflies of Britain and Ireland, with geographical and type information (Diptera: Simuliidae). *Entomologist's Gazette* **42**: 206–17.
Provides an updated checklist to the British simuliid fauna, also a table summarising name usage in the principal works on British Simuliidae compared with current nomenclature.

Crosskey, R.W. 1996. Names of British Simuliids. *British Simuliid Group Bulletin* **7**: 6.
Gives all name changes for British simuliids subsequent to Davies' (1968) handbook.

Davies. L. 1968. A key to the species of Simuliidae (Diptera) in the larval, pupal and adult stages. *Scientific Publications of the Freshwater Biological Association* **24**: 126pp.
The most comprehensive published keys to British blackflies, but needs revision, especially the sections covering the immature stages; a revised key is being prepared.

Disney, R.H.L. 1975 A key to the British Dixidae. *Scientific Publications of the Freshwater Biological Association* **31**: 78pp.
Keys to adults and immature stages, also distribution maps, although the latter have subsequently been updated by Goldie-Smith (1989).

Edwards, F.W. 1926. On the British biting midges (Diptera, Ceratopogonidae). *Transactions of the Entomological Society of London* **74**: 389–426.
Contains keys to genera and descriptions of British species; now somewhat out of date but the most comprehensive key to this family currently available.

Edwards, F.W. 1929. A revision of the Thaumaleidae (Dipt.). *Zoologischer Anzeiger* **82**: 121–42.
A world revision, with descriptions of each species.

Goldie-Smith, E.K. 1989. Distribution maps for Dixidae in Great Britain and Ireland. *Dipterist's Digest* **3**: 8–26.
Updated maps with several additional records compared to those in Disney (1975).

Kettle, D.S. & Lawson, J.W.H. 1952. The early stages of British biting midges *Culicoides* Latreille (Diptera: Ceratopogonidae) and allied genera. *Bulletin of Entomological Research* **43**: 421–67.
Keys to fourth-instar larvae, and pupae.

Pinder, L.C.V. 1978. A key to adult males of the British Chironomidae (Diptera). *Scientific Publications of the Freshwater Biological Association* **37**: 169pp.
Significant updating of key by Coe *et al.* (1950), with several additional species, better illustrations and up-to-date nomenclature.

Snow, K.R., Rees, A.T. & Brooks, J.A.H. 1997. *A revised bibliography of the mosquitoes of the British Isles.* Occasional Publication of the British Mosquito Group, University of East London Press.

A comprehensive listing of literature on the biology, ecology, distribution, identification, disease relationships and control of British mosquitoes: this revision cites an additional 150 references to the first edition published in 1993.

Wiederholm, T. (Ed.) 1983. Chironomidae of the Holarctic region. Keys and diagnoses. Part 1 Larvae. *Entomologica Scandinavica* (suppl.) **19**: 1–457.

Wiederholm, T. (Ed.) 1986. Chironomidae of the Holarctic region. Keys and diagnoses. Part 2 Pupae. *Entomologica Scandinavica* (suppl.) **28**: 1–482.

Wiederholm, T. (Ed.) 1989. Chironomidae of the Holarctic region. Keys and diagnoses. Part 3 Adult males. *Entomologica Scandinavica* (suppl.) **34**: 1–532.

These three volumes contain a compilation of papers by miscellaneous authors, providing keys to larvae, pupae and adult males of all Holarctic species.

Suborder Brachycera

This group contains what are generally considered to be the most advanced Diptera. Until recently the species placed here were usually included in two suborders, Brachycera and Cyclorrhapha, but current thinking is that they should all be considered as Brachycera. Many members of the infraorders Tabanomorpha and Asilomorpha show relatively primitive features such as complex wing venation, increased abdominal segmentation, a multi-segmented antennal flagellum, and in the immature stages a distinct larval head capsule and well-sclerotised pupal case with, except in a few cases, the last larval skin cast aside and not used as a protective case for the pupa (puparium). However, in the infraorder Muscomorpha, and in a few of the most advanced Asilomorpha, abdominal and antennal segmentation is reduced (in the latter only one flagellar segment is present bearing a bristle-like appendage, the arista) and the wing venation also reduced. The larvae are characterised by having a much reduced, unsclerotised and retracted head capsule, the most obvious feature at the head end being a pair of sclerotised mouth-hooks attached to fused internal sclerotised rods, forming a 'cephalopharyngeal skeleton'. Pupation occurs within the sclerotised last larval skin, which forms a protective case (puparium) for the fragile pupa.

General

References covering more than one infraorder within Brachycera.

Drake, C.M. 1991 *Provisional atlas of the larger Brachycera (Diptera) of Britain and Ireland.* Biological Records Centre, Huntingdon.

Covers Acroceridae, Asilidae, Bombyliidae, Rhagionidae, Scenopinidae, Stratiomyidae, Tabanidae, Therevidae, Xylomyidae and Xylophagidae. Provides notes on the status of each species, a table showing occurrence of each species by vice-county and selected distribution maps.

Oldroyd, H. 1969 Diptera Brachycera Section (a) Tabanoidea and Asiloidea. *Handbooks for the identification of British insects* 9(7): 74pp.
Keys to British species of Acroceridae, Asilidae, Bombyliidae, Rhagionidae, Scenopinidae, Stratiomyidae, Tabanidae, Therevidae, Xylomyidae and Xylophagidae. A new book covering these families is being prepared.

Infraorder Xylophagomorpha

The larvae of the single family, the Xylophagidae, which includes just three British species, are predators or saprophages in rotting wood, but the feeding habits of the adults are not well known. For identification of adults see Oldroyd (1969), above.

Brindle, A. 1961. Taxonomic notes on the larvae of British Diptera. 2. The genus *Xylophagus* Meigen (*Erinna* Meigen) (Rhagionidae). *The Entomologist* 94: 144–8.
Key to larvae of two of the three British species now placed in Xylophagidae.

Infraorder Tabanomorpha

Larvae of this primitive infraorder occur in wet soil, or tree-holes where they are predators of other invertebrates or even occasionally cannibalistic (Tabanidae), aquatic, among stones on river-beds (Athericidae), either predators of other soft-bodied invertebrates in soil or litter, or feeding on decaying plant matter (Rhagionidae), in rotting vegetable matter, moss, or aquatic (Stratiomyidae or soldier-flies), and predators or saprophages in rotting wood (Xylomyidae). Adult horse-flies (Tabanidae) have piercing mouthparts and females suck blood to obtain protein for egg production, while the males are chiefly nectar-feeders. Adult Athericidae feed mainly on honeydew; the feeding habits of Xylomyidae and Rhagionidae (snipe-flies) are generally not well known, although a few species of the latter family are known to suck the blood of mammals. Stratiomyidae are most often seen resting on vegetation and visit flowers to feed on nectar.

Brindle, A. 1959. Notes on the larvae of the British Rhagionidae and Stratiomyidae with a key to genera. *Entomologist's Record and Journal of Variation* 71: 126–33.
Keys to British genera, also including species now placed in Xylomyiidae, Xylophagidae and Athericidae.

Brindle, A. 1961. Taxonomic notes on the larvae of British Diptera. 3. The genus *Solva* Walker (*Xylomyia* Rondani: *Subula* Meigen) (Stratiomyiidae). *The Entomologist* 94: 201–5.
Key to larvae of the three British species now placed in Xylomyiidae.

Brindle, A. 1961. Taxonomic notes on the larvae of British Diptera. 4. The genus *Atherix* Meigen (Rhagionidae). *The Entomologist* 94: 218–20.
Key to larvae of British species of this genus, now placed in Athericidae.

Brindle, A. 1962. Taxonomic notes on the larvae of British Diptera. 12. The genus *Rhagio* F. Rhagionidae. *The Entomologist* 95: 311–15.
  Keys to larvae and pupae of four British species.
Chvala, M., Lyneborg, L. & Moucha, J. 1972. *The horse flies of Europe (Diptera, Tabanidae)*. Entomological Society of Copenhagen.
  A comprehensive account of the European species of Tabanidae.
Rozkosny, R. 1982. *A biosystematic study of the European Stratiomyidae (Diptera)*. Vol. 1, *Introduction, Beridinae, Sarginae and Stratiomyinae*. Dr W. Junk, The Hague.
Rozkosny, R. 1983. *A biosystematic study of the European Stratiomyidae (Diptera)*. Vol. 2, *Clitellariinae, Hermetiinae, Pachygasterinae and bibliography*. Dr W. Junk, The Hague.
  These two volumes provide a comprehensive account of the European Stratiomyidae.

Infraorder Asilomorpha

The larvae in this group are mostly predaceous on other small invertebrates, except those of Bombyliidae, which are parasitic or predaceous on the immature stages of other insects, and Acroceridae, which parasitise spiders. However, the immature stages are generally still poorly known in much of this group. Adult Therevidae feed on nectar and insect secretions; asilids, empids (and closely related families), and dolichopodids are mainly predators of other insects, although at least some empids also feed on nectar or pollen; adult Acroceridae appear not to feed at all, while many Bombyliidae have an elongate proboscis that enables them to feed on the nectar of tubular flowers. This group contains families such as the robber-flies (Asilidae), bee-flies (Bombyliidae) and dance-flies (Empididae).

Brindle, A. 1962. Taxonomic notes on the larvae of British Diptera. 10. The Asilidae. *The Entomologist* 95: 241–7.
  Keys to larvae of 15 British species.
Brindle, A. 1964. Taxonomic notes on the larvae of British Diptera. 18. The Hemerodromiinae (Empididae). *The Entomologist* 97: 162–5.
  Key to larvae of a few British species (also including Clinocerinae).
Brindle, A. 1969. Taxonomic notes on the larvae of British Diptera. 27. The Hemerodromiinae (Empididae). *The Entomologist* 102: 35–9.
  Key to pupae of a few British species (also including Clinocerinae).
Chvala, M. 1975. The Tachydromiinae (Dipt. Empididae) of Fennoscandia and Denmark. *Fauna Entomologica Scandinavica* 3: 1–336.
Chvala, M. 1983. The Empidoidea (Diptera) of Fennoscandia and Denmark. II General Part. The families Hybotidae, Alelestidae and Microphoridae. *Fauna Entomologica Scandinavica* 12: 1–279.
  See Chvala (1994).

Chvala, M. 1989. Monograph of northern and central European species of
Platypalpus (Diptera, Hybotidae), with data on the occurrence in
Czechoslovakia. *Acta Universitatis Carolinae* (Biologica) **32**: 209–376.
Key to species, including all British species.

Chvala, M. 1994. The Empidoidea (Diptera) of Fennoscandia and Denmark. III.
Genus *Empis*. *Fauna Entomologica Scandinavica* **29**: 1–192.
**Although not yet complete, this series of volumes will provide a comprehensive account
of these groups (formerly lumped as Empididae) in NW Europe. Most of the British
species are covered, including some not in Collin (1961).**

Chvala, M. 1992. Key to British species of *Empis* s. str. (Diptera, Empididae).
Dipterist's Digest **12**: 7–15.
Keys British and potentially British species.

Cole, J.H. 1989. Two species of *Medetera* Fischer (Diptera, Dolichopodidae) new to
Britain. *British Journal of Entomology and Natural History* **2**: 115–18.
Includes modification of Fonseca's (1978) key to incorporate the additional species.

Collin, J.E. 1961. *British flies*. Vol. VI, Empididae. Cambridge University Press.
**A dated but comprehensive account of the British fauna of Empididae (s. str.); also
includes species now placed in the families Atelestidae, Hybotidae and Microphoridae.
Issued in three parts, all in 1961 and with continuous pagination.**

Fonseca, E.C.M.d'Assis 1978. Diptera: Orthorrhapha, Brachycera, Dolichopodidae.
Handbooks for the identification of British insects **9**(5): 90pp.
Keys and checklist to the British species.

Meuffels, H.J.G. & Grootaert, P. 1990. The identity of *Sciapus contristans*
(Wiedemann, 1817) (Diptera: Dolichopodidae), and a revision of the species
group of its relatives. *Bulletin de l'Institut Royal des Sciences Naturelles de
Belgique* **60**: 161–78.
**Provides revised and improved key to *Sciapus* covering all the British species, including
S. basilicus: also gives much-needed clarification of species identities in the *contristans*
species-group.**

Pollet, M. 1990. Phenetic and ecological relationships between species of the subgenus *Hercostomus* (*Gymnopternus*) in western Europe with the description of two
new species (Diptera: Dolichopodidae). *Systematic Entomology* **15**: 359–82.
**Provides revised and improved key to western European species of this subgenus includ-
ing two that are new to science, both of which occur in Britain.**

Pollet, M. 1996. Systematic revision and phylogeny of the Palaearctic species of the
genus *Achalcus* Loew (Diptera: Dolichopodidae) with the description of four
new species. *Systematic Entomology* **21**: 353–86.
**Includes four British species additional to those in Fonseca's (1978) key, and provides
revised key to the Palaearctic species of this genus.**

Infraorder Muscomorpha

This infraorder contains the groups that were formerly usually placed in
the suborder Cyclorrhapha. Generally considered to contain the most

advanced Diptera. Muscomorpha is divided into two groups, the Schizo-phora and the Aschiza, the former distinguished from the latter by having a suture on the front of the head from which an inflatable sac, the ptilinum, is expressed during emergence from the puparium to rupture the puparial case.

Aschiza

Represented by just six families in Britain, all of which are morphologically very distinct from each other. The most well known of these are the hover-flies (Syrphidae), but the small and more unobtrusive scuttle flies (Phoridae) include almost as many species. The Conopidae have been included here as usual, but some authorities now consider them to be more closely related to the acalyptrates. The larvae occur among decaying vegetable matter in soil or litter (Lonchopteridae), are fungivorous (Platypezidae), predaceous or para-sitic on a variety of other invertebrates, or saprophagous (Phoridae), and endoparasitic on Homoptera (Pipunculidae) or bees and wasps (Conopidae). Syrphidae biology is very varied; examples are predators of soft-bodied insects, especially aphids; fungivores; phytophages attacking roots and bulbs, leaf-miners and stem-borers; saprophages in rotting wood, sap-runs, dung, damp compost or aquatic 'rat-tailed maggots' with an elon-gated respiratory siphon enabling them to breathe while submerged, and scavengers or predators in the nests of wasps or bees. Adult Syrphidae are among the commonest flower-feeding Diptera, but the feeding habits of many other adult Aschiza are still poorly known.

There are good keys available covering the British fauna of all these groups, although those to the Pipunculidae are now somewhat in need of revision, being partly updated by Ackland (1993).

Ackland, D.M. 1993. Notes on British *Cephalops* Fallen, 1810 with description of a new species, and *Microcephalops* De Meyer, a genus new to Britain (Dipt., Pipunculidae). *Entomologist's Monthly Magazine* **129**: 95–105.
    Includes revised key to British *Cephalops* and description of a new species.
Brindle, A. 1961. Taxonomic notes on the larvae of British Diptera. 5. The Clythiidae (Platypezidae). *The Entomologist* **94**: 274–8.
    Includes key to genera covering larvae of British species of *Callomyia*, *Agathomyia*, *Plesioclythia*, *Platypeza* and *Orthovena*.
Chandler, A.E.F. 1968. A preliminary key to the eggs of some of the commoner aphi-dophagous Syrphidae (Diptera) occurring in Britain. *Transactions of the Royal Entomological Society of London* **120**: 199–218.
    Covers some common species in the tribes Syrphini, Melanostomatini and Pipizini.

Chandler, P.J. 1973. The flat-footed flies (Diptera, Aschiza-Platypezidae) known to occur in Kent with a key to the genera and species so far recorded from the British Isles. *Transactions of the Kent Field Club* **5**: 15–44.
    Includes key to all species recorded from Britain at the time, plus more detailed information on those occurring in Kent, also includes species now placed in the families Atelestidae and Opetiidae, which are now considered to be more closely related to Empididae.

Chandler, P.J. 1974. Additions and corrections to the British list of Platypezidae (Diptera), incorporating a revision of the Palaearctic species of *Callomyia* Meigen. *Proceedings and Transactions of the British Entomological and Natural History Society* **7**: 1–32.
    Adds a further six species to the British fauna, and provides revised keys to the genera where changes occur.

Coe, R.L. 1966. Diptera Pipunculidae. *Handbooks for the identification of British insects* **10**(2c): 83pp.
    Includes keys to British species.

Disney, R.H.L. 1983. Scuttle flies. Diptera, Phoridae (except *Megaselia*). *Handbooks for the identification of British insects* **10**(6): 81pp.
    Includes keys and checklist to British species.

Disney, R.H.L. 1989. Scuttle flies. Diptera, Phoridae Genus *Megaselia*. *Handbooks for the identification of British insects* **10**(8): 155pp.
    Includes keys and checklist to British species.

Disney, R.H.L. 1994. *Scuttle flies: the Phoridae.* Chapman & Hall, London.
    Includes detailed accounts of adult and immature stages, and of phorid ecology.

Rotheray, G.E. 1993. Colour guide to hoverfly larvae (Diptera, Syrphidae). *Dipterist's Digest* **9**: 1–155.
    Keys to all currently known larvae of British species; illustrations include several excellent colour photographs.

Smith, K.G.V. 1969. Diptera Lonchopteridae. *Handbooks for the identification of British insects* **10**(2ai): 9pp.
    Includes key to British species.

Smith, K.G.V. 11969. Diptera Conopidae. *Handbooks for the identification of British insects* **10**(3a): 19pp.
    Includes key to British species.

Stubbs, A.E. 1996. *British hoverflies. Second supplement.* British Entomological and Natural History Society.
    Updates information on biology, conservation and reference list from original guide (Stubbs & Falk, 1983), and includes keys to facilitate identification of a further 15 species and females of difficult genera.

Stubbs, A.E. & Falk, S. 1983. *British hoverflies, an illustrated identification guide.* British Entomological and Natural History Society.
    Comprehensive guide to British syrphids, with keys to species level, biological and ecological information, and guides to collecting techniques. Illustrations include some excellent colour plates.

Schizophora: Calyptratae

These are distinguished from the other group of Schizophora, the acalyptrates, by having a complete mesonotal suture on the thorax, a complete dorsal 'seam' on the antennal pedicel, and usually by having well developed calypters (or squamae). Many are rather bristly, and the group includes such familiar insects as house-flies (Muscidae) and blowflies (Calliphoridae). Several larvae in the superfamily Oestroidea are parasitic: endoparasitic on other insects, especially larvae of Lepidoptera (Tachinidae), or on woodlice (Rhinophoridae), kleptoparasites in the nests of solitary wasps; endoparasites of insects such as Orthoptera, of molluscs, earthworms or spiders' egg sacs (many Sarcophagidae and Calliphoridae), endoparasites of amphibians or blood-sucking ectoparasites of bird nestlings, and sometimes feeding in external wounds on living mammals (some Calliphoridae); developing in the nasopharyngeal cavities of sheep or deer, or under the skin of cattle, horses or deer, forming 'warbles' (Oestridae), or in the alimentary canal of horses (Gasterophilidae); others are saprophagous in carrion or dung (some Sarcophagidae and Calliphoridae). In the superfamily Hippoboscoidea (Hippoboscidae and Nycteribiidae) larvae are retained in the mother's body after hatching and are eventually deposited when fully grown, when they immediately pupate. In the superfamily Muscoidea larvae are mainly phytophagous, but a few are predators living in dung or rotting seaweed (Scathophagidae); sometimes phytophagous, including a few pests of commercial crops; or saprophages in rotting vegetable matter, dung, seaweed or fungi; or kleptoparasitic on aculeate Hymenoptera, feeding on stored pollen (Anthomyiidae); saprophages in carrion, dung, nests of mammals, birds or insects, or in fungi (Fanniidae); in soil, fungi, wasps' nests, dung or other decaying organic matter where they may be predaceous or saprophagous, and a few are aquatic (Muscidae). Although many adult calyptrates feed on nectar, pollen or honeydew there are notable exceptions: the Hippoboscoidea are blood-sucking ectoparasites of mammals or birds, and many of them have the wings reduced or absent, while others have wings that break off once they have found a suitable host; the stomoxyines (Muscidae) suck the blood of large mammals; other muscids feed on animals' body secretions, or various forms of decaying organic matter, scathophagids are predators on other invertebrates, and oestrids and gasterophilids have reduced mouthparts and apparently do not feed in the adult stage.

Most of the British fauna are well covered by the literature, a notable exception being the Anthomyiidae, although there are reasonably up-to-date keys by Hennig (1966–76) including British species of this family available in *Die Fliegen der palaearktischen Region.*

Belshaw, R. 1993. Tachinid flies. Diptera: Tachinidae. *Handbooks for the identification of British insects* **10**(4ai): 169pp.
Keys to British species, with illustrations clearly indicating important identification characters.

Collin, J.E. 1958. A short synopsis of the British Scathophagidae (Diptera). *Transactions of the Society for British Entomology* **13**: 37–56.
Keys to all British species.

Dusek, J. 1969–70. Praeimaginale stadiens mitteleuropaischer Anthomyiiden (Diptera) I-II. *Prirodovedne Prace ustavu Ceskoslovenske Akademie Ved v Brne (N.S.)* **3**(2): 1–37; **4**(1): 1–29.
(In German.) Keys to some larvae of European Anthomyiidae, including the agriculturally important species.

Emden, F.I. van 1954. Diptera Cyclorrhapha. Calyptrata (I) section (a). Tachinidae and Calliphoridae. *Handbooks for the identification of British insects* **10**(4a): 133pp.
Keys to British species: to a large extent, but not entirely, superseded by more recent works; as well Tachinidae and Calliphoridae, includes species now placed in Oestridae, Rhinophoridae and Sarcophagidae.

Fonseca, E.C.M. d'Assis 1956. A review of the British subfamilies and genera of the family Muscidae. *Transactions of the Society for British Entomology* **12**(4): 113–28.
Provides keys to genera also including those now placed in Anthomyiidae and Fanniidae, but nomenclaturally out of date.

Fonseca, E.C.M. d'Assis 1968. Diptera Cyclorrhapha Calyptrata Section (b) Muscidae. *Handbooks for the identification of British insects* **10**(4b): 119pp.
Keys to species, also includes species now placed in Fanniidae.

Hennig, W. 1966–1976. 63a. Anthomyiidae. *In* Lindner, E. (Ed.) *Die Fliegen der palaearktischen Region* **7**(1): 974pp. (In German.)
Keys to Palaearctic species, still reasonably comprehensive and up-to-date, unfortunately no comprehensive keys to the British fauna are available in English for this large and important family.

Hutson, A.M. 1984 Keds, flat-flies and bat-flies. Diptera, Hippoboscidae and Nycteribiidae. *Handbooks for the Identification of British Insects* **10**(7): 40pp.
Keys and checklist to British species.

Pape, T. 1987. The Sarcophagidae (Diptera) of Fennoscandia and Denmark. *Fauna Entomologica Scandinavica* **19**: 1–203.
Keys to Fennoscandian species, but covers nearly all of the British fauna; uses more reliable characters, is better illustrated, and more up-to-date nomenclaturally than Emden's (1954) key.

Rognes, K. 1991. The Calliphoridae (Diptera) of Fennoscandia and Denmark. *Fauna Entomologica Scandinavica* **24**: 1–272.

Keys to Fennoscandian species, but includes all of the British fauna; uses more reliable characters, is better illustrated and more nomenclaturally up-to-date than Emden's (1954) key.

Skidmore, P. 1985. The biology of the Muscidae of the world. *Series Entomologica* **29**: 1–550.

Includes descriptions, keys and illustrations of all known muscid larvae.

Schizophora: Acalyptratae

These are mostly small, compact flies distinguished by their lack of a thoracic suture and reduced calypters (or squamae), although a few, such as the long-legged and slender Micropezidae, and the wingless bee ectoparasites in the Braulidae, are strikingly different. Some families include several species with strikingly patterned wings. Among the families included in Acalyptratae are the fruit-flies (Tephritidae), leaf-mining flies (Agromyzidae), and vinegar flies (Drosophilidae). Larvae of many of this large group are saprophagous, occurring in rotting vegetable matter, dung, birds' nests, carrion, decaying wood, decaying seaweed, or fungi. There are however several notable exceptions, examples including: phytophages in fruits, flower-heads, leaves, stems and roots of flowering plants (Tephritidae), also phytophagous, but mostly attacking roots (Psilidae), mining in leaves and stems (Agromyzidae), on grasses (several Chloropidae, Lonchaeidae, Opomyzidae, Anthomyzidae), predators of aphids and coccids (Chamaemyiidae), or of other insect larvae under bark (Pallopteridae, Megamerinidae, some Lonchaeidae), mollusc endoparasites (Sciomyzidae), ectoparasitic blood-suckers of bird nestlings (Piophilidae – *Neottiophila*), mainly aquatic or semi-aquatic larvae which may be phytophagous, parasitic or predaceous, some capable of tolerating high levels of temperature or salinity (Ephydridae), or feeding on stored pollen in bees' nests (Braulidae). Adult acalyptrates utilise flowers to a much lesser degree than most other Diptera, although many Tephritidae, Sepsidae, Chloropidae and Milichiidae regularly visit flowers. Decaying organic matter and sap runs also attract acalyptrates to feed, while fermenting fruit juices are particularly attractive to drosophilids. Many acalyptrates however are unobtrusive, occurring mainly among low vegetation, where they can be collected by sweep netting. Adult Ephydridae mainly feed on unicellular micro-organisms, Carnidae frequent birds' nests where they feed on the birds' bodily secretions, dung or carrion, and the unusual Braulidae are wingless ectoparasites of honey bees.

Unfortunately this group is in general relatively poorly covered by the Royal Entomological Society's series of identification handbooks, with some notable exceptions, and although keys to species for all families are available in *Die Fliegen der palaearktischen Region,* and to a lesser extent covered by some published works of J.E. Collin, many of these are now significantly out of date.

Andersson, H. 1971. The European species of *Limnellia* (Dipt., Ephydridae). *Entomologica Scandinavica* 2: 53–9.
Key to European *Limnellia* and description of a new British species.
Andersson, H. 1976. Revision of the *Anthomyza* species of Northwest Europe (Diptera: Anthomyzidae) I. The *gracilis* group. *Entomologica Scandinavica* 7: 41–52.
Notes on species in this group, illustration of genitalia.
Andersson, H. 1977. Taxonomic and phylogenetic studies of Chloropidae (Diptera) with special reference to Old World genera. *Entomologica Scandinavica* (suppl.) 8: 1–200.
Includes keys to Old World genera.
Andersson, H. 1984. Revision of the *Anthomyza* species of Northwest Europe (Diptera: Anthomyzidae) II. The *pallida* group. *Entomologica Scandinavica* 15: 15–24.
Notes on species in the group and figures of genitalia.
Basden, E.B. 1954. The distribution and biology of Drosophilidae (Diptera) in Scotland, including a new species of *Drosophila. Transactions of the Royal Society of Edinburgh* 62: 603–54.
Provides key to Scottish species of Drosophilidae, and detailed notes on distribution and biology, including laboratory cultures.
Becker, T. 1926. 56a. Ephydridae. *In* Lindner, E.(Ed.) *Die Fliegen der palaearktischen Region* 6(1): 1–115.
The only comprehensive work on this group for this region, but difficult to use, lacks keys for many genera, and includes dubious synonymies. (In German.)
Brindle, A. 1965. Taxonomic Notes on the larvae of British Diptera. No. 21 – The Piophilidae. *The Entomologist* 98: 158–60.
Keys to larvae of four of the eleven British species.
Chandler, P.J. 1975. Notes on the British status of three unusual Acalyptrate flies (Diptera). *Proceedings and Transactions of the British Entomological and Natural History Society* 8: 66–72.
Notes on *Rainieria calceata* (Micropezidae), *Tanypeza longimana* (Tanypezidae) and *Megamerina loxocerina* (Megamerinidae), with distribution maps provided for latter two species, which are the only British representatives of their respective families.
Chandler, P.J. 1978. A revision of the British Asteiidae (Diptera) including two additions to the British list. *Proceedings and Transactions of the British Entomological and Natural History Society* 11: 23–34.
Provides keys to British species.
Clements, D.K. 1990. Provisional keys to the Otitidae and Platystomatidae of the British Isles. *Dipterist's Digest* 6: 32–41.
Keys to species, couplets enhanced by line drawings illustrating diagnostic characters.

Collin, J.E. 1930. Some species of the genus *Meonura* (Diptera). *Entomologist's Monthly Magazine* **66**: 82–9.
Contains key to British species of the largest British genus of Carnidae, and descriptions of new species.

Collin, J.E. 1933. Five new species of Diptera. *Entomologist's Monthly Magazine* **69**: 272–5.
Includes key to British Camillidae.

Collin, J.E. 1943. The British species of *Psilopa* Fln. and *Discocerina* Mcq. (Dipt., Ephydridae). *Entomologist's Monthly Magazine* **79**: 145–51.
Includes keys to British species of these genera, all British Discocerinini included here in *Discocerina*.

Collin, J.E. 1943. The British species of Helomyzidae (Diptera). *Entomologist's Monthly Magazine* **79**: 234–51.
Includes keys to British species.

Collin, J.E. 1944. The British species of Psilidae (Diptera). *Entomologist's Monthly Magazine* **80**: 214–24.
Includes keys to British species.

Collin, J.E. 1944. The British species of Anthomyzidae (Diptera). *Entomologist's Monthly Magazine* **80**: 265–72.
Keys to British species, also includes *Stenomicra* (now placed in Stenomicridae).

Collin, J.E. 1945. British Micropezidae (Diptera). *Entomologist's Record and Journal of Variation* **57**: 115–19.
Keys to British species, includes species now placed in Tanypezidae.

Collin, J.E. 1946. The British genera and species of Oscinellinae (Diptera, Chloropidae). *Transactions of the Royal Entomological Society of London* **97**: 117–48.
Includes keys to British species.

Collin, J.E. 1948. A short synopsis of the British Sapromyzidae (Diptera). *Transactions of the Royal Entomological Society of London* **99**: 225–42.
Includes keys to British species of this family, now known as Lauxaniidae.

Collin, J.E. 1949. The Palaearctic species of the genus *Aphaniosoma* Beck. (Diptera, Chiromyiidae). *Annals and Magazine of Natural History* (12)**2**: 127–47.
Key and notes, including the two British species of this genus.

Collin, J.E. 1951. The British species of the genus *Palloptera* Fallen (Diptera). *Entomologist's Record and Journal of Variation* (suppl.) **63**: 1–6.
Includes key to most British Pallopteridae, except the single British species not in this genus.

Collin, J.E. 1951. British Helomyzidae (Diptera) – additions and corrections. *Journal of the Society for British Entomology* **4**: 37–9.
Two species added to the British fauna, plus correction of a previous misidentification.

Collin, J.E. 1952. On the European species of the genus *Odinia* R.-D. (Diptera, Odiniidae). *Proceedings of the Royal Entomological Society of London* (B) **21**: 110–16.
Key includes British Odiniidae.

Collin, J.E. 1953. A revision of the British (and notes on other) species of Lonchaeidae (Diptera). *Transaction of the Society for British Entomology* 11: 181–207.
Includes keys to British species.

Collin, J.E. 1960. British Tethinidae (Diptera). *The Entomologist* 93: 191–3.
Includes keys to British species.

Collin, J.E. 1964. The British species of *Ephydra* (Dipt., Ephydridae). *Entomologist's Monthly Magazine* 99: 147–52.
Includes keys to British species of *Ephydra* and *Setacera*.

Collin, J.E. 1966. A contribution towards the knowledge of the male genitalia of species of *Hydrellia* (Diptera, Ephydridae). *Bollettino del Museo Civico di Storia Naturale di Venezia* 16: 7–18.
Includes key to British species of this genus and illustrations of male genitalia.

Collin, J.E. 1966. A revision of the Palaearctic species of *Tethina* and *Rhicnoessa*. *Bollettino del Museo Civico di Storia Naturale di Venezia* 16: 19–32.
Revised keys to species covering all British Tethinidae apart from the two British species of *Pelomyiella*.

Collin, J.E. 1966. The British species of *Chamaemyia* Mg. (*Ochthiphila* Fln.) (Diptera). *Transactions of the Society for British Entomology* 17: 121–8.
Includes key to British species of this genus of Chamaemyiidae.

Czerny, L. 1927. 53c. Chiromyidae. *In* Lindner, E. (Ed.) *Die Fliegen der palaearktischen Region* 5(1): 51–4.
Keys to Palaearctic species. (In German.)

Czerny, L. 1928. Clusiidae. *In* Lindner, E. (Ed.) *Die Fliegen der Palaearktischen Region* 6(1): 112.
Keys to Palaearctic species, including those now placed in Acartophthalmidae. (In German.)

Dahl, R.G. 1959. Studies on Scandinavian Ephydridae (Diptera Brachycera). *Opuscula Entomologica. Supplementum.* 15: 1–224.
Mainly ecological data, but also includes descriptions of Scandinavian species, many of which occur in Britain, with mention of diagnostic characters.

Drake, C.M. 1993. A review of the British Opomyzidae. *British Journal of Entomology and Natural History* 6(4): 159–76.
Provides key to British species.

Duda, O. 1934. 58a. Periscelididae. *In* Lindner, E. (Ed.) *Die Fliegen der palaearktischen Region* 6(1): 1–13. (In German.)
Keys to Palaearctic species.

Duda, O. 1934. 58e. Diastatidae. *In* Lindner, E. (Ed.) *Die Fliegen der palaearktischen Region* 6(1): 1–15. (In German.)
Keys to Palaearctic species.

Egglishaw, H.J. 1960. Studies on the family Coelopidae. *Transactions of the Royal Entomological Society of London* 112: 109–40.
Notes on identification, biology and laboratory studies; also includes *Orygma* (now Sepsidae) and species now placed in Helcomyzidae.

Fonseca, E.C.M.d'Assis 1965. A short key to the British Drosophilidae (Diptera)

including a new species of *Amiota*. *Transactions of the Society for British Entomology* **16**: 233–44.
Key to British species.

Hennig, W. 1937. 60a. Milichiidae et Carnidae. *In* Lindner, E. (Ed.) *Die Fliegen der palaearktischen Region* **6**(1): 1–91.
Keys to Palaearctic species.

Imms, A.D., 1942. On *Braula coeca* Nitzsch and its affinities. *Parasitology* **34**: 88–100.
Detailed study of biology, and morphology of immature stages, of the sole British species of Braulidae.

Irwin, A.G. 1982. A new species of *Stenomicra* Coquillett (Diptera, Aulacigastridae) from Anglesey, North Wales. *Entomologist's Monthly Magazine* **118**: 235–8.
Description of new species, now placed in Stenomicridae, the second of this genus recorded from Britain.

McAlpine, J.F. 1962. A revision of the genus *Campichoeta* Macquart (Diptera: Diastatidae). *Canadian Entomologist* **94**: 1–10.
World revision of this genus.

McAlpine, J.F. 1977. A revised classification of Piophilidae, including 'Neottiophilidae' and 'Thyreophoridae' (Diptera: Schizophora). *Memoirs of the Entomological Society of Canada* **103**: 1–66.
World revision of family, with keys to species.

Olafsson, E. 1991. Taxonomic revision of western Palaearctic species of the genera *Scatella* R.-D. and *Lamproscatella* Hendel, and studies on their phylogenetic positions within the subfamily Ephydrinae (Diptera, Ephydridae). *Entomologica Scandinavica* (suppl.) **37**: 1–100.
Descriptions and phylogenetic analysis of western Palaearctic species, and discussion of relationships between western Palaearctic genera of the subfamily Ephydrinae.

Pitkin, B.R. 1988. Lesser dung flies. Diptera: Sphaeroceridae. *Handbooks for the Identification of British Insects* **10**(5e): 175pp.
Keys and checklist to species, illustrations include many of immature stages, especially puparia.

Pont, A.C. 1979. Sepsidae Diptera Cyclorrhapha, Acalyptrata. *Handbooks for the Identification of British Insects* **10**(5c): 35pp.
Keys and checklist to species.

Robinson, I. 1953. The postembryonic stages in the life-cycle of *Aulacigaster leucopeza* (Meigen) (Diptera: Cyclorrhapha: Aulacigastridae). *Proceedings of the Royal Entomological Society of London* (A)**28**: 77–84.
Description of immature stages.

Rozkosny, R. 1984. The Sciomyzidae (Diptera) of Fennoscandia and Denmark. *Fauna Entomologica Scandinavica* **14**: 1–224.
Keys to species, includes all British fauna apart from *Pherbellia knutsoni*.

Shorrocks, B. 1972. *Invertebrate types: Drosophila*. Ginn & Co., London.
Includes keys to the commoner British species in the genus, with some further information on them, especially on ecology and genetics.

Skidmore, P. 1993. Notes on the taxonomy of the puparia of British Sphaeroceridae. *Dipterist's Digest* 13: 6–22.

Gives diagnostic characters for known British sphaerocerid puparia.

Smith, K.G.V. 1963. A short synopsis of British Chamaemyiidae (Dipt.). *Transactions of the Society for British Entomology* 15 (6): 103–15.

Includes keys to British genera, and 'tentative' keys to most of the species.

Spencer, K.A. 1972 Diptera Agromyzidae. *Handbooks for the Identification of British Insects* 10(5g): 136pp.

Keys to British species.

Spencer, K.A. 1976. The Agromyzidae (Diptera) of Fennoscandia and Denmark. *Fauna Entomologica Scandinavica* 5(1): 1–304;(2): 305–606.

Keys to Fennoscandian species with detailed accounts of each; relevant for most of the British fauna.

Steyskal, G.C. 1957. A revision of the family Dryomyzidae (Diptera, Acalyptratae). *Papers from the Michigan Academy of Sciences, Arts and Letters* 42: 55–68.

World revision of family, with key to species.

Stubbs, A.E. 1982. An identification guide to British Clusiidae. *Proceedings and Transactions of the British Entomological and Natural History Society* 15: 89–92.

Keys genera and species, including some species not yet recorded from Britain.

White, I.M. 1988. Tephritid flies. Diptera: Tephritidae. *Handbooks for the Identification of British Insects* 10(5a): 134pp.

Keys and checklist to species, illustrations including photographs of diagnostic wing patterns; a host-plant list is also provided.

Wirth, W.W. 1951. A revision of the dipterous family Canaceidae. *Occasional Papers of the Bernice Pauahi Bishop Museum* 20: 245–75.

World revision of this family, with keys to species.

Wirth, W.W. 1975. A revision of the brine flies of the genus *Ephydra* of the Old World. *Entomologica Scandinavica* 6: 11–44.

Provides key to Old World *Ephydra*.

Withers, P. 1987. The British species of the genus *Suillia* (Diptera, Heleomyzidae), including a species new to science. *Proceedings and Transactions of the British Entomological and Natural History Society* 20: 91–104.

Improved key compared with that by Collin (1943), includes male genitalic characters, and notes including a synthesis of recently discovered biological data.

Siphonaptera: the fleas

(57 species in 6 families)

THERESA M. HOWARD

There are 57 species of flea currently recorded from the British Isles, although more may await discovery, especially associated with small mammals such as voles, shrews, moles and the rarer species of bat. Three species have unique distributions: *Ornithopsylla laetitiae*, found on puffins and storm petrels and only on the islands around the west coast of England and Wales and off the coast of Ireland; *Actenophthalmus integella*, found as a winter parasite of voles in Scotland, and *Euctenophthalmus congener congener*, occurring on bank voles in England only. Sometimes when mainland hosts such as voles, moles and the common shrew are absent, as in Ireland, their fleas are found on other hosts such as woodmice, which may be an indicator as to their ancient original host. Ireland does have one shrew, the pygmy shrew, and its flea (*Palaeopsylla soricis soricis*) is also found on the common and pygmy shrews on mainland British Isles.

Fleas are important for their ability to transmit diseases such as plague and typhus, as a carrier of the dog tapeworm (*Dipylidium caninum*), as pests of poultry, and general nuisance value. Of the species found in the British Isles only seven commonly come into contact with people.

Fleas are small, 1–8 mm long (the largest recorded flea species in the British Isles is the mole flea (*Hystrichopsylla talpae*), which reaches about 6 mm), wingless and usually heavily 'armoured', hairy, shiny, and light red-brown to black in colour. They live as ectoparasites on mammals and birds and are distinguished from other parasitic groups by their ability to jump and being laterally flattened. Fleas are parasitic in the adult stage only, sucking blood from mammals and birds. Host blood is usually needed as a stimulus to breeding, for ovulation and for larval development. An adult flea can survive for several months without taking a blood meal. The larvae are legless, elongate, eyeless, with fairly sparse but strong bristles, and have biting mouthparts; they are not usually parasitic, but feed on organic matter which they find in the nest or home of the host and on faecal blood from adult fleas. Larvae cannot be identified to family or species level. The pupae are contained in cocoons.

An adult female flea can produce about 25 eggs a day for at least three or four weeks. Eggs are laid in the host's nest or regular sleeping place. Hatching occurs after about five days. The larval stage lasts for two or three weeks and includes two moults; the fully grown larvae spin cocoons of silk from their salivary glands and transform into pupae. Duration of the pupal stage is dependent on temperature but is usually two weeks. Adult fleas need a mechanical stimulus to start their emergence from the cocoon and this is generally in the form of vibrations caused by the movements of the host.

The identification of fleas is not easy to family or species level and, partivcularly because of their size, a good stereo microscope is essential. The most recent comprehensive key to families, species and hosts of fleas of the British Isles is by Smit (1957); although now out of print it can be used by specialist and amateur alike. A simple key to the European families can be found in Chinery (1993).

The six families in Britain are: Histrichopsyllidae, Leptopsyllidae, Ischnopsyllidae, Ceratophyllidae, Vermipsyllidae, Pulicidae.

References

Chinery. M. 1993. Order Siphonaptera – Fleas. *In: A field guide to the insects of Britain and northern Europe,* pp. 165–9. Collins, London.
Provides a simple key to the European families of fleas, and a general account of their life history.
Smit, F.G.A.M. 1957. Siphonaptera. *Handbooks for the Identification of British Insects* 1(16): 95pp.
The only complete guide to British fleas and their hosts, with complete glossary of terms and fully illustrated, easy to use by a specialist, more demanding for an amateur. Sadly now out of print.
Smit, F.G.A.M. 1976. Siphonaptera. *In* Kloet, G.S. & Hincks, C.D. (Eds.): A check list of British insects. Part 5, Diptera and Siphonaptera. *Handbooks for the Identification of British Insects* 11(5): 121–3.
The most up-to-date British list.

Hymenoptera: the bees, wasps and ants

(over 7000 species in 57 families)

JOHN S. NOYES, MIKE G. FITTON, DONALD L.J. QUICKE,
DAVID G. NOTTON, GEORGE R. ELSE, NIGEL D.M. FERGUSSON,
BARRY BOLTON, SUZANNE LEWIS AND LARAINE C. TAREL

The Hymenoptera is the largest and most diverse order of insects in the British Isles with a little over 7000 recorded species, yet large parts of it remain among the least studied groups of British animals. In some ways this is surprising because many species of the order are familiar as pollinators of flowers and as pests in the garden and home. However, these groups constitute only a very small proportion of the order as a whole, with by far the largest number of species belonging to groups that are largely unknown even to professional entomologists.

A characteristic of the order as a whole is the mechanism of sex determination, which is now known as haplodiploidy. In general, males are produced from unfertilised (haploid) eggs and females are produced from fertilised (diploid) ones. Thus the ovipositing female has the ability to choose the sex of her offspring, enabling her to invest more resources in the sex that is of the greatest benefit. For example, queen bees lay fertilised eggs during most of the year to build up worker numbers to provision and generally care for the hive, and only lay unfertilised eggs at certain times of the year in order to produce drones (males), which will mate with other queens. This type of parthenogenesis is termed arrhenotoky. Quite a large number of Hymenoptera exhibit a different form of parthenogenesis called thelytoky. In these species, females are produced from unfertilised eggs by a variety of mechanisms, and sometimes even by the presence of a *Rickettsia* type of micro-organism, *Wolbachia*. In some sawflies and gall-wasps, unfertilised eggs can develop into individuals of either sex. This is termed deuterotoky. In at least some cases the sex of the offspring is temperature-dependent, suggesting that sex determination may also be under the control of *Wolbachia*.

As might be expected of such large group of insects, the order as whole exhibits a great variety of biologies and life-histories. These will be outlined briefly in the sections dealing with the individual groups below.

The order is traditionally broken down into two subgroups, the Symphyta (or sawflies, *ca.*490 species) and the Apocrita, which is further subdivided into the Aculeata (stinging Hymenoptera, *ca.*590 species) and Parasitica (parasitic Hymenoptera, *ca.*6000 species). The division of the order is artificial and cannot be supported from a phylogenetic standpoint but is one of the most convenient ways in which it can be subdivided (see discussion in Gauld & Bolton, 1988).

The sawflies are structurally the most primitive hymenopterans. This is reflected in the phytophagous habit of their immature stages, which feed externally on plant tissue or internally by forming galls, mining leaves or boring into stems or timber.

The aculeate Hymenoptera, as the name implies, possess an ovipositor that is modified into a sting and does not function primarily as a means of placing the egg. The Aculeata includes species whose immature stages develop as parasitoids (e.g. some Pompilidae, Chrysididae, Mutillidae and Dryinidae), free-living solitary species (e.g. some bees and wasps) and truly social species (some bees, wasps and all ants).

The groups of parasitic Hymenoptera are the least well known within the order, yet they account for more than three-quarters of the species. Included are species where the immature stages are not only ecto- or endoparasitic, but also predatory or phytophagous. The phytophagous species are restricted to the superfamilies Chalcidoidea and Cynipoidea and feed on highly nutritious parts of the plant such the endosperm of seeds or by forming galls within the plant. As the name implies, the vast majority are parasitoids of other insects, and also of some arachnids, mites or even nematodes. The adult stages are free-living, either not feeding as adults or feeding on nectar, insect honeydew or host tissues (often haemolymph exuding from the puncture made by insertion of the ovipositor). The parasitic Hymenoptera are important as natural enemies and are extensively used in the biological control of insect pests of agriculture, horticulture and forestry.

Higher classification of the British Hymenoptera

Symphyta: Xyelidae, Pamphiliidae, Blasticotomidae, Cimbicidae, Argidae, Diprionidae, Tenthredinidae, Siricidae, Xiphydriidae, Orussidae, Cephidae

Apocrita Aculeata
Superfamily Vespoidea: Tiphiidae, Mutillidae, Sapygidae, Scoliidae, Formicidae, Pompilidae, Vespidae

Superfamily Apoidea: Sphecidae, Apidae

Superfamily Chrysidoidea: Dryinidae, Embolemidae, Bethylidae, Chrysididae

Apocrita Parasitica
Superfamily Trigonaloidea: Trigonalidae
Superfamily Evanioidea: Evaniidae, Aulacidae, Gasteruptiidae
Superfamily Ichneumonoidea: Ichneumonidae, Braconidae
Superfamily Cynipoidea: Cynipidae, Ibaliidae, Figitidae
Superfamily Proctotrupoidea: Diapriidae, Heloridae, Proctotrupidae
Superfamily Platygastroidea: Platygastridae, Scelionidae
Superfamily Ceraphronoidea: Ceraphronidae, Megaspilidae
Superfamily Chalcidoidea: Chalcididae, Eurytomidae, Torymidae, Ormyridae, Eucharitidae, Perilampidae, Pteromalidae, Eupelmidae, Encyrtidae, Signiphoridae, Aphelinidae, Elasmidae, Tetracampidae, Eulophidae, Trichogrammatidae, Mymaridae
Superfamily Mymarommatoidea: Mymarommatidae

General references

As the order is so large and diverse, the various taxonomic groups are treated separately in the following pages, generally by superfamily. There have been a number of publications dealing with the order as whole, the following references generally being considered the best as a means of introduction.

Askew, R.R. 1971. *Parasitic insects.* Heinemann, London.
    A very readable book and an excellent introduction to the biology of parasitic insects, although now somewhat dated.

Askew, R.R. 1984. The biology of gall wasps. *In* Ananthakrishnan, T.N. (Ed.) *Biology of gall insects,* pp. 223–71. Oxford & IBH Publishing Co., New Delhi, Bombay and Calcutta
    A very useful summary of the biology of gall-wasp communities, including their parasitoids.

Clausen, C.P. 1940. *Entomophagous insects.* McGraw Hill, New York; London.
    A classic book summarising information relating to entomophagous insects, which draws together much published information on their biologies.

Fitton, M.G., Graham, M.W.R. de V., Boucek, Z.R.J., Fergusson, N.D.M., Huddleston, T., Quinlan, J. & Richards, O.W. 1978. Kloet and Hincks: A check list of British insects. (2nd edn) (completely revised). Part 4, Hymenoptera. *Handbooks for the Identification of British Insects* 11(4): 159pp.
    Synopsis of the classification of the order with lists of taxa recorded as British. Will probably be most useful for those studying the lesser-known groups such as Chalcidoidea and Proctotrupoidea. Many sections now in need of revision.

Gauld, I.D. & Bolton, B. (Eds) 1988. *The Hymenoptera.* British Museum (Natural History) and Oxford University Press. [Reprinted 1996, with minor alterations and additions.]

An excellent volume, which summarises everything that the serious student of Hymenoptera would wish to know, from the systematics of the group to morphology and information about biology. Includes keys to all the families of the order recorded from the British Isles.

Godfray, D. 1993. *Parasitoids: behavioural and evolutionary ecology.* Princeton University Press, New Jersey.

An excellent review of different aspects of the biology of parasitoids and provides a good synthesis of theories relating to the evolution and other aspects of parasitism.

Goulet, H. & Huber, J.T. (Eds). 1993. *Hymenoptera of the world: an identification guide to the families.* Canada Communication Group – Publishing, Ottawa.

Well-illustrated keys to all families of Hymenoptera with some additional notes on their recognition and useful taxonomic publications. Limited notes on biology, etc.

Hanson, P. & Gauld, I.D. (Eds) 1995. *The Hymenoptera of Costa Rica.* Oxford University Press.

Do not be put off by the title – definitely the best book currently available on the order as a whole and includes excellent summaries of all that is known on the biologies of the different families. A must for any serious student of the Hymenoptera.

Quicke, D.L.J. 1997. *Parasitic wasps.* Chapman & Hall, London.

A very readable up-to-date review of the biology of parasitic wasps and includes an extensive guide to the literature. Very little that will aid the identification of taxa, but includes a summary of the present hypotheses relating to the classification of the order.

Richards, O.W. 1977. Hymenoptera. Introduction and key to families (2nd edn). *Handbooks for the Identification of British Insects* 6(1): 100pp.

Includes a summary of the morphology and classification of the order with keys to the families. Rather outdated by Gauld and Bolton (1988).

Symphyta

The Symphyta includes the sawflies and woodwasps, the groups of Hymenoptera that are structurally most primitive. Historically the Symphyta has been accorded the status of a suborder, but in evolutionary terms it is not a monophyletic unit. The majority of sawflies are rather weak-flying, clumsy insects and are often associated with moist habitats. With the exception of the Orussidae the larvae of Symphyta are phytophagous. Eggs are laid singly or in groups in slits or borings made in the host plant tissue by the female's ovipositor. It is the saw-like nature of this organ that gives the common name for the group. Many sawfly larvae feed exposed on the vegetation, or are partly concealed in leaf rolls or webs. Others may feed concealed in the reproductive parts of plants or bore in stems or timber. A few mine leaves or cause galls. The majority of free-feeding larvae are caterpillar-like with abdominal prolegs, but unlike those of the Lepidoptera these organs are never furnished with small hooks.

The essential works for identifying adult British Symphyta are the Royal Entomological Society's *Handbooks* (Benson, 1951, 1952, 1958; Quinlan & Gauld, 1981) and Liston's *Compendium* (Liston, 1995, 1996). Liston's work gives modern up-to-date classification and nomenclature and facilitates access to the recent European literature and details of additions to the British fauna made since publication of the *Handbooks*. Some modern papers relevant to particular families are cited in the text below, but Liston should be consulted for further European key works. Another work, giving a valuable alternative key to symphytan genera in Britain, is the AIDGAP key by Wright (1990). For identification of larvae the work of Lorenz & Kraus (1957) has not really been superseded, although the keys and descriptions (for families and for subfamilies of Tenthredinidae) by Smith (in Stehr, 1987) are the best starting point.

Family Xyelidae

Only two species of *Xyela* occur in the British Isles. Females of these peculiar primitive sawflies oviposit in staminate cones of pines as they begin to push out through the surrounding bracts in the spring, and the larvae feed on the developing sporophylls. Keys: Benson (1951); Quinlan & Gauld (1981).

Family Pamphiliidae

The Pamphiliidae is a small family with 19 species in four genera found in Britain. (The related family Megalodontesidae was once recorded as occurring in Britain (Benson, 1951), but Quinlan & Gauld (1981) could find no authenticated British material and deleted it from the British list.) Pamphiliid larvae feed on trees, especially of Pinaceae, Rosaceae, Betulaceae and Salicaceae. The young larvae make silken webs on the underside of the leaves or needles, and on angiosperms they later generally move to the leaf margin and make a characteristic leaf roll. Several species are forestry pests in Europe. Identification: adults, Quinlan & Gauld (1981) and van Achterberg & van Aartsen (1986); larvae, Lorenz & Kraus (1957).

Family Blasticotomidae

A single species of this very small family occurs in Britain. The larvae are stem-borers in ferns. Identification: adult and larva, Quinlan & Gauld (1981).

Family Cimbicidae

Fourteen species in four genera are recorded from Britain. Cimbicines are striking insects, which fly rapidly, making a loud buzzing sound. The larvae of most species feed on woody angiosperms, particularly trees of the families Rosaceae, Betulaceae and Salicaceae. They are usually solitary and when disturbed often curl up. The cocoon may be in an underground chamber or spun attached to the host plant. Members of the subfamily Abiinae are smaller and generally less common and their larvae feed on climbing Caprifoliaceae or herbs of the family Dipsacaceae. Identification: adults, Quinlan & Gauld (1981); larvae, Lorenz & Kraus (1957).

Family Argidae

The Argidae is the second largest family of Symphyta and with over 800 described species worldwide. Unlike most of the other families, it is most diverse in tropical habitats. Only 15 species in three genera extend to Britain. Adults are slow-moving, heavily built insects not uncommonly encountered feeding on the flowers of Umbelliferae. Most British species are associated with woody plants, especially Salicaceae, Rosaceae and Betulaceae, but the larvae of one genus feed on herbaceous Leguminosae. Identification: adults, Quinlan & Gauld (1981); larvae, Lorenz & Kraus (1957).

Family Diprionidae

This is a small family mainly occurring in the northern temperate coniferous forests. Seven species in four genera are recorded as British. Their larvae develop on conifers (Pinaceae and Cupressaceae). Identification: adults, Quinlan & Gauld (1981); larvae, Lorenz & Kraus (1957) and Wong & Szlabey (1986) (genera).

Family Tenthredinidae

The Tenthredinidae is very large, with more species than all other symphytan families combined. Just under 400 species occur in Britain. Most tenthredinid larvae feed externally on leaves, although some species of several of the subfamilies are miners in leaves, shoots, catkins and fruits. A few nematines make leaf rolls and some cause galls.

Fitton *et al.* (1978) (see above) adopted the classification proposed by

Benson (1952, 1958), but it seems sensible now to use a more modern system (see Smith, 1979a, b) and recognise seven subfamilies: Selandriinae, Dolerinae, Nematinae, Heterarthrinae, Blennocampinae, Allantinae and Tenthredininae.

Selandriine larvae are mainly feeders on ferns, although a number are associated with the monocotyledonous families Gramineae, Cyperaceae and Juncaceae. Most are external feeders but the larva of *Heptamelus* bores in the rachis of ferns and may be a minor pest in gardens.

Several species of dolerine can be extremely common in spring in grassland or damp, marshy habitats. The adults may be found feeding on the blossom of trees, or on catkins of birches, bordering wet meadows. The larvae feed on low vegetation in wet areas. Most species are associated with Gramineae and Equisetaceae, but a number consume Cyperaceae or Juncaceae. The mature larvae overwinter in an earthen cell in the ground.

The Nematinae is a large and diverse subfamily and forms an increasingly large proportion of the symphytan fauna the further north one goes, until in the Arctic they are virtually the only sawflies present. This is the largest subfamily in Britain, where it is represented by nearly 200 species in 21 genera. Several genera are very large and contain complexes of species that are extremely difficult to identify. The majority of nematines are associated with trees, especially of the families Salicaceae, Betulaceae and Pinaceae, and with arborescent Rosaceae. Their larvae may be exposed feeders on leaves, but many are leaf-rollers, miners or gall-formers.

The Heterarthrinae is represented by 25 species in Britain. All are small blackish sawflies and, as larvae, are angiosperm leaf feeders. Most attack woody plants, some are external feeders, but others are leaf miners. The larvae of *Caliroa* are slug-like.

The Blennocampinae is a rather heterogeneous group of species. Most are confined to the humid temperate and tropical parts of the world. The limits and classification of the group have changed considerably since Benson (1952) published his British keys. Smith (1969a, 1979a) can be consulted for details of the presently accepted groupings. Fifteen genera are represented in Britain by 26 species. The larvae of most species are external feeders on the foliage of various angiosperms, particularly Rosaceae

The subfamily Allantinae was included in the Blennocampinae by Benson (1952). It is represented in Britain by eleven genera and 47 species. The larvae of allantines are all external feeders on foliage. Several species are notorious pests.

The Tenthredininae contains most of the large, colourful sawflies commonly encountered in spring and early summer. About 60 species are found in Britain. Tenthredinine larvae mainly feed exposed on the leaves of herbaceous plants, though many are nocturnally active and remain concealed during the day. Of all sawfly subfamilies the Tenthredininae has the broadest spectrum of larval food-plants, embracing 28 vascular plant families.

Identification:

Selandriinae: Smith (1969b) (Nearctic species but a valuable generic key); larvae, Lorenz & Kraus (1957).

Dolerinae: adults, Benson (1952); larvae, Lorenz & Kraus (1957).

Nematinae: Benson (1958, 1960); larvae, Lorenz & Kraus (1957) and Nigitz (1974) (European species on spruce).

Heterarthrinae: Benson (1952); Smith (1967, 1971) (Nearctic species, but includes some Holarctic ones); Smith (1976) (world genera); larvae- Lorenz & Kraus (1957).

Blennocampinae: Benson (1952); Smith (1969a) (generic key and review of classification); larvae, Lorenz & Kraus (1957).

Allantinae: Benson (1952); Smith (1979a) (supraspecific keys); Taeger (1986) (European genera); larvae, Lorenz & Kraus (1957).

Tenthredininae: Benson (1952, 1965) (*Rhogogaster* species); larvae, Lorenz & Kraus (1957).

Family Siricidae

Most species of siricid woodwasp occur in the temperate forests of the northern hemisphere. The subfamily Siricinae is associated with Pinaceae, whereas the Tremicinae attack angiosperm trees, especially those of the families Aceraceae, Fagaceae and Ulmaceae. Some of the siricines are probably native to Britain, but the Tremicinae is represented in Britain by a single North American species which is introduced with imported timber. Probably only five of the eleven so-called British siricids have reproducing populations in Britain. Identification: adults, Quinlan & Gauld (1981).

Family Xiphydriidae

Xiphydriid woodwasp larvae bore in deciduous trees of the families Salicaceae, Betulaceae, Aceraceae and Ulmaceae. Only three species occur in Britain. Identification: adults, Shaw & Liston (1985); larvae, Quinlan & Gauld (1981).

Family Orussidae

The one reputedly British species has not been collected in the UK for over 150 years and it may be extinct. Adult orussids are associated with wood-boring insects, and there have been conflicting reports about their developmental biology. However, a number of observations clearly show that some do develop as parasitoids of the larvae of buprestid beetles and siricid wood-wasps. Identification can be effected by using any of the keys to families, Benson (1951) or Quinlan & Gauld (1981).

Family Cephidae

This is a small family, but with more than 40 species in Europe, and twelve in five genera in Britain. The larvae of cephids are stem-borers. Those of the tribe Hartigiini bore twigs of woody plants or stems of herbaceous Rosaceae, whereas larval Cephini bore in stems of various Gramineae. Several species of cephid attack cereal crops and one in particular, *Cephus pygmeus*, was formerly regarded as a serious pest. Identification: adults, Quinlan & Gauld (1981).

References

Achterberg, C. van & Aartsen, B. van 1986. The European Pamphiliidae (Hymenoptera: Symphyta), with special reference to The Netherlands. *Zoologischen Verhandelingen* **234**: 1–98.
  Includes keys to species.
Benson, R.B. 1951. Hymenoptera Symphyta. Section A. *Handbooks for the Identification of British Insects* **6**(2a): 49pp.
  Keys to species of all families of Symphyta except Tenthredinidae. Out of print and superseded by new edition (Quinlan & Gauld, 1981), but some users prefer the layout of keys in the first edition. This and Benson (1952, 1958) are basic identification guides, but taxonomy needs modifying to take account of more recent work (Benson, 1960, 1965; Liston, 1995, 1996; Smith, 1967–79).
Benson, R.B. 1952. Hymenoptera Symphyta. Section B. *Handbooks for the Identification of British Insects* **6**(2b): 51–137.
  See note above. Out of print. Covers Tenthredinidae except subfamily Nematinae.
Benson, R.B. 1958. Hymenoptera Symphyta. Section C. *Handbooks for the Identification of British Insects* **6**(2c): 139–252.
  See note above. Covers Tenthredinidae, subfamily Nematinae.
Benson, R.B. 1960. Studies in *Pontania* (Hym. Tenthredinidae). *Bulletin of the British Museum (Natural History)* (Entomology) **8**: 367–87.
  Supplements Benson (1958).

Benson, R.B. 1965. The classification of *Rhogogaster* Konow. *Proceedings of the Royal Entomological Society of London* (B) **34**: 105–12.
  Supplements Benson (1952).
Liston, A.D. 1995. *Compendium of European sawflies.* 190pp. Chalastos Forestry, Gottfrieding.
  A checklist of European sawflies, with brief sections on various topics including identification literature.
Liston, A.D. 1996. *Compendium of European sawflies: Supplement.* 16pp. Chalastos Press, Gottfrieding.
  See note above.
Lorenz, H. & Kraus, M. 1957. Die larvalsystematik der Blattwespen. *Abhandlungen zur Larvalsystematik der Insekten* **1**: 1–339.
  In German. Now dated, but not superseded. See also Stehn (1987) for key to higher taxa.
Nigitz, H.P. 1974. Über die Fichten-Nematinen (Hym., Tenthredinidae) der Steiermark. *Zeitschrift für Angewandte Entomologie* **75**: 264–84.
  In German. Supplements Benson (1958).
O'Connor, J.P., Liston, A.D. & Speight, M.C.D. 1997. A review of the Irish sawflies (Hymenoptera: Symphyta) including a checklist of species. *Bulletin of the Irish Biogeographical Society* **20**: 2–99.
  Useful review of the 272 Irish species, with information on nomenclature and distribution.
Quinlan, J. & Gauld, I.D. 1981. Symphyta (except Tenthredinidae). *Handbooks for the Identification of British Insects* **6**(2a): 1–67.
  See note under Benson (1951).
Shaw, M.R. & Liston, A.D. 1985. *Xiphydria longicollis* (Geoffroy) (Hym. Xiphydriidae) new to Britain. *Entomologist's Gazette* **36**: 233–5.
  Supplements Quinlan & Gauld (1981) and Benson (1951).
Smith, D.R. 1967. A review of the subfamily Heterarthrinae in North America. (Hymenoptera: Tenthredinidae). *Proceedings of the Entomological Society of Washington* **69**: 277–84.
  Includes keys, covers species with Holarctic distribution. See also Smith (1971).
Smith, D.R. 1969a. Nearctic Sawflies I. Blennocampinae: adults and larvae (Hymenoptera Tenthredinidae). *U.S. Department of Agriculture Technical Bulletin* **1397**: 1–179.
  Gives modern classification.
Smith, D.R. 1969b. Nearctic Sawflies II. Selandriinae: adults (Hymenoptera Tenthredinidae). *U.S. Department of Agriculture Technical Bulletin* **1398**: 1–48.
  Valuable generic key.
Smith, D.R. 1971. Nearctic Sawflies III. Heterarthrinae: adults and larvae (Hymenoptera Tenthredinidae). *U.S. Department of Agriculture Technical Bulletin* **1420**: 1–84.
  See note under Smith (1967).
Smith, D.R. 1976. World genera of the leaf-mining sawfly tribe Fenusini (Hymenoptera: Tenthredinidae). *Entomologica Scandinavica* **7**: 253–60.
  Includes key.

Smith, D.R. 1979a. Nearctic Sawflies IV. Allantinae: adults and larvae (Hymenoptera Tenthredinidae). *U.S. Department of Agriculture Technical Bulletin* **1595**: 1–172.
Useful key to genera.

Smith, D.R. 1979b. Symphyta. *In* Krombein, K.V., Hurd, P.D., Smith, D.R. & Burks, B.D. (Eds) *Catalog of Hymenoptera in America north of Mexico* **1**: 3–137.
A modern classification. Mainly useful for changes in Tenthredinidae since Benson (1951, 1952, 1958).

Stehr, F.W. (Ed.) 1987. *Immature insects*, 1. Kendall Hunt, Dubuque, Iowa. 754pp.
The chapter on Hymenoptera (pp. 597–710) includes a section on Symphyta by D.R. Smith (pp. 602–4, 618–49).

Taeger, A. 1986. Beiträg zur Taxonomie und Verbreitung paläarktischer Allantinae (Hymenoptera, Symphyta). *Beiträge zur Entomologie* **36**: 107–18.
In German. See also Smith (1979a).

Wong, H.R. & Szlabey, D.L. 1986. Larvae of North American genera of Diprionidae (Hymenoptera: Symphyta). *Canadian Entomologist* **118**: 577–87.
Although dealing with the Nearctic fauna, a useful adjunct to Lorenz & Kraus (1957).

Wright, A. 1990. British sawflies (Hymenoptera: Symphyta): a key to adults of the genera occurring in Britain. *Field Studies* **7**: 530–93.
Practical keys. Poor illustrations.

Apocrita: Aculeata

Superfamily Vespoidea

Family Tiphiidae

The family contains about 1500 species placed in seven subfamilies, of which only two occur in the British Isles: the Tiphiinae and the Methochinae.

The Tiphiinae is represented in Britain by two species of *Tiphia*. Both have largely black bodies (in the female *T. femorata* the legs are reddish) and are fully winged. Neither is particularly common, although they may occur locally in large numbers flying over low vegetation or on blossoms, particularly those of umbellifers. The larvae of both species develop on those of scarabaeid beetles in their cells beneath the soil surface. The female wasp reaches the host larva by digging down into the soil, stinging to paralysis the host before laying a single egg on it. Very little, however, is known about the biology of the British species.

Only one species of Methochinae occurs in the British Isles: *Methocha ichneumonides*. Sexual dimorphism is extreme: the male is slender with a black body and is fully winged, whereas the female has a shiny black head and gaster and a red thorax and legs. It is entirely wingless and ant-like in appearance. *M. ichneumonides* is a rather rare species, mainly confined to sandy soils in southern Britain. The female seeks out tiger beetle (Cicindelidae) larvae in their open burrows in the soil. The beetle larva is paralysed by being stung several times in the thorax, before an egg is laid on the body. The wasp seals the burrow when it leaves. On hatching the wasp larva feeds externally on the beetle larva.

Champion, H.G. & Champion, R.J. 1914. Observations on the life-history of *Methocha ichneumonides* Latr. *Entomologist's Monthly Magazine* **50**: 266–70.
  Observations under artificial conditions of the development on tiger beetle larvae.
Richards, O.W. 1980. Scolioidea, Vespoidea and Sphecoidea (Hymenoptera, Aculeata). *Handbooks for the Identification of British Insects* **6**(3b): 118pp.
  Illustrated keys and brief species notes to the British species of *Tiphia* and *Methocha*.

Family Mutillidae

The Mutillidae comprises about 8000 species worldwide, placed in seven subfamilies. Only two of the subfamilies are found in the British Isles: the Mutillinae and the Myrmosinae.

In the British Isles, the Mutillidae is represented by three species, *Mutilla europaea, Smicromyrme rufipes* and *Myrmosa atra*. These wasps are rather rare and mainly inhabitants of light, sandy soils, particularly in heathland and coastal dunes. There is strong sexual dimorphism: females lack wings and are rather ant-like in appearance, with the body clad in bands of adpressed silvery hairs (hence the collective, common name for the family, 'velvet ants'). The male is fully winged but is otherwise similar in appearance to the female (except in *M. atra*, where the entirely black-bodied male strongly contrasts with the reddish brown, and usually smaller, female). Females seek out the nests of other aculeates: bumblebees (*Bombus* species) (*M. europaea*); probably small pompilid and sphecid wasps' nests, and mining bee nests (*S. rufipes* and *M. atra*). Their eggs are probably laid on the mature host larva or pupa, which is eventually eaten by the ectoparasitic wasp larva. When disturbed both sexes of *M. europaea* are capable of stridulating (audible to the human ear). This is achieved by rubbing a finely striated median area (the plectron) at the base of the third gastral tergite against a transverse ridge of the underside of the apex of the second tergite.

Richards, O.W. 1980. Scolioidea, Vespoidea and Sphecoidea (Hymenoptera, Aculeata). *Handbooks for the Identification of British Insects*. **6**(3b): 118pp.
  Illustrated keys and brief species notes to the British species of *Mutilla, Smicromyrme* and *Myrmosa*.

Family Sapygidae

The Sapygidae is a small family containing about 80 species worldwide, placed in two subfamilies, the Fedtschenkiinae and the Sapyginae. Only the latter occurs in Britain. This is represented by two species: *Sapyga clavicornis* and *S. quinquepunctata*. These are solitary, brightly coloured wasps, generally with black-and-yellow banded gasters (the female *S. quinquepunctata* has a gaster marked with yellow, red and black). The larvae of both species are kleptoparasites in the nests of certain bees in the subfamily Megachilinae. The female *Sapyga* inserts her ovipositor through the cell cap and lays an egg. The first-instar wasp larva has a pair of large, sickle-shaped mandibles with which it destroys the host egg or young larva (and other *Sapyga* if present). Later instars have short, simple mandibles, which are used to devour the stored pollen and nectar. Unlike that of most aculeates, the sting retains its original purpose as an ovipositor.

Richards, O.W. 1980. Scolioidea, Vespoidea and Sphecoidea (Hymenoptera, Aculeata). *Handbooks for the Identification of British Insects.* **6**(3b): 188pp.
  Includes a key to both British species and illustrates *Sapyga clavicornis.*

Family Scoliidae

A predominantly tropical family containing about 300 species worldwide, placed in two subfamilies, of which only the Scoliinae occurs in the United Kingdom (this contains a single species, *Scolia sexmaculata*, found in Jersey, Channel Islands, only). These are solitary wasps in which the species are mainly large and robust, the integument hairy, and the abdomen black, usually with conspicuous yellow or red markings. Females in particular have strong legs, each with numerous large spines. The two pairs of fully developed wings are generally iridescent, with dense, fine, longitudinal wrinkles near the apices.

The larvae of all species are ectoparasitoids of soil-dwelling larvae of Scarabaeoidea and Curculionoidea. The female wasp digs into the soil to reach a larva and stings it to paralysis; the resulting wasp larva feeds externally on the host. The fully grown scoliid larva spins a dense silken cocoon in which it pupates; the adult eventually emerges through a circular cap at the anterior end.

Only one species occurs in the United Kingdom, where it is confined to the Channel Islands. It is omitted from identification keys to the British wasp fauna.

Family Formicidae: the ants

All ant species are contained in a single aculeate family of worldwide distribution, the Formicidae. There are 16 subfamilies with over 9500 described extant species (Bolton, 1995), with about half that number again awaiting description. The British fauna is depauperate, with only 50 native species (representing 4 subfamilies) and about 10 common introductions that are mostly confined to heated buildings, but despite the small number of British species they present a remarkable range of biologies. The family has been fully defined in Gauld & Bolton (1988) and in Bolton (1994), where diagnoses of the subfamilies can also be found.

In general ants may be recognised by their perennial colonies and presence of a wingless worker caste, but in a few species the workers have been secondarily lost. In female castes (queens and workers) the head is prognathous. The antennae have 4–12 segments in female castes and 9–13 segments in males. Antennae are geniculate between the long basal segment (scape) and the remaining funicular segments. The second abdominal segment is reduced and forms a node or scale (the petiole), isolated from the alitrunk in front and from the remaining abdominal segments behind. The third abdominal segment is frequently also reduced and isolated (the postpetiole). The wings of alate queens are deciduous and are shed after mating. There is usually a metapleural gland present on the alitrunk that opens above the metacoxa.

Because of their eusocial organisation, enormous diversity and extreme ecological amplitude a great deal has been written about ants. The book by Hölldobler & Wilson (1990) gives a superb introduction to all the many subdisciplines that constitute myrmecology, the scientific study of ants.

Most British ant species nest directly in the soil, in the soil beneath stones, or in rotten wood. A few build more complex mound-nests and one species lines its galleries with carton. There are two dulotic (slave-making) species in the British fauna and a single species that is xenobiotic in the nests of much larger ants. Most queens of British species are capable of commencing new colonies alone, but a number are temporarily or permanently socially parasitic (list in Gauld & Bolton, 1988), requiring the parasite queen to usurp the pre-existing nest of a host species in order to begin her own colony.

Baroni Urbani, C. & Collingwood, C.A. 1977. The zoogeography of ants in Northern Europe. *Acta Zoologica Fennica* **152**: 1–34.
    Analysis of ant distribution in Northern Europe; contains many useful maps.

Barrett, K.E.J. 1979. *Provisional atlas of the insects of the British Isles* 5. Hymenoptera, Formicidae. European Invertebrate Survey. Biological Records Centre, Monks Wood.
Distribution maps of British ants, based on 10 km squares of Ordnance Survey and Irish National Grids.

Bolton, B. 1994. *Identification guide to the ant genera of the world.* Harvard University Press, Cambridge, Massachusetts.
Contains identification keys to world subfamilies and genera, subfamily diagnoses, synoptic classifications, references to papers for species-rank identifications and other critical taxonomic references.

Bolton, B. 1995. A taxonomic and zoogeographical census of the extant ant taxa. *Journal of Natural History* 29: 1037–56.
Enumeration of the world's living ant species, analysed both taxonomically and by distribution.

Bolton, B. 1995. *A new general catalogue of the ants of the world.* Harvard University Press, Cambridge, Massachusetts.
A fully referenced catalogue of all the world's ants, including the fossil forms, providing a taxonomic history of all the names ever used in the family.

Bolton, B. & Collingwood, C. A. 1975. Hymenoptera: Formicidae. *Handbooks for the Identification of British Insects* 6(3c): 34pp.
Nomenclature somewhat out of date but keys still functional for all the common species. Checklist in this publication used as base-line for more recent additions and corrections to British fauna.

Brian, M.V. 1977. *Ants.* Collins New Naturalist, London.
Introduction to the biology of the more common British ant species.

Collingwood, C.A. 1979. The Formicidae of Fennoscandia and Denmark. *Fauna Entomologica Scandinavica* 8: 174pp.
Keys to species include all those found in Britain; includes addition/correction to British fauna.

Collingwood, C.A. 1991. New records for British ants. *Entomologist's Record and Journal of Variation* 103: 92–3.
Addition to British ant fauna.

Donisthorpe, H.St.J.K. 1927. *The guests of British ants.* Routledge & Sons, London.
Much out of date but still the only synoptic account of other arthropods that inhabit British ants' nests.

DuBois, M.B. 1993. What's in a name? A clarification of *Stenamma westwoodi, S. debile,* and *S. lippulum. Sociobiology* 21(3): 299–334.
Addition to British ant fauna.

Dumpert, K. 1981. *The social biology of ants.* Pitman, London.
English translation from the original German edition of 1978 (Verlag Paul Parey). A good general introduction to ant social biology.

Hölldobler, B. & Wilson, E.O. 1990. *The ants.* Harvard University Press, Cambridge, Massachusetts.
The definitive introduction to all aspects of modern myrmecology and a superbly readable book. If you intend to buy a single introductory book on ants, this is it.

Seifert, B. 1988. A revision of the European species of the ant subgenus *Chthonolasius. Entomologische Abhandlungen. Staatliches Museum für Tierkunde, Dresden* **51**: 143–80.

Addition to British ant fauna.

Seifert, B. 1988. A taxonomic revision of the *Myrmica* species of Europe, Asia Minor, and Caucasia. *Abhandlungen und Berichte des Naturkundemuseums Görlitz* **62**(3): 1–75.

Updates on all British species of *Myrmica*.

Seifert, B. 1992. A taxonomic revision of the Palaearctic members of the ant subgenus *Lasius* s. str. *Abhandlungen und Berichte des Naturkundemuseums Görlitz* **66**(5): 1–67.

Additions to British ant fauna.

Seifert, B. 1993. Taxonomic description of *Myrmica microrubra* n. sp. – a social parasitic ant so far known as the microgyne of *Myrmica rubra* (L.). *Abhandlungen und Berichte des Naturkundemuseums Görlitz* **67**(5): 9–12.

Addition to British ant fauna.

Skinner, G.J. & Allen, G.W. 1996. Ants. *Naturalists' Handbooks* No. 24. Richmond Publishing Co., Slough. 83pp.

Introduction to British ants and keys for identification of species. Nomenclature of some species updated.

Ward, P.S., Bolton, B., Shattuck, S.O. & Brown, W.L. Jr. 1996. A Bibliography of Ant Systematics. *University of California Publications. Entomology* **16**: 417pp.

A complete bibliography of all publications on ant systematics and related subjects from 1758 to 1995.

Family Pompilidae

Commonly known as 'spider-hunting wasps' or simply as 'spider wasps' the Pompilidae comprises about 5000 species worldwide, divided into three subfamilies. Of these 41 species (in 14 genera) occur in the British Isles, with a further 3 species (and one additional genus) in the Channel Islands. These are very active, solitary wasps, which are generally encountered running erratically over the ground and taking short flights. They also constantly flick their wings. These habits have earned them the colloquial name 'Hymenoptera neurotica'. Most British species have a black-and-red marked abdomen, although in a few others this is entirely black or black with reduced cream markings. The legs are long and the body rather slender.

Pompilids provision their nests with paralysed adult spiders, one spider per cell. This is in contrast to those Sphecidae that also prey on spiders, as these generally provision each cell with several immature spiders. Nests consist of a simple burrow excavated in the soil, or a cluster of earthen pots built from numerous mud pellets (as in *Auplopus carbonarius*), or the web of the spider itself (as, for example, in *Aporus unicolor* and *Homonotus sanguinolentus*). The three British *Evagetes* are kleptoparasites of some other soil-nesting pompilid species. They seek out and dig down to the host nest and apparently eat the egg of the original provisioner before laying one of their own. The two British *Ceropales* also have kleptoparasitic larvae. In these species the egg is concealed in the lung book of the spider, sometimes being done so as the spider is dragged along the ground by the host pompilid. The kleptoparasite's larva emerges before that of its host, the young larva seeking out and destroying the latter's egg. It then devours the paralysed spider prey. Most pompilid species spin dense silken cocoons in which the mature larva (prepupa) overwinters. A few species, however, pass the winter as adults, emerging early the following spring.

Day, M.C. 1988. Spider Wasps (Hymenoptera: Pompilidae). *Handbooks for the Identification of British Insects* 6(4): 60pp.
Contains a detailed diagnosis of the family, with a further section on biology and evolution. The bulk of the handbook consists of an illustrated key to all the species recorded from the British Isles and the Channel Islands.

Richards, O.W. & Hamm, A.H. 1939. The biology of the British Pompilidae (Hymenoptera). *Transactions of the Society for British Entomology* 6: 51–114.
Describes the nesting biologies of all species reported from the British Isles.

Family Vespidae

Until recently the Vespidae and Eumenidae were each treated as a separate family. However, it has now been accepted that these should be combined as a single family, the Vespidae. This now includes six subfamilies: Euparagiinae, Masarinae, Eumeninae, Stenogastrinae, Polistinae and Vespinae. Only the Eumeninae and Vespinae occur in the British Isles (the Polistinae is occasionally found there but is not considered to be native). The family is large, with about 4000 species worldwide, of which 34 species (in 13 genera) occur in the United Kingdom. These include one species that occurs in the Channel Islands but not on the British mainland.

The Eumeninae are all solitary, whereas the Vespinae are mainly social (a few — only *Vespula austriaca* in the British Isles — are social parasites of other vespid species). Adult Vespidae visit flowers for nectar and also, for example, leaves to imbibe honeydew. The Eumeninae provision their nests with arthropod prey, mostly the larvae of Lepidoptera and Coleoptera (mainly leaf beetles and weevils). The nests, according to the species, are built of mud either in cavities (such as hollow stems and dead wood generally) or on exposed surfaces of rocks and soil. The potter wasps, *Eumenes coarctatus* and *E. papillarius,* suspend their single-celled, pot-like nests from the branches of low shrubs, such as heather. Many of these wasps are quite rare, most being restricted to southern England. The entrances to the nest burrows of *Odynerus* species are often surmounted by a delicate tube or chimney, constructed from mud, brought to the site as numerous, individual pellets held in the female's mandibles. Most eumenid nests contain only a few cells, often less than a dozen, and the cells are mass-provisioned. In contrast the social Vespinae build large, mainly spherical nests, entirely from wood fibres mixed with saliva to form 'wasp paper'. Such nests are initiated in the early summer by an over-wintered queen and, a little later, by her worker offspring. The nest consists (when fully developed) of an outer, multi-layered covering or envelope, almost entirely surrounding a series of transverse sheets of comb, each containing several hundred downwardly directed cells. The combs are interconnected by a series of short pedicels. The nest entrance is at the bottom of the envelope. *Vespa crabro* (the hornet) and the four *Vespula* species tend to nest in cavities, such as holes in the ground, rot-holes in tree trunks, and in attics. The four *Dolichovespula* species and *Polistes dominulus* (the latter not native to Britain, though occasionally found) often suspend their nests from the branches of trees and bushes. *Polistes* nests consist of a single comb, which is not surrounded by an envelope. The *Dolichovespula* species include *D. media* and *D. saxonica,* which have both colonised Britain since 1980. Vespid larvae are progressively fed on masticated insect and spider prey. *Vespula austriaca* queens invade the nests of *V. rufa.* The queen of the latter is killed by that of the usurper and the *V. austriaca* brood (queens and males; there is no worker brood) is subsequently reared by the *V. rufa* workers.

Archer, M.E. 1978. *Provisional atlas of the insects of the British Isles.* Part 9, *Hymenoptera Vespidae.* Institute of Terrestrial Ecology, Abbots Ripton, Huntingdon. [Second edition with printing errors corrected: published 1979.]
  Distribution maps to seven species of social Vespidae (excludes *Dolichovespula media* and *D. saxonica).*

Edwards, R. 1980. *Social wasps.* Rentokil Ltd, East Grinstead.
  Excellent account of the biology and habits of British social wasps. Contains an identification key, though *Dolichovespula media* and *D. saxonica* omitted (these had not been reported from Britain at the time of publication). Numerous figures and photographs (both black and white and colour) included.

Edwards, R. (Ed.) 1997. *Provisional atlas of the aculeate Hymenoptera of Britain and Ireland.* Part 1, Bees, Wasps and Ants Recording Society. Biological Records Centre, Huntingdon.
  Annotated distribution maps to species in the genera *Odynerus, Symmorphus, Vespa,* and *Dolichovespula* (part).

Else, G.R. 1994. Identification. Social wasps. *British Wildlife* 5: 304–11.
  Species profiles and an illustrated key to the identification of the nine species of social
  vespid recorded from the United Kingdom.

Falk, S. 1991. A review of the scarce and threatened bees, wasps and ants in Great
  Britain. *Research and Survey in Nature Conservation* 35. Nature Conservancy
  Council, Peterborough.
  Detailed accounts to all nationally threatened Vespidae recorded from the United
  Kingdom. Updates those accounts presented by D.B. Shirt (see below).

Guichard, K.M. 1991. The occurrence of *Eumenes papillarius* (Christ) (Hymenoptera,
  Eumenidae) in England. *Entomologist's Monthly Magazine* 127: 71.
  Species added to the British wasp list. Characters to distinguish it from *Eumenes coarcta-
  tus* presented.

Hammond, P.M., Smith, K.G.V., Else, G.R. & Allen, G.W. 1989. Some recent additions
  to the British insect fauna. *Entomologist's Monthly Magazine* 125: 95–102.
  Short accounts (complete with colour photographs) of the occurrence in Britain of
  Dolichovespula media and *D. saxonica*.

Richards, O.W. 1979. The Hymenoptera Aculeata of the Channel Islands. *Report and
  Transactions of the Société Guernesiaise* 20: 389–424.
  All Channel Island records of Vespidae listed with short accounts of their habits.

Richards, O.W. 1980. Scolioidea, Vespoidea and Sphecoidea (Hymenoptera,
  Aculeata). *Handbooks for the Identification of British Insects* 6 (3b): 118pp.
  Illustrated keys and brief species notes to all Vespidae recorded from the United
  Kingdom, excluding *Dolichovespula media* and *D. saxonica*.

Shirt, D. (Ed.) 1987. Insects. *British Red Data Books. 2.* Nature Conservancy Council,
  Peterborough.
  Includes species accounts of nationally threatened Vespidae.

Spradbery, J.P. 1973. *Wasps.* Sidgwick & Jackson, London.
  Useful introduction to the biologies and habits of Vespidae found in the British Isles. Keys
  to species and provisional distribution maps included.

Stelfox, A.W. 1927. A list of the Hymenoptera Aculeata (*sensu lato*) of Ireland.
  Proceedings of the Royal Irish Academy (B) 69(22): 201–355.
  Irish records of Vespidae listed under county headings.

Yeo, P.F. & Corbet, S.A. 1995. *Solitary wasps.* Naturalists' Handbooks No. 3. Richmond
  Publishing Company, Slough.
  A good introduction to the study and identification of British solitary wasps. Key to iden-
  tification of species included. Several species illustrated in colour.

Superfamily Apoidea

Family Sphecidae

About 8000 species of sphecid wasps are known worldwide and are placed in 11 subfamilies of which six (Sphecinae, Pemphredoninae, Astatinae, Crabroninae, Nyssoninae and Philanthinae) occur in the British Isles and collectively include 129 species (in 31 genera). Six of these species occur in the Channel Islands but have not been found on the British mainland.

Most Sphecidae, in common with many other aculeates, visit flowers for nectar, extra-floral nectaries, and leaves for honeydew. They provision their nests with a variety of arthropod prey, in the British Isles these being (according to the species of sphecid) Ephemeroptera, Orthoptera (grasshopper nymphs), Psocoptera, Thysanoptera, Trichoptera, Hemiptera (aphids and other plant bugs, including the nymphs of heteropteran bugs), the larvae of Lepidoptera and Symphyta, adult Diptera, adult Apidae and Coleoptera, and immature spiders. Sphecid larvae feed externally on the prey species. Nests are usually burrows in dead stems, rotten wood and the soil. Prey is generally immobilised by stinging and nest cells usually contain several prey items. Mass provisioning is mainly universal, though in Britain *Ammophila pubescens* provisions progressively. Indeed this species often provisions more than one nest at a time. *Ammophila* species are often tool-users, using a small stone held in the mandibles to tamp down the soil when a nest burrow is being filled in with soil. Some sphecid genera are kleptoparasites of other Sphecidae. All European species are solitary, each female being responsible for her own nest; there is rarely co-operation between individuals.

Allen, G.W. 1987. *Stigmus pendulus* Panzer (Hymenoptera: Sphecidae) new to Britain. *Entomologist's Gazette* **38**: 214.
  Species formally added to the British faunal list. Species compared and contrasted with its close relative *S. solskyi.*

Bohart, R.M. & Menke, A.S. 1976. *Sphecid wasps of the world, a generic revision.* Berkeley, California: University of California Press.
  Major monograph of the Sphecidae of the world. Contains subfamily and generic descriptions, catalogues for each genus (listing all species and their synonyms), key to genera and an exhaustive bibliography.

Dollfuss, H. 1983. The taxonomic value of male genitalia of *Spilomena* Shuckard, 1838, from the palaearctic region (excl. Japan) (Hymenoptera, Sphecidae). *Entomofauna* **4**: 349–70.
  Dollfuss reports that *Spilomena vagans* is probably synonymous with *S. troglodytes.*

Dollfuss, H. 1991. Bestimmungsschlüssel der Grabwespen Nord- und Zentraleuropas (Hymenoptera, Sphecidae) mit speziellen Angaben zur Grabwespenfauna Österreichs. *Stapfia* No. 24: 247pp.
  Good illustrated key to European Sphecidae. (In German.)

Edwards, M. 1993. *Chrysis gracillima* Forster (Hym., Chrysididae) and *Nitela spinolae* Latreille (Hym., Sphecidae) in West Sussex. *Entomologist's Monthly Magazine* **129**: 198.
  The occurrence of *Nitela* in West Sussex (*N. spinolae* misidentified and represents an undescribed species).

Else, G.R. 1993. Recent records of *Chrysis gracillima* Forster (Hym., Chrysididae) and *Nitela spinolae* (Hym., Sphecidae) in southern England. *Entomologist's Monthly Magazine* 129: 198.

Review of the known records of the genus *Nitela* known from Britain.

Else, G.R. & Felton, J.C. 1994. *Mimumesa unicolor* (Vander Linden) (Hymenoptera: Sphecidae), a wasp new to the British list, with observations on related species. *Entomologist's Gazette* 45: 107–14.

Recent addition to the British faunal list. Paper includes an identification key distinguishing *Mimumesa unicolor* from its close relative *M. littoralis*. Species notes for the five British species in the genus also presented.

Falk, S. 1991. A review of the scarce and threatened bees, wasps and ants in Great Britain. *Research and Survey in Nature Conservation* 35. Nature Conservancy Council, Peterborough.

Detailed species accounts for all threatened and vulnerable British Sphecidae.

Felton, J.C. 1987. The genus *Nitela* Latreille (Hym., Sphecidae) in southern England. *Entomologist's Monthly Magazine* 123: 235–8.

A review of the first British records of *Nitela borealis* and *N. spinolae* (the latter was misidentified and currently represents an undescribed species).

Felton, J.C. 1988. The genus *Trypoxylon* Latreille (Hym., Sphecidae) in Kent, with a first British record for *T. minus* de Beaumont. *Entomologist's Monthly Magazine* 124: 221–4.

***Trypoxylon minus* formally added to the British list.**

Lomholdt, O. 1984. The Sphecidae of Fennoscandia and Denmark. *Fauna Entomologica Scandinavica* 4: 452pp.

Detailed account and illustrated identification keys for the Sphecidae (includes key to larvae) recorded from Fennoscandia. Includes most species found in the British Isles. An update of the original 1976–77 edition (which was published in two volumes). In English.

Pulawski, W.J. 1984. The status of *Trypoxylon figulus* (Linnaeus, 1758), *medium* de Beaumont, 1945, and *minus* de Beaumont, 1945 (Hymenoptera: Sphecidae). *Proceedings of the California Academy of Sciences* 43: 123–40.

Contains an illustrated key to the three species of the *Trypoxylon figulus* group recorded from the British Isles.

Richards, O.W. 1979. The Hymenoptera Aculeata of the Channel Islands. *Report and Transactions of the Société Guernesiaise* 20: 389–424.

Short species accounts of all Sphecidae recorded from the Channel Islands.

Richards, O.W. 1980. Scolioidea, Vespoidea and Sphecoidea (Hymenoptera, Aculeata). *Handbooks for the Identification of British Insects* 6(3b): 118pp.

Illustrated keys to and brief species accounts of the Sphecidae reported from the British Isles (Channel Island species omitted).

Shirt, D. (Ed.) 1987. Insects. *British Red Data Books.* 2. Nature Conservancy Council, Peterborough.

Includes species accounts of threatened and vulnerable British Sphecidae; rare species also listed.

Stelfox, A.W. 1927. A list of the Hymenoptera Aculeata (*sensu lato*) of Ireland. *Proceedings of the Royal Irish Academy* (B) 37(22): 201–355.

All records of Irish Sphecidae listed under county headings.

Stelfox, A.W. 1929. Report on recent additions to the Irish fauna and flora. *Proceedings of the Royal Irish Academy* (B) **39**(1): 11–15.

A further sphecid added to the Irish list.

Stelfox, A.W. 1933. Some recent records for Irish aculeate Hymenoptera. *Entomologist's Monthly Magazine* **69**: 47–53.

Additions to the Irish list or verification of old and doubtful records.

Yeo, P.F. & Corbet, S.A. 1995. *Solitary Wasps.* Naturalists' Handbooks No. 3. Richmond Publishing Company, Slough.

Good introduction to British aculeate wasps. Includes identification key and some colour illustrations.

Family Apidae

The family Apidae contains all the bees, comprising about 25000 species in 400 genera worldwide. Of the seven subfamilies six occur in the British Isles: the Colletinae, Andreninae, Halictinae, Melittinae, Megachilinae, and Apinae, all formerly treated as distinct families. The greatest concentration of species is found in the tropics and subtropics (particularly in the deserts of both the Old and New Worlds). A total of 267 species (31 genera) have been recorded from the British Isles, of which 11 have been reported from the Channel Islands but not elsewhere in the United Kingdom.

The bees are basically sphecid wasps that provision their nests with a mixture of nectar or honey (rarely floral oils) and pollen, instead of paralysed arthropods. Most of the morphological characteristics of bees are a result of adaptations to the collection of these plant products. Some bees are so specialised that they forage for pollen from only a single species of plant (monolectic species), others from several closely related species (oligolectic), whereas others visit many unrelated species (polylectic). The females of all three of these groups may visit many other plant species for nectar. The females of most bees possess discrete areas of specially modified body hairs for storing pollen while in flight. Such hairs commonly occur as brushes or pouches on the hind femora and tibia, the lateral surfaces of the propodeum or the abdominal sterna. These structures are known as scopae, although the pollen baskets on the hind tibiae of bumblebees and honeybees are termed corbiculae. Bees that lack scopae (or corbiculae) include most species of *Hylaeus* (Hylaeinae), which ingest pollen along with nectar and regurgitate both products into their cells, and all kleptoparasitic and socially parasitic bees, which, as they do not provision the nests they invade, have secondarily lost such pollen-collecting hairs. Most plant visits by bees, are to angiosperms, although there are some records of gymnosperm pollen being collected.

Kleptoparasitic bees lay their eggs in the cells of their host species (the egg is generally secreted in the pollen loaf or in the cell wall (very rarely the host egg is replaced by that of the inquiline)). The offspring of such species destroy the host egg or larva and feed on the provision intended for the latter. Social parasitic bees (confined in the British Isles to the cuckoo bumblebees) usurp a nest of the bumblebee host, subdue or kill the host *Bombus* queen, and use the worker bumblebees to raise their own brood of queens and males (there is no worker caste).

Most bees are solitary species. The females of these build and provision their nests without the co-operation of other individuals. However, as these bees may nest in large aggregations, they *appear* to have a social organisation. A few female offspring in the nests of some other species continue to enlarge the nest, adding more cells and provisioning these. These workers do not, however, mate, and rarely lay eggs. At the other end of the scale, many species of the subfamily Apinae (e.g. honeybees and many bumblebees) have nest populations numbered in hundreds or thousands of individuals. Most of these are sterile females (workers), all the daughters of their mother (the queen).

Bees exploit all manner of cavities and surfaces in or on which to build their nests. Many excavate the pith of dead stems and arrange their cells end to end. Others utilise empty snail shells, beetle exit holes in dead wood, and similar niches; still others dig burrows in the soil, constructing their cells at the distal end of the burrow and often at the end of lateral burrows, which branch from the main one. In most species the cells have distinct walls; often these consist of a secretion of the abdominal Dufour's gland, which impregnates the surrounding substrate, such as pith or soil particles, and prevents the ingress of water and microorganisms. Others (depending on the species) bring into the nest products collected from outside, such as sections of neatly cut leaf (as in *Megachile*), masticated leaves (leaf mastic), petals, sap, leaf hairs, resin and mud. Cells are mass-provisioned, there rarely being contact between the mother and her offspring. Bumblebees and honeybees construct their nests from wax and nest mainly in large cavities (*Apis mellifera*) or often in the disused nests of rodents and small birds (*Bombus* species).

Baker, D.B. 1964. Two bees new to Britain (Hym., Apoidea). *Entomologist's Monthly Magazine* **100**: 279–86.

Melitta dimidiata (as *Pseudocilissa dimidiata*) and *Eucera nigrescens* (as *tuberculata*) added to the British list. Species diagnosis and accounts included.

Baker, D.B. 1978. Changes of name affecting Apoidea (Hym.) on the British list. *Entomologist's Monthly Magazine*. **113**: 137–8.

Name changes for a number of bees listed.

Baker, D.B. 1994. On the nomenclature of two sibling species of the *Andrena tibialis* (Kirby, 1802) group (Hymenoptera, Apoidea). *Entomologist's Gazette* **45**: 281–90.

Key characters presented for distinguishing *Andrena nigrospina* from *A. spectabilis* (the former species an addition to the British bee fauna). Species accounts (habits and distributions) omitted.

Danks, H.V. 1971. Biology of some stem-nesting aculeate Hymenoptera. *Transactions of the Royal Entomological Society of London* **122**: 323–99.

Summary of various species of Hymenoptera reared from dead plant stems (particularly bramble). Includes a useful key to such stem-nests.

Ebmer, A.W. 1969. Die Bienen des Genus *Halictus* Latr. *s.l.* im Grossraum von Linz (Hymenoptera, Apidae). Pt. 1. *Naturkundliches Jahrbuch der Stadt Linz* (1969): 133–83.

Key includes *Halictus* species recorded from the British Isles. (In German.)

Ebmer, A.W. 1970. Die Bienen des Genus *Halictus* Latr. *s.l.* im Grossraum von Linz (Hymenoptera, Apidae). Pt. 2. *Lasioglossum (Lasioglossum)* Curt. *Naturkundliches Jahrbuch der Stadt Linz* (1970): 19–82.

Key includes certain *Lasioglossum* species recorded from the British Isles. (In German.)

Ebmer, A.W. 1971. Die Bienen des Genus *Halictus* Latr. *s.l.* im Grossraum von Linz (Hymenoptera, Apidae). Pt. 3. *Lasioglossum (Evylaeus* Rob.). *Naturkundliches Jahrbuch der Stadt Linz* (1971): 63–156.

Key includes remaining *Lasioglossum* species recorded from the British Isles. (In German.)

Ebmer, A.W. 1976. Liste der Mitteleuropaischen *Halictus*- und *Lasioglossum*-arten. *Linzer Biologische Beitrage* **8**: 393–405.

List includes those species reported from the British Isles. (In German.)

Edwards, R. (Ed.) 1997. *Provisional atlas of the aculeate Hymenoptera of Britain and Ireland. Part 1.* Bees, Wasps and Ants Recording Society. Biological Records Centre, Huntingdon.

Distribution maps and accounts for species in the genera *Colletes* (part), *Hylaeus* (part), *Anthidium, Stelis* (part), *Osmia* (part) and *Ceratina*.

Else, G.R. & Edwards, M. 1996. Observations on *Osmia inermis* (Zetterstedt) and *O. uncinata* Gerstaecker (Hym., Apidae) in the central Scottish Highlands. *Entomologist's Monthly Magazine* **132**: 291–8.

Includes information on the habits of *Osmia uncinata*, a species only recently recognised from the British Isles (where it is confined to the central Scottish Highlands).

Falk, S. 1991. A review of the scarce and threatened bees, wasps and ants in Great Britain. *Research and Survey in Nature Conservation* **35**. Nature Conservancy Council, Peterborough.

Detailed species accounts for all threatened and vulnerable species of bee in Britain. Other rare species listed.

Gardner, W. 1901. *Coelioxys mandibularis*, Nyl., an addition to the British list of aculeates. *Entomologist's Monthly Magazine* **37**: 166–7.

First found in Britain in north-west England (has since been recorded from Wales and east Kent).

Guichard, K.M. 1971. A bee new to Britain from Wiltshire – *Andrena lathyri* Alfken (Hym., Apidae). *The Entomologist* **104**: 40–2.
> Identification characters of this bee presented. (A much older record from Somerset was not known at the time of publication.)

Guichard, K.M. 1974. *Colletes halophila* Verhoeff (Hym., Apidae) and its *Epeolus* parasite at Swanscombe in Kent, with a key to the British species of *Colletes* Latreille. *Entomologist's Gazette* **25**: 195–9.
> Key largely based on characters associated with the mouthparts (galea) and abdominal sternites. No figures presented. *Epeolus variegatus* shown to be a kleptoparasite of *C. halophilus* in Kent.

Hammond, P.M., Smith, K.G.V., Else, G.R. & Allen, G.W. 1989. Some recent additions to the British insect fauna. *Entomologist's Monthly Magazine* **125**: 95–102.
> Short account of *Osmia uncinata* in Britain presented, plus a colour photograph of a female specimen of this species.

Koster, A. 1986. Het genus *Hylaeus* in Nederland (Hymenoptera, Colletidae). *Zoologische Bijdragen*. No. 36.
> Key includes all *Hylaeus* found in the British Isles. (Text in Dutch, key in English.)

Malyshev, S.I. 1935. The nesting habits of solitary bees. A comparative study. *Eos* **11**: 201–309.
> An introduction to the nesting habits of bees (mainly in Europe).

Michener, C.D. 1953. Comparative morphological and systematic studies of bee larvae with a key to the families of hymenopterous larvae. *University of Kansas Science Bulletin* **35**: 987–1102.
> One of the very few keys to bee larvae (at generic level), worldwide.

Morice, F.D. 1908. *Coelioxys afra* Lep. – a bee new to Britain – from the New Forest. *Entomologist's Monthly Magazine* **44**: 178–80.
> Only record from Britain (although the species occurs in the Channel Islands).

O'Toole, C. 1974. A new subspecies of the vernal bee, *Colletes cunicularius* (L.) (Hymenoptera: Colletidae). *Journal of Entomology* (B) **42**(2): 163–9.
> Account of British populations of *Colletes cunicularius* which are recognised as subspecifically distinct on morphological and biological grounds.

O'Toole, C. & Raw, A. 1991. *Bees of the world*. Blandford.
> An excellent general introduction to bees worldwide. Numerous colour photographs of bees at work and rest.

Perkins, R.C.L. 1895. On two apparently undescribed British species of Andrenidae. *Entomologist's Monthly Magazine* **31**: 39–40.
> *Lasioglossum* (as *Halictus*) *angusticeps* formally added to the British list.

Perkins, R.C.L. 1919. The British species of *Andrena* and *Nomada*. *Transactions of the Entomological Society of London* **1919**: 218–319.
> The most recent published key to British *Andrena* and *Nomada*. Out of date but still quite useful. Contains good species accounts, particularly for *Andrena*.

Perkins, R.C.L. 1922. The British species of *Halictus* and *Sphecodes*. *Entomologist's Monthly Magazine* **58**: 46–52, 94–101, 167–74.
> Good key to British *Halictus* (most in this paper currently included in the genus *Lasioglossum*). No illustrations and text largely out of date. See Perkins (1931) for correction of an error.

Perkins, R.C.L. 1922. *Sphecodes scabricollis* Wesm. in Somerset, and description of female of *S. kershawi* Perk. *Entomologist's Monthly Magazine* **58**: 89–91.
Characters for *Sphecodes scabricollis* described.

Perkins, R.C.L. 1925. An *Andrena* new to Britain. *Entomologist's Monthly Magazine* **60**: 277–8.
Andrena congruens formally added to the list of British fauna. Identification characters presented.

Perkins, R.C.L. 1925. The British species of *Megachile*, with descriptions of some new varieties from Ireland, and of a species new to Britain in F. Smith's collection. *Entomologist's Monthly Magazine* **61**: 95–101.
Key to British species of leafcutter bees. Still relevant in the absence of a more up-to-date British key.

Perkins, R.C.L. 1931. Notes on some Aculeate Hymenoptera, with corrections of errors. *Entomologist's Monthly Magazine* **67**: 20.
Megachile lapponica formally added to the British list.

Perkins, R.C.L. 1935. A note on some British species of *Halictus*. *Entomologist's Monthly Magazine* **71**: 104–6.
Key for the identification of a group of four metallic green *Lasioglossum* species.

Prŷs-Jones, O. & Corbet, S.A. 1991. *Bumblebees*. Naturalists' Handbooks No. 6. Richmond Publishing Company, Slough.
Most recent summary and illustrated key to the bumblebees of the British Isles, with distribution maps. All species illustrated in colour.

Richards, O.W. 1937. A study of the British species of *Epeolus* Latr. and their races, with a key to the species of *Colletes* (Hymen., Apidae). *Transactions of the Society for British Entomology* **4**: 89–130.
Includes key characters for separating the two species of British *Epeolus*.

Richards, O.W. 1979. The Hymenoptera Aculeata of the Channel Islands. *Report and Transactions of the Société Guernesiaise* **20**: 389–424.
Short species accounts for those bees recorded from the Channel Islands.

Ruttner, F. 1988. *Biogeography and taxonomy of honeybees*. Springer-Verlag, Berlin.
Biogeography of honeybees worldwide. Includes useful sections on *Apis mellifera*, the species native to the British Isles.

Saunders, E. 1896. *The Hymenoptera Aculeata of the British Islands*. Reeve, London.
A classic work on the bees (plus wasps and ants) of the British Isles. Long out of date but still the only work that includes identification keys to all bees known at the time of publication.

Saunders, E. 1901. *Coelioxys mandibularis*, Nyl. *Entomologist's Monthly Magazine* **37**: 167.
Species formally added to the list of British fauna. Identification characters listed.

Saunders, E. 1910. On four additions to the list of British Hymenoptera. *Entomologist's Monthly Magazine* **46**: 10–12.
Includes the occurrence of *Dufourea halictula* in Britain.

Schmidt, K. & Westrich, P. 1993. *Colletes hederae* n. sp., eine bischer unerkannte, auf Efeu (*Hedera*) spezialisierte Bienenart (Hymenoptera: Apoidea). *Entomologische Zeitschrift* **103**: 89–93.
Species formerly misidentified on the Channel Islands (where, in the United Kingdom,

this species is only known) as *Colletes halophilus*. Identification characters listed (in German).

Shirt, D. (Ed.) 1987. Insects. *British Red Data Books. 2.* Nature Conservancy Council, Peterborough.
Includes species accounts of threatened and vulnerable British bees.

Sladen, F.W.L. 1897. *Cilissa melanura,* a species new to the British list, and other bees at St. Margaret's Bay. *Entomologist's Monthly Magazine* 33: 229–30.
Melitta tricincta formally added to the British faunal list (author apparently overlooked the fact that the species was described from British specimens by W. Kirby in 1802).

Snodgrass, R.E. 1956. *Anatomy of the honey bee.* Constable and Company Ltd, London.
Useful guide to the internal and external anatomy of *Apis mellifera*.

Spooner, G.M. 1929. *Halictus angusticeps* Perk. in Dorset. *Entomologist's Monthly Magazine* 65: 233–4.
Report of first discovery of the female of *Lasioglossum angusticeps* in Britain (see Perkins (1885) for details of the male).

Spooner, G.M. 1946. *Nomada errans* Lep., a bee new to Britain. *Entomologist's Monthly Magazine* 82: 105–6.
Species formally added to the British faunal list. Identification characters listed.

Stelfox, A.W. 1927. A list of the Hymenoptera Aculeata (*sensu lato*) of Ireland. *Proceedings of the Royal Irish Academy* 37 (Section B, No. 22): 201–355.
All Irish bees listed under county headings.

Stelfox, A.W. 1929. Report on recent additions to the Irish fauna and flora. Hymenoptera. *Proceedings of the Royal Irish Academy* 39 (Section B, No. 1): 11–15.
Records of a further two species of Irish bee.

Stelfox, A.W. 1933. Some recent records for Irish aculeate Hymenoptera. *Entomologist's Monthly Magazine* 69: 47–53.
Additions to the Irish list and verification of old and doubtful records.

Stephen, W.P., Bohart, G.E. & Torchio, P.F. 1969. *The biology and external morphology of bees, with a synopsis of the genera of northwestern America.* Agricultural Experiment Station, Oregon State University, Corvallis, Oregon.
Excellent introduction to the biology and anatomy of bees worldwide (out of print).

Westrich, P. 1989. *Die Wildbienen Baden-Württembergs.* Ulmer, Stuttgart.
A very important monograph (in two volumes) on the bees of a German province. Includes profiles on most British species, many illustrated with stunning colour photographs. No identification keys. Out of print. (In German.)

Yarrow, I.H.H. 1955. *Andrena combinata* Christ (Hym., Apidae): a bee new to Britain. *Entomologist's Monthly Magazine* 92: 234–5.
Andrena lepida (misidentified by Yarrow as *A. combinata)* formally added to the British bee list.

Yarrow, I.H.H. 1968. Recent additions to the British bee-fauna, with comments and corrections. *Entomologist's Monthly Magazine* 104: 60–4.
Includes *Eucera nigrescens* (as *tuberculata)* and *Melitta dimidiata*.

Yarrow, I.H.H. 1970. *Hoplitis claviventris* (Thomson 1872) (= *Osmia leucomelana* Auctt. Nec Kirby) and the identity of *Apis leucomelana* Kirby 1802 (Hymenoptera, Megachilidae). *The Entomologist* 103: 62–9.

Includes morphological characters to separate *Hoplitis claviventris* from *H. leucome-lana*.

Yarrow, I.H.H. & Guichard, K.M. 1941. Some rare Hymenoptera Aculeata, with two species new to Britain. *Entomologist's Monthly Magazine* 77: 2–13.
Includes *Andrena floricola* new to Britain.

Superfamily Chrysidoidea (= Bethyloidea)

This aculeate superfamily contains 7 families worldwide, of which 4 occur in Britain, the Dryinidae, Embolemidae, Bethylidae and Chrysididae. The chrysidoid fauna of Britain consists of 79 native species and a few well-established bethylid introductions, mostly associated with stored products. The Bethylidae is the largest family occurring in Britain with some 13 endemic British species and a further 7 introductions.

In the Chrysidoidea, the number of antennal segments is the same in both sexes, whereas in the other aculeate superfamilies (Vespoidea and Apoidea) the antennal segments usually exhibit sexual dimorphism.

Family Dryinidae

A recent world revision of the family (Olmi, 1984) recognised ten subfamilies, made up of approximately 850 species. Five subfamilies occur in Britain: Anteoninae, Aphelopinae, Bocchinae, Dryininae and Gonatopodinae, which contain in total 32 species.

Males are always winged whereas females can be winged or wingless or have reduced wings. The antennae are 10–segmented in both sexes and the antennal insertions are close to the posterior clypeal margin. Females of all species, except those of the genus *Aphelopus*, have the fore tarsi chelate, the chela composed of a strongly projecting fifth tarsal segment, which is opposed by a much enlarged apical claw.

Adult dryinids are active from April to September but are most common in the summer months. British species have one or two generations each year and usually overwinter in the pupal stage. The prey of dryinid larvae consists of nymphal or more rarely adult auchenorrhynchan Homoptera belonging to the subfamilies Cicadelloidea and Fulgoroidea. The host is gripped by the female dryinid by the strong chelae or fore and middle legs and stung in the abdomen. This results in the temporary paralysis of the victim, during which an egg is being laid between two of the host's abdominal sclerites. The dryinid hatches into a first-instar larva, which may be an ecto- or endoparasitoid. Later instars of the dryinid larva develop outside the host in a sac or cyst, which projects from the host's abdomen and is now actively feeding on the host's tissues. On completion of its development the dryinid larva has consumed most or all of the host's internal organs and thus seriously injured or killed it. The larva emerges from its development sac and pupates in the soil or on the food plant of the host. The dryinid cocoon is typically of very dense silk and has a close-fitting inner lining. Dryinid cocoons are often parasitised by Hymenoptera of the families Diapriidae and Encyrtidae.

Family Embolemidae

Olmi (1995) recognises two genera worldwide: *Ampulicomorpha*, with six extant and one fossil species, and *Embolemus*, with ten extant and one fossil species.

Of this small and rare family, *Embolemus ruddii* is the only species known to occur in Britain. It is widely distributed throughout England, but has been found only once in Scotland and once in Wales. There is little sexual dimorphism known in this species, but males are always fully winged whereas females are wingless. Both sexes have ten-segmented antennae, which arise high on the head and are therefore widely separated from the clypeus. They are typically found by sweeping between August and October. Females are thought to overwinter as they have also been found in April and May, and in all the winter months.

Little is known about the biology of embolemids, but their life histories are thought to be similar to that of dryinids. The host of *E. ruddii* is unknown but the North American embolemid species *Ampulicomorpha confusa* is known to be an internal parasitoid of the nymphs of both *Epiptera floridae* and *E. pallida* (Homoptera: Achilidae). According to Olmi (1995), females of *Embolemus ruddii* have occasionally been found in nests of ants and moles in or near forests where there were probably rotten trees with colonies of Achilidae close by and so this family is suggested as a likely host. This association with ants is another character of Embolemidae that is common to Dryinidae.

Family Bethylidae

The world fauna of Bethylidae consists of about 2000 species divided into four subfamilies. Three of these, Bethylinae, Eryrinae and Pristocerinae, occur in Britain.

Males are almost always winged; rarely they have reduced wings. However, females may be winged, wingless or with reduced wings. The head is often elongate and depressed. In both males and females the antennae have 12–13 segments. The antennae are situated close to the clypeus. The forewing has reduced venation.

Bethylids are specialist predators of coleopterous or lepidopterous larvae, both as larvae and as adults. The coleopterous or lepidopterous larvae are often attacked by bethylids in cryptic situations such as rolled leaves or under bark, within pieces of rotten wood or in earth cells. Often the prey chosen is larger than the bethylid female, in consequence bethylid stings are efficient and powerful, and many are painful to humans. When the female has found appropriate prey she attacks and stings it once or several times, inducing rapid paralysis or even killing it. The prey may be dragged or carried into a secluded place. The female will often feed on the haemolymph that oozes from the sting site of the victim. One or more eggs are laid on the prey. The bethylid larvae that hatch are external predators on the coleopterous or lepidopterous larva.

Family Chrysididae

A recent world revision of chrysidids (Kimsey & Bohart, 1990), recognises around 3000 species arranged in 84 genera and 4 subfamilies. Two subfamilies occur in Britain: Cleptinae and Chrysidinae.

Chrysidids are commonly termed cuckoo wasps, because of their klepto-parasitic habits, and ruby-tailed, jewel or gold wasps, because of their metallic coloration. The thick and sculptured cuticle is usually brilliantly coloured with various combinations of red, green, blue, purple and gold. Sexual dimorphism is minimal, the two sexes are not easily discriminated on external characters. In all British species both sexes are winged and have 13 antennal segments with the sockets positioned low on the face, on the dorsal clypeal margin. The number of visible gastral segments is reduced in chrysidids, 3–4 visible in females and 3–5 visible in males.

The adults fly in bright sunlight from late April to early September and are frequently seen flying around investigating holes and crevices in their search for nests of the hosts for their larvae, which consists in Britain of other Hymenoptera. All adult chrysidids feed at flowers and extrafloral nectaries, and all except the cleptines are able to adopt a rolled-up defensive posture when threatened.

To identify British species, keys in Morgan, (1984) should be used and then the validity of the names checked by using Kimsey and Bohart (1990).

References

Jervis, M.A. 1977. A new key for the identification of the British species of *Aphelopus* (Hym: Dryinidae). *Systematic Entomology* 2: 301–3.
   Useful key.

Jervis, M.A. 1979. Parasitism of *Aphelopus* species (Hymenoptera: Dryinidae) by *Ismarus dorsiger* (Curtis) (Hymenoptera: Diapriidae). *Entomologist's Gazette* 30: 127–9.
Useful host records.

Kimsey, L.S. & Bohart, R.M. 1990. *The chrysidid wasps of the world*. Oxford University Press.
Up to date; contains keys to all genera and comprehensive biological information. Useful because of its complete coverage.

Kieffer, J.J. 1914. Bethylidae. *Das Tierreich* 41. Berlin.
A monograph of the [then] world species. Unreliable for identifications, but useful because of its completeness.

Kunz, P.X. 1994. Die Goldwespen (Chrysididae) Baden-Wurttembergs. *Beihefte zu den Veröffentlichungen für Naturschutz und Landschaftspflege in Baden-Wurttemberg* 77: 1–188.
New but covers German fauna only. Includes keys, host lists, colour forms of species and colour photographs.

Morgan, D. 1984. Hymenoptera: Chrysididae. *Handbooks for the Identification of British Insects* 6(5): 37pp.
Keys to species in Britain.

Olmi, M. 1984. A revision of the Dryinidae (Hymenoptera). *Memoirs of the American Entomological Institute* 37(2): 1913pp.
Revision of world Dryinidae including keys to Palaearctic species. An extensive work, which has been updated by the author in various latter papers: see Olmi (1995b) for a full reference list.

Olmi, M. 1995a. A revision of the world Embolemidae (Hymenoptera Chrysidoidea). *Frustula Entomologia* n.s. XVIII (XXXI): 85–146.
Keys and descriptions of all species.

Olmi, M. 1995b. Contribution to the knowledge of the world Dryinidae (Hymenoptera: Chrysidoidea). *Phytophaga* 6: 3–54.
Includes keys to Palaearctic species.

Perkins, J.F. 1976. Hymenoptera: Bethyloidea (excluding Chrysididae). *Handbooks for the Identification of British Insects* 6(3a): 38pp.
Keys to British species.

Richards, O.W. 1932. Observations on the genus *Bethylus* Latr. (= *Perisemus* Foerst.) (Hymenoptera: Bethylidae). *Transactions of the Entomological Society of the South of England* 8: 35–40.
Notes on localities, habitats and biology of two *Bethylus* species.

Richards, O.W. 1936. The Cameron collection of Dryinidae. *Annals and Magazine of Natural History* (18) 10: 463–7.
Important observations and lectotype designations are made in this short paper.

Richards, O.W. 1939. The British Bethylidae. *Transactions of the Royal Entomological Society of London* 89: 185–344.
Includes keys to genera and species.

Richards, O.W. 1948. New records of Dryinidae and Bethylidae (Hymenoptera). *Proceedings of the Royal Entomological Society of London* (A) 23: 14–18.
Limited use for identifications.

Richards, O.W. 1953. The classification of the Dryinidae (Hym.) with the descriptions of new species. *Transactions of the Royal Entomological Society of London* **104**: 51–70.

Includes keys.

Richards, O.W. 1977. Hymenoptera: introduction and key to families. *Handbooks for the identification of British Insects* 6(1) (2nd edn): 100pp.

Useful introduction.

Apocrita: Parasitica

Superfamilies Trigonaloidea and Evanioidea

These two superfamilies include one and three families, respectively, with a total of only nine species in Britain.

Family Trigonalidae

The family (and superfamily) name has usually been misspelled as Trigonalyidae. Worldwide, about 100 species are known, but only one, *Pseudogonalos hahnii* (recently mistakenly attributed to *Trigonalis*), occurs in Britain. Identification can be effected by using any of the keys to families (see general references) and Oehlke (1984).

Trigonalids have a very unusual biology. They are parasitoids and most species have two successive hosts. The female trigonalid lays large numbers of minute eggs on leaves. The eggs do not hatch unless ingested by a lepidopterous caterpillar or a sawfly larva, but in many species there is no further development unless one of two things happens. In the first case, if the primary host is parasitised by an endoparasitoid, such as an ichneumonid, braconid or tachinid, the trigonalid larva seeks out and enters the body of the parasitoid larva and develops there as a hyperparasitoid. The second mode of development occurs if the primary host is consumed by a predaceous wasp of the family Vespidae. If the primary host, or part of it, containing the trigonalid first-instar larva is fed by a worker wasp to a larva then the trigonalid develops as an endoparasitoid of the vespid. The pattern of development in the British species is not known.

Family Evaniidae

Evaniids have a uniform, very characteristic appearance and although they have no common name in Britain they are known as ensign-flies elsewhere because of the way in which the metasoma is held up like a flag. Evaniids develop as parasitoids in the oothecae of cockroaches. The family is mainly tropical and only two species occur in Britain, one of which is cosmopolitan and associated with domestic cockroaches. The British native species, *Brachygaster minutus*, attacks the three British species of *Ectobius* (Dictyoptera: Ectobiidae). The standard identification work is Crosskey (1951).

Family Aulacidae

Only one species, *Aulacus striatus*, occurs in Britain. The female *Aulacus* locates the egg-shaft of the woodwasp host (*Xiphydria camelus* (Xiphydriidae)) in a crevice in alder bark, inserts her ovipositor down it and lays single eggs in as many of the batch of host eggs as she can reach. Not until almost a year later, when the woodwasp larva is fully grown and has tunnelled up to just below the surface, does the parasitoid larva complete feeding and kill the host. *Aulacus* can be identified by using the keys to families (see general references) or Crosskey (1951).

Family Gasteruptiidae

This distinctive family comprises some 500 species. It is almost worldwide in distribution but is particularly well represented in Australia, where the most primitive extant forms occur. Only *Gasteruption* is found in Britain and of the seven species keyed by Crosskey (1951) two were almost certainly included as British in error (Fitton *et al.*, 1978). Despite some statements indicating other hosts, it seems clear that *Gasteruption* develop as secondary kleptoparasites on the food stored in the cells of only solitary bees.

References

Crosskey, R.W. 1951. The morphology, taxonomy, and biology of the British Evanioidea (Hymenoptera). *Transactions of the Royal Entomological Society of London* **102**: 247–301.
   Includes keys.
Oehlke, J. 1984. Beitrage zur Insektenfauna der DDR: Hymenoptera – Evanioidea, Stephanoidea, Trigonalyoidea. *Faunistiche Abhandlungen Staatliches Museum fur Tierkunde in Dresden* **11**: 161–90.
   In German. Includes keys and covers virtually all of NW European fauna.

Superfamily Ichneumonoidea

The species of this superfamily, comprising the Ichneumonidae and Braconidae, account for almost half of the British Hymenoptera fauna, with records of approximately 3300 species. On average, adult ichneumonoids are larger than other parasitic wasps but, despite their size, their numbers, and the spectacular appearance of some, they have failed to attract much attention from naturalists. None has a common, English name: not even 'ichneumon-fly' has general currency, as it is sometimes understood to apply to all members of the superfamily, or sometimes just to the Ichneumonidae. Probably it is best to say ichneumonid wasp or braconid wasp, as appropriate. Ichneumonoids are all parasitoids of insects or, in a few cases, other arthropods.

A few ichneumonoids are nocturnal and are attracted to lights. A short cut to their identification is provided by Huddleston & Gauld (1988). The other useful key to a group not defined taxonomically is that to Hymenoptera associated with spiders (Fitton *et al.* 1987).

Family Ichneumonidae

The Ichneumonidae is an extremely large family with an estimated 60 000 extant species worldwide (Townes, 1969) of which over 2100 species in 35 subfamilies occur in Britain. Ichneumonids are common insects in most terrestrial habitats, but the group is most species-rich in the temperate regions and the humid tropics. The family is currently divided into 37 subfamilies, and only two of these are not represented in Britain.

Most ichneumonids attack larvae or pupae of holometabolous insects, particularly those of the Lepidoptera and Hymenoptera-Symphyta. Although described as parasitoids, ichneumonid larvae exhibit a wide diversity of biological habits and range from highly modified endoparasitoids, to mobile predators consuming spiders' eggs in silken sacs or bee larvae in a series of cells. However, the great majority of ichneumonids are intimately associated with a single host individual. Although most are solitary parasitoids, small numbers of gregarious species occur in several subfamilies. In Britain Lepidoptera are the hosts of almost two-thirds of the ichneumonid species, a further quarter utilise sawflies. Relatively few ichneumonid groups have adapted to parasitise the great diversity of Coleoptera and Diptera potentially available as hosts. Some Neuroptera and Mecoptera are parasitised as larvae or pupae by a few species of Cryptinae and Campopleginae, and a few species of Trichoptera serve as hosts for a few Cryptinae and for the very small subfamily Agriotypinae. The Hymenoptera-Apocrita do not escape attack and because various ichneumonids behave as hyperparasitoids even the Ichneumonoidea are exploited as hosts. Species of Mesochorinae are obligatory hyperparasitoids, their larvae developing solitarily inside the bodies of primary parasitoid larvae. The groups most prone to attack by mesochorines are the various braconid, ichneumonid and tachinid endoparasitoids of growing caterpillars and sawfly larvae. Large numbers of small cryptines and some pimplines are pseudohyperparasitoids. Species of *Pseudorhyssa* (Poemeniinae) and some *Temelucha* (Cremastinae) are known to be specialist multiparasitoids. Spiders are the hosts of ectoparasitic polysphinctines (Pimplinae) and their egg sacs are attacked by some other pimplines and cryptines. One genus of Cryptinae parasitises the egg nests of a pseudoscorpion. A few ichneumonids achieve synchronisation with their hosts by overwintering as adults within the cocoon, whereas others overwinter as a first-instar larva within a diapausing host. However, in temperate areas like Britain the majority pass the winter as full-grown larvae within their cocoons. A few species, particularly ichneumonines, hibernate as fertilised adult females in grass tussocks, leaf litter, etc. Some bivoltine species utilise different host species at different times of the year.

Understanding ichneumonid classification, getting to know the group, accessing the inadequate and complex literature, and identifying specimens are all difficult. Much of the British and European fauna is still extremely poorly known and much of the useful literature is fragmentary and scattered. Unfortunately, there are few taxonomists working on ichneumonids. Perhaps more than any other group of insects identification still relies, to a large extent, on access to advice from specialists and extensive and reliably named collections. The few collections in museums in Britain and Ireland that are useful are those in the Natural History Museum, London; the Royal Museum of Scotland, Edinburgh; the Manchester Museum; the Ulster Museum, Belfast; the National Museum of Ireland, Dublin; the Oxford University Museum of Natural History; and the Horniman Museum, London. Not all of these collections are useful or up-to-date for all subfamilies.

The best general introductions to the family are in Gauld & Bolton (1988), Gauld (1991), Hanson & Gauld (1995) and Goulet & Huber (1993). Only the first of these deals specifically with the British fauna. The current British check list (Fitton, in Fitton *et al.*, 1978) requires considerable revision. The best general source of information is the recently published catalogue of the Ichneumonidae of the world (Yu & Horstmann, 1997). Yu has intentions of publishing this in the form of a database on CD-ROM, which will make it a very powerful tool.

For identification the first requirement is to get to subfamily. The best recent key is that by David Wahl (in Goulet & Huber, 1993). The standard work, which includes keys to subfamilies and to genera for all subfamilies (except Ichneumoninae) on a world basis is by Townes (1969, 1970a,b, 1971).

However, Townes' work is not always easy to use: his keys are not illustrated (although there are habitus figures for each genus), his nomenclature is idiosyncratic and the subfamily classification has changed considerably in the past twenty-odd years. Other useful keys to subfamilies, although not applying strictly to the British fauna, include Gauld (1991) and Kasparyan (1981). The latter work, in Russian, has keys to the subfamilies, tribes, genera, and (for many groups) species in the European USSR. A new key to subfamilies on a world basis is in preparation (Notton, Fitton & Quicke). Perkins' (1959) key to subfamilies cannot be recommended (although the keys to Ichneumoninae in the same publication can). Older general works, such as those of Morley (1903–15) are of very limited value, even to those who have a good knowledge of the family. All the keys so far mentioned deal only with adults, and the need to identify ichneumonid larvae is limited. However, the cast skin of the final larval-instar is found associated with the host remains after the ichneumonid adult has emerged. It is often possible to identify these cast skins, using features mainly of the head sclerites, to subfamily and sometimes genus. Short (1978) and Finlayson & Hagen (1977) are useful compendia. Wahl's papers (1986 onwards) should also be consulted.

The synopsis below shows the current subfamily and tribal classification of the Ichneumonidae occurring in Britain (the first column), followed by the names and groupings used by Townes, Wahl and a few other authors. The column headed Townes reflects the classification and nomenclature given in Townes' treatment of world genera (1969–71) [and takes into account changes made by Townes in the 'Errata' in the 1971 volume]. The column headed Wahl is the classification used by Wahl in Goulet & Huber (1993). The column headed Others includes a few names, some widely used in early works, which might cause confusion and some different ranks given to a few groups by some contemporary workers. A dash in a column indicates that the name and grouping are (more or less) as in the first column and a blank means that the group is not mentioned in the work concerned.

For each subfamily the approximate number of species in Britain and the papers that are of value in identifying the genera and species are given in the list below (in alphabetical order of subfamily name). Sufficient notes are given in the references section for the user to be able to judge the relevance and utility of particular works. The listings are not meant to be fully comprehensive. For example, if a genus is represented by only one species in Britain, a paper dealing with the European fauna of several species may not be included. For recently and currently active authors citations are given in the form 'Wahl, 1986 onwards' and abstracting publications, such as *Zoological Record*, should be consulted for details of their subsequent papers.

Current classification	Townes	Wahl	Others
ACAENITINAE	—	—	
Coleocentrini	—		
Acaenitini	—		
ADELOGNATHINAE	—	—	
AGRIOTYPINAE	—	—	AGRIOTYPIDAE
ALOMYINAE	part of ICHNEUMONINAE	part of ICHNEUMONINAE	
ANOMALONINAE	ANOMALINAE	—	
Anomalonini	Anomalini		ANOMALINAE
Gravenhorstiini	Gravenhorstiini + Podogastrini + Theriini		THERIINAE
BANCHINAE	—	—	
Banchini	—		
Glyptini	—		
Lissonotini	—	Atrophini	
CAMPOPLEGINAE	PORIZONTINAE	—	
Campoplegini	Campoplegini + Porizontini		Porizontini + Cymodusini + Macrini/Limneriini
COLLYRIINAE	—	—	
CREMASTINAE	—	—	
CRYPTINAE	GELINAE	PHYGADEUONTINAE	
Aptesini	Echthrini		Hemigastrini/INAE
Cryptini	Mesostenini		MESOSTENINAE
Phygadeuontini	Gelini		PHYGADEUONTINAE
CTENOPELMATINAE	SCOLOBATINAE	—	
Ctenopelmatini			
Euryproctini	Euryproctini + part of Westwoodiini		
Mesoleiini	—		
Olethroditini	—		
Perilissini	—		
Pionini	—		
Scolobatini	Scolobatini + part of Westwoodiini		
CYLLOCERIINAE	part of MICROLEPTINAE	—	
DIACRITINAE	tribe of EPHIALTINAE	—	

Current classification	Townes	Wahl	Others
DIPLAZONTINAE	—	—	Bassini
EUCEROTINAE	tribe of TRYPHONINAE	—	
HELICTINAE	part of MICROLEPTINAE	part of ORTHOCENTRINAE	
ICHNEUMONINAE	—	—	
Eurylabini			
Goedartiini			
Heresiarchini	Ichneumonini		Protichneumonini
Ichneumonini	Joppini		
Listrodromini			
Phaeogenini	part of Alomyini	part of Alomyini	
Platylabini			
Trogini			
Zimmeriini			
LYCORININAE	—	—	
MESOCHORINAE	—	—	
METOPIINAE	—	—	
MICROLEPTINAE	part of MICROLEPTINAE	—	
NEORHACODINAE	—	—	
OPHIONINAE	—	—	
ORTHOCENTRINAE	—	part of ORTHOCENTRINAE	
ORTHOPELMATINAE	—	—	
OXYTORINAE	part of MICROLEPTINAE	—	
PAXYLOMMATINAE	not in ICHNEUMONIDAE	—	HYBRIZONTIDAE/INAE
PHRUDINAE			
PIMPLINAE	part of EPHIALTINAE		
Ephialtini	Pimplini + some Theroniini		some Delomeristini
Polysphinctini	—		
Pimplini	Ephialtini + some Theroniini		Echthromorphini
POEMENIINAE	tribe of EPHIALTINAE	—	Neoxoridini
RHYSSINAE	tribe of EPHIALTINAE	—	

Current classification	Townes	Wahl	Others
STILBOPINAE	tribe of BANCHINAE	—	
TERSILOCHINAE	—	—	Porizonini
TRYPHONINAE	—	—	
Exenterini	Cteniscini		
Idiogrammatini	—		
Oedemopsini	Eclytini		Eclytini + Oedemopsini
Phytodietini	—		
Sphinctni	—		
Tryphonini	—		
XORIDINAE	—	—	

Acaenitinae (7 species): Fitton (1981); Shaw (1986).

Adelognathinae (about 15 species): Fitton et al. (1982).

Agriotypinae (1 species): Perkins (1960).

Alomyinae (2 species): Perkins (1960).

Anomaloninae (about 40 species): Gauld & Mitchell (1977a).

Banchinae (about 110 species): Aubert (1978a); Fitton (1985a, 1987); Townes (1970b).

Campopleginae (more than 250 species): Barron & Walley (1983); Dbar (1983, 1984, 1985); Hinz (1963, 1977); Horstmann (1969 onwards); Horstmann & Shaw (1984); Sanborne (1984 onwards); Shaw & Horstmann (1997); Townes (1970b).

Collyriinae (2 species): Fitton (1984).

Cremastinae (about 13 species): Fitton & Gauld (1980).

Cryptinae (more than 450 species): Fitton et al. (1987); Horstmann (1967 onwards); Jussila (1979 onwards); van Rossem (1966 onwards); Sawoniewicz (1980 onwards); Schwarz (1988 onwards); Townes (1970a, 1983).

Ctenopelmatinae (more than 200 species): Gauld & Mitchell (1977b); Idar (1979, 1981); Hinz (1975, 1976, 1980, 1985, 1991); Townes (1970b).

Cylloceriinae (about 5 species): van Rossem (1981, 1987); Townes (1971); Wahl (1990).

Diacritinae (1 species): Fitton, Shaw & Gauld (1988).

Diplazontinae (about 50 species): Beirne (1941); Diller (1969); Fitton & Boston (1988); Fitton & Rotheray (1982); Rotheray (1990); Wahl (1990).

Eucerotinae (3 species): Fitton (1984).

Helictinae (about 50 species): van Rossem (1981–90); Townes (1971).

Ichneumoninae (about 340 species): Hilpert (1992); Perkins (1959, 1960); Selfa & Diller (1994).

Lycorininae (1 species): Perkins (1960).

Mesochorinae (over 50 species): Fitton (1985b); Lawton (1981); Wahl (1993).

Metopiinae (over 60 species): Aeschlimann (1973 onwards); Perkins (1936); Townes (1971).

Microleptinae (3 species): Schwarz (1991).

Neorhacodinae (1 species): Fitton (1984).

Ophioninae (about 23 species): Brock (1982); Gauld (1973 onwards).

Orthocentrinae (over 40 species): Aubert (1978b, 1981); Horstmann (1994a); Townes (1971).

Orthopelmatinae (2 species): Gauld & Mitchell (1977a).

Oxytorinae (2 species): van Rossem (1981, 1987).

Paxylommatinae (2 species): Watanabe (1984).

Phrudinae (6 species): Gauld & Fitton (1980); Shaw (1991).

Pimplinae (about 100 species): Fitton, Shaw & Austin (1987); Fitton, Shaw & Gauld (1988).

Poemeniinae (5 species): Fitton, Shaw & Gauld (1988).

Rhyssinae (2 species): Fitton, Shaw & Gauld (1988).

Stilbopinae (4 species): Fitton (1984).

Tersilochinae (over 50 species): Horstmann (1971a, 1981).

Tryphoninae (about 160 species): Delrio (1975); Fitton (1975); Fitton & Ficken (1990); Kasparyan (1973, 1994); Kerrich (1952); Townes (1969).

Xoridinae (about 14 species): Gauld & Fitton (1981).

References

Aeschlimann, J.P. 1973a. Révision des espèces Ouest-Paléarctiques du genre *Trieces* (Hym., Ichneumonidae). *Annales de la Société Entomologique de France* (N.S.) **9**: 975–88.
  In French. Includes key to species.

Aeschlimann, J.P. 1973b. Révision des espèces Ouest-Paléarctiques du genre *Triclistus* Foerster (Hymenoptera: Ichneumonidae). *Mitteilungen der Schweizerischen Entomologischen Gesellschaft* **46**: 219–52.
  In French. Includes key to species.

Aeschlimann, J.P. 1975. Révision des espèces Ouest-Paléarctiques du genre *Chorinaeus* (Hym., Ichneumonidae). *Annales de la Société Entomologique de France* (N.S.) **11**: 723–44.
  In French. Includes key to species.

Aeschlimann, J.P. 1989. Révision des espèces ouest-paléarctiques du genre *Hypsicera* Latreille (Hymenoptera, Ichneumonidae). *Annales de la Société Entomologique de France* (N.S.) **25**: 33–9.
  In French. Includes key to species.

Aubert, J.F. 1978a. *Les Ichneumonides ouest-paléarctiques et leurs hôtes.* 2. (Banchinae et suppl. aux Pimplinae). 318pp. O.P.I.D.A., Echauffeur.
  In French. Includes keys to species of *Glypta*, *Lissonota* and *Apophua*.

Aubert, J.F. 1978b. Révision préliminaire des Ichneumonides Orthocentrinae européenes (Hym. Ichneumonidae). *Eos* 52: 7–28.

In French. Includes key to species of *Orthocentrus*.

Aubert, J.F. 1981. Révision des Ichneumonides *Stenomacrus* sensu lato. *Mitteilungen der Münchener Entomologischen Gesellschaft* 71: 139–59.

In French. Includes key to species.

Barron, J.R. & Walley, G.S. 1983. Revision of the Holarctic genus *Pyracmon* (Hymenoptera: Ichneumonidae). *Canadian Entomologist* 115: 227–41.

Includes key to species.

Beirne, B.P. 1941. British species of Diplazonini (Bassini auctt.) with a study of the genital and postgenital sclerites in the male (Hym.: Ichneum.). *Transactions of the Royal Entomological Society of London* 91: 661–712.

Useful, but cannot be regarded as providing very reliable means of identification. In addition there are many changes in taxonomy to be taken into account (see Fitton & Rotheray, 1982).

Brock, J.P. 1982. A systematic study of the genus *Ophion* in Britain. *Tijdschrift voor Entomologie* 125: 57–97.

Includes key. A more detailed treatment than Gauld (1978) and with some differences in taxonomy.

Dbar, R.S. 1983. Parasitic wasps of the genus *Bathypiesta* Aubert (Hymenoptera, Ichneumonidae). *Entomological Review* 62(4): 103–9.

In English, originally in Russian in *Entomolocheskoe Obozrenie* 62. Includes key to species.

Dbar, R.S. 1984. Review of the Palaearctic species of the genus *Cymodusa* Holmgren (Hymenoptera, Ichneumonidae). I. *Entomological Review* 63(4): 127–38.

In English, originally in Russian in *Entomolocheskoe Obozrenie* 63.

Dbar, R.S. 1985. Review of the Palaearctic species of the genus *Cymodusa* Holmgren (Hymenoptera, Ichneumonidae). II. *Entomological Review* 65(1): 5–17.

In English, originally in Russian in *Entomolocheskoe Obozrenie* 64. See also Dbar (1984). Includes key to species.

Delrio, G. 1975. Revision des espèces ouest-paléarctiques du genre *Netelia* Gray (Hym., Ichneumonidae). *Studi Sassaresi. Annali della Facolta di Agraria dell'Universita di Sassari* 23: 1–126.

In French. Includes key to species.

Diller, E.H. 1969. Beitrag zur Taxonomie der Gattung *Syrphoctonus* Foerster mit Beschreibung einer neuen Holarktischen Art (Hymenoptera, Ichneumonidae). *Acta Entomologica Musei Nationalis Pragae* 38: 545–52.

In German. Includes a key to species. The genus covered is *Woldstedtius* not *Syrphoctonus* (see Fitton & Rotheray (1982) for nomenclature).

Finlayson, T. & Hagen, K.S. 1977. Final-instar larvae of parasitic Hymenoptera. *Pest Management Papers, Simon Fraser University* 10: 1–111.

Keys to families and subfamilies of Ichneumonidae and large bibliography. See also Short (1978).

Fitton, M.G. 1975. A review of the British species of *Tryphon* Fallén (Hym., Ichneumonidae). *Entomologist's Monthly Magazine* 110: 153–71.

Includes a key to species.

Fitton, M.G. 1981. The British Acaenitinae (Hymenoptera: Ichneumonidae). *Entomologist's Gazette* **32**: 185–92.
    Includes a key to species.

Fitton, M.G. 1984. A review of the British Collyriinae, Eucerotinae, Stilbopinae and Neorhacodinae (Hymenoptera: Ichneumonidae). *Entomologist's Gazette* **35**: 185–95.
    Includes keys to species.

Fitton, M.G. 1985a. The ichneumon fly genus *Banchus* (Hymenoptera) in the Old World. *Bulletin of the British Museum (Natural History)* (Entomology) **51**: 1–60.
    Includes a key to species.

Fitton, M.G. 1985b. The British species of *Cidaphus* (Hymenoptera: Ichneumonidae). *Entomologist's Gazette* **36**: 293–7.
    Includes a key to species.

Fitton, M.G. 1987. A review of the *Banchus*-group of ichneumon-flies, with a revision of the Australian genus *Philogalleria* (Hymenoptera: Ichneumonidae). *Systematic Entomology* **12**: 33–45.
    Includes key to genera and note on *Rhynchobanchus* in Britain.

Fitton, M.G. & Boston, M. 1988. The British species of *Phthorima* (Hymenoptera: Ichneumonidae). *Entomologist's Gazette* **39**: 165–70.
    Includes a key to species.

Fitton, M.G. & Ficken, L.C. 1990. British ichneumon-flies of the tribe Oedemopsini (Hymenoptera: Ichneumonidae). *The Entomologist* **109**: 210–14.
    Includes a key to tribes of Tryphoninae in Britain, as well as a key to genera and species of Oedemopsini.

Fitton, M.G. & Gauld, I.D. 1980. A review of the British Cremastinae (Hymenoptera: Ichneumonidae), with keys to the species. *Entomologist's Gazette* **31**: 63–71.
    Includes keys to genera and species.

Fitton, M.G., Gauld, I.D. & Shaw, M.R. 1982. The taxonomy and biology of the British Adelognathinae (Hymenoptera: Ichneumonidae). *Journal of Natural History* **16**: 275–83.
    Includes a key to species. Advances in taxonomy mean that some changes are necessary (see Kasparyan, 1990).

Fitton, M.G. & Rotheray, G.E. 1982. A key to the European genera of diplazontine ichneumon-flies, with notes on the British fauna. *Systematic Entomology* **7**: 311–20.
    Clarifies generic taxonomy.

Fitton, M.G., Shaw, M.R. & Austin, A.D. 1987. The Hymenoptera associated with spiders in Europe. *Zoological Journal of the Linnean Society* **90**: 65–93.
    Includes keys to genera for all groups of Hymenoptera except Pompilidae.

Fitton, M.G., Shaw, M.R. & Gauld, I.D. 1988. Pimpline ichneumon-flies. *Handbooks for the Identification of British Insects* **7**(1): 110pp.
    Includes keys to genera and species. The group has since been divided into four separate subfamilies (Pimplinae, Poemeniinae, Diacritinae and Rhyssinae) and the status of some tribes altered.

Gauld, I.D. 1973. Notes on the British Ophionini (Hym., Ichneumonidae) including a provisional key to species. *Entomologist's Gazette* **24**: 55–65.
    See also Gauld (1974, 1978) and Brock (1982).

Gauld, I.D. 1974. Further notes on the British Ophionini (Hym., Ichneumonidae). *Entomologist's Gazette* 25: 147–8.

A note, and key couplet, on an additional species of *Enicospilus.*

Gauld, I.D. 1978. Notes on the British Ophioninae (Hymenoptera, Ichneumonidae). Part 4. A revised key to species of the genus *Ophion* Fabricius. *Entomologist's Gazette* 29: 145–9.

See also Brock (1982).

Gauld, I.D. 1991. The Ichneumonidae of Costa Rica, 1. *Memoirs of the American Entomological Institute* 47: 1–589.

Although dealing with an exotic fauna this volume includes valuable introductory sections on the family.

Gauld, I.D. & Bolton, B. (Eds) 1988. *The Hymenoptera.* British Museum (Natural History), London, and Oxford University Press, Oxford. [Reprinted 1996, with minor alterations and additions.]

Includes an introductory account of the British ichneumonid fauna, with a paragraph on each subfamily.

Gauld, I.D. & Fitton, M.G. 1980. The British species of Phrudinae (Hym., Ichneumonidae). *Entomologist's Monthly Magazine* 115 (1979): 197–9.

Includes key to genera and species. See also Shaw (1991).

Gauld, I.D. & Fitton, M.G. 1981. Keys to the British xoridine parasitoids of wood-boring beetles (Hymenoptera: Ichneumonidae). *Entomologist's Gazette* 32: 259–67.

Includes keys to genera and species.

Gauld, I.D. & Mitchell, P.A. 1977a. Ichneumonidae, subfamilies Orthopelmatinae and Anomaloninae. *Handbooks for the Identification of British Insects* 7(2b): 32pp.

Includes keys to genera and species.

Gauld, I.D. & Mitchell, P.A. 1977b. Nocturnal Ichneumonidae of the British Isles: the genus *Alexeter* Foerster. *Entomologist's Gazette* 28: 51–5.

Includes a key to the two nocturnal species.

Goulet, H. & Huber, J.T. (Eds) 1993. *The Hymenoptera of the world: an identification guide to families.* Agriculture Canada, Research Branch, Ottawa.

Includes a practical, workable key to ichneumonid subfamilies and has useful subfamily diagnoses.

Hanson, P.E. & Gauld, I.D. (Eds) 1995. *The Hymenoptera of Costa Rica.* Oxford University Press, Oxford.

The section on ichneumonids is a good introduction to the family, although concerned primarily with the fauna of Costa Rica. The key to subfamilies is of limited use for the British fauna.

Hilpert, H. 1992. Zur Systematik der Gattung *Ichneumon* Linnaeus, 1758 in der Westpalaearktis (Hymenoptera, Ichneumonidae, Ichneumoninae). *Entomofauna* (suppl.) 6: 1–389.

In German. Includes keys for males and females.

Hinz, R. 1963. Zur Systematik und Ökologie der Ichneumoniden III. *Deutsche Entomologische Zeitschrift* (N.F.) 10: 116–21.

In German; includes a key to a group of species of *Dusona.*

Hinz, R. 1975. Die Arten der Gattung *Glyptorhaestus* Thomson (Hymenoptera, Ichneumonidae) *Zeitschrift der Arbeitsgemeinschaft Österreichischer Entomologen* **27**: 39–46.
    In German. Includes a key to species [Ctenopelmatinae].

Hinz, R. 1976. Zur Systematik und Ökologie der Ichneumoniden V. *Deutsche Entomologische Zeitschrift* (N.F.) **23**: 99–105.
    In German. Includes a key to species of *Lethades* [Ctenopelmatinae].

Hinz, R. 1977. Über einige Arten der Gattung *Dusona* Cameron (Hymenoptera, Ichneumonidae). *Nachrichtenblatt der Bayerischen Entomologen* **26**: 47–54.
    In German; includes a key to two groups of species of *Dusona*.

Hinz, R. 1980. Die europaischen Arten der Gattung *Trematopygodes* Aubert (Hymenoptera, Ichneumonidae). *Nachrichtenblatt der Bayerischen Entomologen* **29**: 89–93.
    In German. Includes a key to species [Ctenopelmatinae].

Hinz, R. 1985. Die paläarktischen Arten der Gattung *Trematopygus* Holmgren (Hymenoptera, Ichneumonidae). *Spixiana* **8**: 265–76.
    In German. Includes a key to species [Ctenopelmatinae].

Hinz, R. 1991. Die paläarktischen Arten der Gattung *Sympherta* Förster (Hymenoptera, Ichneumonidae). *Spixiana* **14**: 27–43.
    In German. Includes a key to species [Ctenopelmatinae].

Horstmann, K. 1967. Untersuchungen zur Systematik einiger *Phygadeuon*-Arten aus der Verwandtschaft des *P. vexator* Thunberg und des *P. fumator* Gravenhorst (Hymenoptera, Ichneumonidae). *Opuscula Zoologica* **98**: 1–22.
    In German. Includes keys to the species of the two species-groups [Cryptinae].

Horstmann, K. 1968. Revision einiger Arten der Gattungen *Mesostenus* Gravenhorst, *Agrothereutes* Foerster und *Ischnus* Gravenhorst [Hymenoptera, Ichneumonidae]. *Entomophaga* **13**: 121–33.
    In German. Includes key to four species of *Agrothereutes* [Cryptinae].

Horstmann, K. 1969. Typenrevision der europäischen Arten der Gattung *Diadegma* Foerster (syn. *Angitia* Holmgren). *Beiträge zur Entomologie* **19**: 413–72.
    In German. Includes a key to species [Campopleginae].

Horstmann, K. 1971a. Revision der europäischen Tersilochinen 1. (Hymenoptera: Ichneumonidae). *Veröffentlichungen der Zoologischen Staatssammlung München* **15**: 45–138.
    In German. See also Horstmann, 1981. The two papers together cover the entire European fauna of this subfamily, including keys to genera and species.

Horstmann, K. 1971b. Revision der europäischen Arten der Gattung *Lathrostizus* Foerster (Hymenoptera, Ichneumonidae). *Mitteilungen der Deutschen Entomologischen Gesellschaft* **30**: 8–12, 16–18.
    In German. Includes key to species [Campopleginae].

Horstmann, K. 1973a. Revision der Gattung *Nepiesta* Foerster (mit einer Übersicht über die Arten der Gattung *Leptoperilissus* Schmiedeknecht) (Hymenoptera, Ichneumonidae). *Polskie Pismo Entomologiczne* **43**: 729–41.
    In German. Includes a key to species [Campopleginae].

Horstmann, K. 1973b. Revision der westpaläarktischen Arten der Gattung *Nemeritis* Holmgren (Hymenoptera, Ichneumonidae). *Opuscula Zoologica* 125: 1–14.
In German. See also Horstmann (1975, 1994b) [Campopleginae].

Horstmann, K. 1973c. Revision der europäischen Arten der Gattung *Dichrogaster* Doumerc (Hym. Ichneumonidae). *Entomologica Scandinavica* 4: 65–72.
In German. See also Horstmann (1976) [Cryptinae].

Horstmann, K. 1974a. Revision der westpaläarktischen Arten der Schlupfwespen-Gattungen *Bathyplectes* und *Biolysia* (Hymenoptera: Ichneumonidae). *Entomologica Germanica* 1: 58–81.
In German. Includes keys to species. See also Dbar (1983) [Campopleginae].

Horstmann, K. 1974b. Typenrevision der von E. Zilahi-Kiss beschriebenen Hemitelinen mit Bemerkungen zu den Gattungen *Hemiteles* Grav. (s.str.), *Gnotus* Foerst. und *Xiphulcus* Townes (Hymenoptera, Ichneumonidae). *Annales Historico-Naturales Musei Nationalis Hungarici* 66: 339–46.
In German. Includes keys to species of the three genera [Cryptinae].

Horstmann, K. 1975. Neuearbeitung der Gattung *Nemeritis* Holmgren (Hymenoptera, Ichneumonidae). *Polskie Pismo Entomologiczne* 45: 251–65.
In German. Includes key to European species. See also Horstmann (1973b, 1994b) [Campopleginae].

Horstmann, K. 1976. Nachtrag zur Revision der europaischen *Dichrogaster*-Arten (Hymenoptera, Ichneumonidae). *Zeitschrift der Arbeitsgemeinschaft Österreichischer Entomologen* 28: 55–61.
In German. Includes key. See also Horstmann (1973c) [Cryptinae].

Horstmann, K. 1978. Revision der Gattungen der Mastrina Townes (Hymenoptera, Ichneumonidae, Hemitelinae). *Zeitschrift der Arbeitsgemeinschaft Österreichischer Entomologen* 30: 65–70.
In German. Includes key to genera of this subtribe [Cryptinae].

Horstmann, K. 1979a. Revision der europäischen Arten der Gattung *Ceratophygadeuon* Viereck (Hymenoptera, Ichneumonidae). *Zeitschrift der Arbeitsgemeinschaft Österreichischer Entomologen* 31: 41–8.
In German. Includes key to species [Cryptinae].

Horstmann, K. 1979b. Revision der von Kokujev beschriebenen Campopleginae-Arten (mit Teiltabellen der Gattungen *Venturia* Schrottky, *Campoletis* Förster und *Diadegma* Förster) (Hymenoptera: Ichneumonidae). *Beiträge zur Entomologie* 29: 195–9.
In German. Includes keys to species of *Campoletis*, *Venturia deficiens* species-group and *Diadegma trochanteratum* species-group [Campopleginae].

Horstmann, K. 1980a. Revision der europäischen Arten der Gattung *Rhimphoctona* Förster (Hymenoptera, Ichneumonidae). *Nachrichtenblatt der Bayerischen Entomologen* 29: 17–24.
In German. Includes a key to species [Campopleginae].

Horstmann, K. 1980b. Revision der europäischen Arten der Gattung *Aclastus* Foerster (Hymenoptera, Ichneumonidae). *Polskie Pismo Entomologiczne* 50: 133–58.
In German. Includes a key to species [Cryptinae].

Horstmann, K. 1981. Revision der europäischen Tersilochinae 2. (Hymenoptera: Ichneumonidae) *Spixiana* Supplement **4**: 1–76.
See note under Horstmann (1971a).

Horstmann, K. 1983. Die westpaläarktischen Arten der Gattung *Chirotica* Foerster, 1869 (Hymenoptera, Ichneumonidae). *Entomofauna* **4**: 1–33.
In German. Includes a key to species [Cryptinae].

Horstmann, K. 1984. Revision der paläarktischen Arten der Gattung *Hidryta* Foerster (Hymenoptera, Ichneumonidae). *Zeitschrift der Arbeitsgemeinschaft Österreichischer Entomologen* **35**: 113–17.
In German. Includes a key to species [Cryptinae].

Horstmann, K. 1985. Revision der mit *difformis* (Gmelin, 1790) verwandten westpaläarktischen Arten der Gattung *Campoplex* Gravenhorst, 1829 (Hymenoptera, Ichneumonidae). *Entomofauna* **6**: 129–63.
In German. Includes a key to species of this species-group [Campopleginae].

Horstmann, K. 1986. Die westpaläarktischen Arten der Gattung *Gelis* Thunberg, 1827, mit macropteren oder brachypteren Weibchen (Hymenoptera, Ichneumonidae). *Entomofauna* **7**: 389–424.
In German. Includes keys to this subset of species. See also Schwarz (1994) [Cryptinae].

Horstmann, K. 1987. Die europäischen Arten der Gattungen *Echthronomas* Förster und *Eriborus* Förster (Hymenoptera, Ichneumonidae). *Nachrichtenblatt der Bayerischen Entomologen* **36**: 57–67.
In German. Includes keys to species [Campopleginae].

Horstmann, K. 1990. Die westpaläarktischen Arten einiger Gattungen der Cryptini (Hymenoptera, Ichneumonidae). *Mitteilungen der Münchener Entomologischen Gesellschaft* **79**: 65–89.
In German. Includes keys to species of *Listrognathus* and *Pycnocryptus* [Cryptinae].

Horstmann, K. 1992. Revision einiger Gattungen der Phygadeuontini (Hymenoptera, Ichneumonidae). *Mitteilungen der Münchener Entomologischen Gesellschaft* **81** (1991): 229–54.
In German. Revised taxonomy of a number of genera. Includes keys to species of *Cremnodes* and *Sulcarius* [Cryptinae].

Horstmann, K. 1994a. Die europäischen Arten von *Picrostigeus* Förster (Hymenoptera, Ichneumonidae, Orthocentrinae). *Zeitschrift der Arbeitsgemeinschaft Österreichischer Entomologen* **46**: 111–20.
In German. Includes a key to species.

Horstmann, K. 1994b. Nachtrag zur Revision der westpaläarktischen *Nemeritis*-Arten (Hymenoptera, Ichneumonidae, Campopleginae). *Mitteilungen der Münchener Entomologischen Gesellschaft* **84**: 79–90.
In German. Includes key to species. See also Horstmann (1973b, 1975) [Campopleginae].

Horstmann, K. 1995. Die europäischen Arten von *Arotrephes* Townes, 1970 und *Pleurogyrus* Townes, 1970 (Hymenoptera, Ichneumonidae, Cryptinae). *Entomofauna* **16**: 261–76.
In German; includes keys to species.

Horstmann, K. & Shaw, M.R. 1984. The taxonomy and biology of *Diadegma chrysostictos* (Gmelin) and *D. fabricianae* sp. n. (Hymenoptera: Ichneumonidae). *Systematic Entomology* 9: 329–37.
  Separation of two cryptic species.

Huddleston, T. & Gauld, I.D. 1988. Parasitic wasps (Ichneumonoidea) in British light-traps. *The Entomologist* 107: 134–54.
  Key to genera of nocturnal ichneumonids and braconids, with notes on further identification.

Idar, M. 1979. Revision of the European species of the genus *Hadrodactylus* Foerster (Hymenoptera: Ichneumonidae). Part 1. *Entomologica Scandinavica* 10: 303–13.
  Includes key. See also Idar (1981). Idar's untimely death in a road accident put an end to his work on ctenopelmatines.

Idar, M. 1981. Revision of the European species of the genus *Hadrodactylus* Foerster (Hymenoptera: Ichneumonidae). Part 2. *Entomologica Scandinavica* 12: 231–9.
  See note on Idar (1979).

Jussila, R. 1979. A revision of the genus *Atractodes* (Hymenoptera, Ichneumonidae) of the western Palearctic region. *Acta Entomologica Fennica* 34: 1–44.
  Includes a key to species.

Jussila, R. 1987. Revision of the genus *Stilpnus* (Hymenoptera, Ichneumonidae) of the western Palaearctic region. *Annales Entomologici Fennici* 53: 1–16.
  Includes a key to species.

Kasparyan, D.R. 1973. Ichneumonidae (subfamily Tryphoninae) tribe Tryphonini. *Fauna SSSR* (New Series) 106 (Hymenoptera 3(1): 1–320).
  In Russian. English translation published in 1981 by Amerind Publishing Co, New Delhi.

Kasparyan, D.R. (Ed.) 1981. Hymenoptera, Ichneumonidae. *Keys to insects of the European parts of the USSR* 3(3): 1–688.
  In Russian, but a very valuable compendium. Includes some original work but is derived mainly from studies published elsewhere.

Kasparyan, D.R. 1990. Ichneumonidae: subfamily Tryphoninae, tribe Exenterini, subfamily Adelognathinae. *Fauna SSSR* (New Series) 141 (Hymenoptera 3(2): 1–340).
  In Russian. Revisionary treatment, including keys to species.

Kasparyan, D.R. 1994. Review of Palearctic species of wasps of the genus Phytodietus Grav. (Hymenoptera, Ichneumonidae). *Entomological Review* 73(7): 56–79.
  In English, originally in Russian in *Entomolocheskoe Obozrenie* 72 (1993). Includes a key to species.

Kerrich, G.J. 1952. A review and a revision in greater part, of the Cteniscini of the Old World (Hym., Ichneumonidae). *Bulletin of the British Museum (Natural History)* (Entomology) 2: 305–460.
  Covers part of the British fauna of this tribe (= Exenterini) of tryphonines.

Lawton, F.D. 1981. An introduction to the Mesochorinae (Hymenoptera, Ichneumonidae). *Proceedings and Transactions of the British Entomological and Natural History Society* 14: 93–7.
  Includes a key to genera. See also Wahl (1993).

Morley, C. 1903–15. *Ichneumonologia Britannica* 1–5. Plymouth & London.
  Morley's five volumes (1903–15) were the standard work on the British fauna. They have many shortcomings in addition to being long out-of-date.

Perkins, J.F. 1936. Notes on British Metopiini (Hym., Ichneumonidae). *Entomologist's Monthly Magazine* 72: 83–6.
Includes key, but covers only the genus *Metopius*. Nomenclature needs updating.

Perkins, J.F. 1959. Ichneumonidae, key to subfamilies and Ichneumoninae-I. *Handbooks for the Identification of British Insects* 7(2ai): 116pp.
Key to subfamilies now superseded. Keys to ichneumonines relatively reliable, but classification and nomenclature need updating.

Perkins, J.F. 1960. Ichneumonidae, subfamilies Ichneumoninae II, Alomyinae, Agriotypinae and Lycorininae. *Handbooks for the Identification of British Insects* 7(2aii): 117–213.
See note on Perkins (1959).

Rossem, G. van 1966. A study of the genus *Trychosis* Foerster in Europe (Hymenoptera, Ichneumonidae, Cryptinae). *Zoologische Verhandelingen* 79: 1–40.
Includes key to species.

Rossem, G. van 1969. A revision of the genus *Cryptus* Fabricius s.str. in the western Palearctic region, with keys to genera of Cryptina and species of *Cryptus* (Hymenoptera, Ichneumonidae). *Tijdschrift voor Entomologie* 112: 299–374.
See also later papers by van Rossem [Cryptinae].

Rossem, G. van 1971. The genus *Buathra* Cameron in Europe (Hymenoptera, Ichneumonidae). *Tijdschrift voor Entomologie* 114: 201–7.
Includes key to species [Cryptinae].

Rossem, G. van 1981. A revision of some western Palearctic Oxytorine genera (Hymenoptera, Ichneumonidae). *Spixiana* (suppl.) 4 (1980): 79–135.
Includes keys. Covers the genera *Microleptes, Hemiphanes, Oxytorus, Hyperacmus, Entypoma, Allomacrus, Apoclima, Cylloceria, Aniseres, Pantisarthrus, Dialipsis, Laepserus, Ephalmator, Symplecis, Catastenus* and *Phosphorus.* The Oxytorinae as recognised by van Rossem covers the subfamilies Oxytorinae, Microleptinae, Cylloceriinae and Helictinae. See also later papers by van Rossem.

Rossem, G. van 1982. A revision of the western Palearctic oxytorine genera. Part II. Genus *Eusterinx* (Hymenoptera, Ichneumonidae). *Spixiana* 5: 149–70.
Includes keys. Also covers the genus *Proeliator* and includes a key to genera (but see van Rossem (1990) for a revised key to genera).

Rossem, G. van 1983. A revision of the western Palearctic oxytorine genera. Part III. Genus *Proclitus* (Hymenoptera, Ichneumonidae). *Contributions of the American Entomological Institute* 20: 153–65.
Includes a key to species.

Rossem, G. van 1983. A revision of the western Palearctic oxytorine genera. Part IV. Genus *Megastylus* (Hymenoptera, Ichneumonidae). *Entomofauna* 4: 121–32.
Includes a key to species.

Rossem, G. van 1985. A revision of the western Palaearctic oxytorine genera Part V Genus *Aperileptus* (Hymenoptera, Ichneumonidae). *Spixiana* 8: 145–52.
Includes a key to species.

Rossem, G. van 1987. A revision of western Palaearctic oxytorine genera. Part VI. (Hymenoptera, Ichneumonidae). *Tijdschrift voor Entomologie* 130: 49–108.
Includes keys. Covers the genera *Hemiphanes, Oxytorus, Apoclima, Cylloceria, Proclitus,*

Pantisarthrus, Plectiscidea, Gnathocorosis, Eusterinx, Helictes, Phosphoriana, Proeliator and *Megastylus.*

Rossem, G. van 1988. A revision of Palaearctic oxytorine genera. Part VII. (Hymenoptera, Ichneumonidae). *Tijdschrift voor Entomologie* **131**: 103–12.
Includes notes on a number of genera, including keys or additional couplets for *Entypoma, Plectiscidea, Symplecis, Eusterinx* and *Microleptes.*

Rossem, G. van 1990. Key to the genera of the Palaearctic Oxytorinae, with the description of three new genera (Hymenoptera: Ichneumonidae). *Zoologische Mededelingen* **63**: 309–23.
The Oxytorinae as recognised by van Rossem covers the subfamilies Oxytorinae, Microleptinae, Cylloceriinae and Helictinae. Also includes a revised key to species of *Eusterinx* subgenus *Eusterinx.*

Rotheray, G.E. 1990. A new species of *Bioblapsis* (Hymenoptera: Ichneumonidae) from Scotland parasitising a mycophagous hoverfly, *Cheilosia longula* (Diptera: Syrphidae). *Entomologica Scandinavica* **21**: 277–80.
Includes a key to the species.

Sanborne, M. 1984. A revision of the world species of *Sinophorus* (Ichneumonidae). *Memoirs of the American Entomological Institute* **38**: 1–403.
Includes key to species.

Sawoniewicz, J. 1980. Revision of European species of the genus *Bathythrix* Foerster (Hymenoptera, Ichneumonidae). *Annales Zoologici* **35**: 319–65.
Includes a key to species.

Sawoniewicz, J. 1985. Revision of European species of the subtribe Endaseina (Hymenoptera, Ichneumonidae). *Annales Zoologici* **39**: 131–46.
Includes key to genera of the subtribe and a key to species of *Glyphicnemis.*

Schwarz, M. 1988. Die europäischen Arten der Gattung *Idiolispa* Foerster (Ichneumonidae, Hymenoptera). *Linzer Biologische Beiträge* **20**: 37–66.
In German. Includes a key to species.

Schwarz, M. 1989a. Ergebnisse von Typenuntersuchungen bei Schlupfwespen (Hymenoptera, Ichneumonidae, Cryptinae). *Entomofauna* **10**: 293–304.
In German. Includes key to European *Apsilops.*

Schwarz, M. 1989b. Revision der Gattung *Enclisis* Townes (Ichneumonidae, Hymenoptera). *Linzer Biologische Beiträge* **21**: 497–522.
In German. Includes a key to species.

Schwarz, M. 1991. Eine neue Art der Gattung *Microleptes* Gravenhorst (Ichneumonidae, Hymenoptera) aus Österreich. *Linzer Biologische Beiträge* **23**: 399–405.
In German. Includes key.

Schwarz, M. 1994. Beitrag zur Systematik und Taxonomie europäischer *Gelis*-Arten mit macropteren oder brachypteren Weibchen (Hymenoptera, Ichneumonidae). *Linzer Biologische Beiträge* **26**: 381–91.
In German. Includes a key to species for brachypterous females. See also Horstmann (1986).

Schwarz, M. 1995. Revision der westpaläarktischen Arten der Gattungen *Gelis* Thunberg mit apteren Weibchen und *Thaumatogelis* Schmiedeknecht (Hymenoptera, Ichneumonidae). Teil 1. *Linzer Biologische Beiträge* **27**: 5–105.

In German. Includes, for the genera of Cryptinae with apterous females, a key to genera and keys to species groups for the *Gelis* and *Thaumatogelis*.

Selfa, J. & Diller, E. 1994. Illustrated key to the western Palaearctic genera of Phaeogenini (Hymenoptera, Ichneumonidae, Ichneumoninae). *Entomofauna* 15: 237–52.

An alternative key to Perkins (1959) for this tribe.

Shaw, M.R. 1986. *Coleocentrus excitator* (Poda) (Hymenoptera: Ichneumonidae) new to Britain. *Entomologist's Gazette* 37: 221–4.

Includes key couplet to add to Fitton (1981).

Shaw, M.R. 1991. *Phrudus badensis* Hilpert (Hym., Ichneumonidae) new to Britain. *Entomologist's Monthly Magazine* 127: 157–8.

Includes key couplet to add to Gauld & Fitton (1980).

Shaw, M.R. & Horstmann, K. 1997. An analysis of host range in the *Diadegma nanus* group of parasitoids in western Europe, with a key to species (Hymenoptera: Ichneumonidae: Campopleginae). *Journal of Hymenoptera Research* 6: 273–96.

Includes definition of the *D. nanus* group (and subgroups), as well as the key to females of the 32 species.

Short, J.R.T. 1978. The final larval instars of the Ichneumonidae. *Memoirs of the American Entomological Institute* 25: 1–508.

Compendium. Includes keys to subfamilies and genera, but see Wahl (1986 onwards). An alternative key to subfamilies is given by Finlayson & Hagen (1977).

Townes, H. 1969. The genera of Ichneumonidae 1. *Memoirs of the American Entomological Institute* 11: 1–300.

Standard work, although now getting out-of-date. Covers world fauna. Volumes 1–4 (1969–71) cover all subfamilies except Ichneumoninae.

Townes, H. 1970a. The genera of Ichneumonidae 2. *Memoirs of the American Entomological Institute* 12: 1–537.

See note on Townes (1969).

Townes, H. 1970b. The genera of Ichneumonidae 3. *Memoirs of the American Entomological Institute* 13: 1–307.

See note on Townes (1969).

Townes, H. 1971. The genera of Ichneumonidae 4. *Memoirs of the American Entomological Institute* 17: 1–372.

See note on Townes (1969).

Townes, H. 1983. Revisions of twenty genera of Gelini (Ichneumonidae). *Memoirs of the American Entomological Institute* 35: 1–281.

Includes a key to Nearctic genera of this tribe which works tolerably well for the European fauna, as well as keys to species of the genera treated (*Agasthenes, Charitopes, Dichrogaster, Ethelurgus, Helcostizus, Lochetica, Lysibia, Megacara, Pygocryptus, Sulcarius* and *Xenolytus* in Britain).

Wahl, D.B. 1986. Larval structures of oxytorines and their significance for the higher classification of some Ichneumonidae (Hymenoptera). *Systematic Entomology* 11: 117–27.

No keys, but essential in conjunction with Short (1978).

Wahl, D.B. 1990. A review of the mature larvae of Diplazontinae, with notes on larvae of Acaenitinae and Orthocentrinae and proposal of two new subfamilies (Insecta: Hymenoptera, Ichneumonidae). *Journal of Natural History* 24: 27–52.
Supersedes Short (1978) for included subfamilies, but no keys are given.

Wahl, D.B. 1993. Cladistics of the genera of Mesochorinae (Hymenoptera: Ichneumonidae). *Systematic Entomology* 18: 371–87.
Includes a revised classification and key to genera.

Watanabe, C. 1984. Notes on Paxylommatinae with review of Japanese species (Hymenoptera, Braconidae). *Kontyû* 52: 553–6.
Includes a key covering the species found in Britain.

Yu, D.S. 1993. *TAXA. A biosystematic data management system. Insecta Hymenoptera Ichneumonidae, 1900–1990.* [A database package for PC-compatible computers.] Yu, Lethbridge, Canada.
Has a number of serious drawbacks and pitfalls for those who do not have a good knowledge of the complexities of ichneumonid taxonomy.

Yu, D.S. & Horstmann, K. 1997. Catalogue of World Ichneumonidae (Hymenoptera). *Memoirs of the American Entomological Institute* 58: 1–1558.
A complete synonymic catalogue and citation index. An absolutely essential reference work.

Family Braconidae

The Braconidae is one of the largest families of insects but much of its diversity is centred on the tropics and the British fauna is not very rich. Currently some 40 subfamilies are recognised and the number may increase still more as some subfamilies are recognised as being polyphyletic. To date this has especially affected those taxa formerly included in the Rogadinae (or sometimes the Doryctinae), which was a heterogeneous assemblage that is now treated as comprising several distinct subfamilies including the Rogadinae *sensu stricto*, Rhysipolinae, Rhyssalinae, Pambolinae, Lysiterminae, Bethylobraconinae, Hormiinae and Exothecinae. In some works the Meteorinae and Euphorinae are treated as synonymous. The Histeromerinae have only recently been shown to be unrelated to the Doryctinae in which they were previously included; the Microgastrinae was often taken as also including Miracinae and Adeliinae, the Blacinae as a tribe of Helconinae, and the Opiinae as including the Gnamptodontinae. As various key works of importance for the identification of the British fauna still use the old system with a heterogeneous Rogadinae, these subfamilies are treated together here. Most of the 40 or so subfamilies that are currently recognised occur in Britain, but absent are the Amicrocentrinae, Apozyginae, Betylobraconinae, Cardiochilinae, Lysiterminae, Masoninae, Meteorideinae, Pambolinae, Pselaphaninae, Trachypetinae and Xiphozelinae, most of which are known from only a few tropical species.

The way to key braconids involves first keying them to subfamily. Therefore the key works listed below are arranged with subfamily keys given first and then keys to each particular subfamily. Identification of most taxa will require a good stereo microscope. Keys that work (i.e. are meaningful) to species are not available for all genera, and particularly problematic at present is the large genus *Bracon* in the Braconinae.

Many books and papers include keys to enable separation of braconids from ichneumonids. Some recent and more thorough ones are Gauld & Bolton (1988) and Goulet & Huber (1993) (see Introduction to the Ichneumonoidea, above).

References

Keys direct to genera

Huddleston, T. & Gauld, I. 1988. Parasitic wasps (Ichneumonoidea) in British light traps. *The Entomologist* **107**: 134–54.

Deals only with nocturnal taxa that regularly turn up at light traps (six genera in total, each now in its own subfamily).

Marsh, P.M., Shaw, S.R. & Wharton, R.A. 1987. An identification manual for the North American genera of the family Braconidae. *Memoirs of the Entomological Society of Washington* **13**: 1–98.

Based on the North American fauna but this will key out many British genera successfully. It is well illustrated with line drawings of venation and SEMs of other characters. However, it must be realised that although there is considerable overlap between British genera and those occurring in the New World North of Mexico, the latter region contains many taxa not present here and Britain has a few not found there.

Tobias, V.I. 1971. A review of the Braconidae (Hymenoptera) of the USSR. *Trudy Vsesoyuznogo Entomologicheskogo Obshchestva* **54**: 156–268.

In Russian: English translation by Amerind, 1975, New Delhi.

Keys to subfamilies

Achterberg, C. van 1976. A preliminary key to the subfamilies of the Braconidae (Hymenoptera). *Tijdschrift voor Entomologie* **119**: 33–78.

Achterberg, C. van 1990. Illustrated key to the subfamilies of the Holarctic Braconidae (Hymenoptera: Ichneumonoidea). *Zoologische Mededelingen* **64**: 1–20.

Achterberg, C. van 1993. Illustrated key to the subfamilies of the Braconidae (Hymenoptera: Ichneumonoidea). *Zoologische Verhandelingen* **283**: 1–189.

Čapek, M. 1973. Key to the final-instar larvae of the Braconidae. *Acta Instituti Forestalis Zvolenensis* **4**: 259–68.

Most subfamilies have been studied, but for some only a few species are known, so it is not certain how well Čapek's larval key will work in general. It is worth a try when only larvae are known.

Gauld, I.D. & Bolton, B. (Eds) 1988. *The Hymenoptera.* British Museum (Natural History), London, and Oxford University Press, Oxford. [Reprinted 1996, with minor alterations and additions.]

Sharkey, M. J. 1993. Family Braconidae. In H. Goulet & J. Huber (Eds) *Hymenoptera of the world: an identification guide to families.* Agriculture Canada, Research Branch, Ottawa.

Shaw, M.R. & Huddleston, T. 1991. Classification and biology of braconid wasps (Hymenoptera: Braconidae). *Handbooks for the Identification of British Insects* 7(11): 126pp.

Subfamily Adeliinae

Very small subfamily of nepticulid parasitoids; in the past usually included in Microgastrinae. No recent key to the four British species exists.

Subfamily Agathidinae

A medium-sized subfamily in Britain.

Nixon, G.E.J. 1986. A revision of the European Agathidinae (Hymenoptera: Braconidae). *Bulletin of the British Museum (Natural History)* (Entomology) **52**: 183–243.

A very good species-level work, but the species placed by Nixon in *Microdus* are now placed in *Bassus* (see Sharkey, 1985).

Sharkey, M.J. 1985. Notes on the genera *Bassus* Fabricius and *Agathis* Latreille, with a description of *Bassus arthurellus* n. sp. (Hymenoptera: Braconidae). *Canadian Entomologist* **117**: 1497–1502.

Provides better character for separation of *Agathis* from *Bassus* (=*Microdus*) and clarifies seniority of *Bassus* over *Microdus*.

Sharkey, M.J. 1992. Cladistics and tribal classification of the Agathidinae (Hymenoptera: Braconidae). *Journal of Natural History* **26**: 425–447.

Contains easy key to tribes (on a worldwide basis), tribal diagnoses and checklists of genera included within each tribe.

Simbolotti, G. & Achterberg, C. van 1992. Revision of the West Palaearctic species of the genus *Bassus* Fabricius (Hymenoptera: Braconidae). *Zoologische Verhandelingen* **281**: 1–80.

Subfamily Alysiinae

This subfamily comprises two tribes, Alysiinae and Dacnusinae, and the literature concerning these is largely separate. In Britain the tribes can be separated easily in winged species by the absence of vein r-m in the forewing of dacnusines. The large dacnusine genera *Dacnusa* and *Chorebus* are mainly parasites of Agromyzidae and are dealt with by Griffiths' many works. Species-level identification in these genera is often difficult.

Achterberg, C. van 1975. A new species of *Chasmodon* Haliday (Hymenoptera, Braconidae, Alysiinae). *Zoologische Mededelingen* **49**: 73–9.

Provides a key to the genera of wingless and brachypterous Alysiinae, although some of these genera have subsequently been sunk into synonymy, and provides a key to separate *C. apterus*, a British species, from a closely related one found in The Netherlands.

Achterberg, C. van 1976. A new species of *Tanycarpa* Foerster from England (Hymenoptera, Braconidae, Alysiinae). *Entomologische Berichten, Amsterdam* **36**: 12–15.

Key to species.

Achterberg, C. van 1976. Revisionary notes on the genus *Coloneura* Foerster with the description of a new subgenus, *Coloneurella*, from the Netherlands (Hymenoptera, Braconidae, Alysiinae). *Entomologische Berichten, Amsterdam* **36**: 186–92.

Keys to subgenera and species of this dacnusine genus.

Achterberg, C. van 1983. Revisionary notes on the genera *Dapsilarthra* auct. and *Mesocrina* Foerster (Hymenoptera, Braconidae, Alysiinae). *Tijdschrift voor Entomologie* **126**: 1–24.

Contains key to species currently and formerly included in *Dapsilarthra* and a new combination affecting the British list.

Achterberg, C. van 1986. The Holarctic genus *Anisocyrta* Foerster (Hymenoptera: Braconidae: Alysiinae). *Zoologische Mededelingen* **60**: 285–97.

Separate keys to Palaearctic and Nearctic species of this genus of Alysiini. Thus far only a single species has been recorded from Britain.

Achterberg, C. van 1988. The genera of the *Aspilota*-group and some descriptions of fungicolous Alysiini from the Netherlands (Hymenoptera: Braconidae: Alysiinae). *Zoologische Verhandelingen* **247**: 1–88.

Diagnoses the *Aspilota* group of Alysiini and provides a key to the genera, keys to the European species of the *globipes* group of *Aspilota*, the species of *Carinthilota* (not yet known from Britain), the fungicolous *Dinotrema* in the Netherlands, the *Dinotrema* with minute occipital tubercles, the Palaearctic species of *Eudinostigma*, the species of *Panerema*, *Pterusa*, and Dutch species of fungicolous Alysiini. The paper includes many new combinations and descriptions of new species affecting the British fauna. Probably, many more species of these genera await discovery in the UK.

Achterberg, C. van 1995. New combinations of names for Palaearctic Braconidae (Hymenoptera). *Zoologische Mededelingen* **69**: 131–8.

New generic and specific synonymies affecting the British list.

Fischer, M. 1962. Das Genus *Synaldis* Förster (Hymenoptera, Braconidae). *Mitteilungen aus dem Zoologischen Museum in Berlin* **38**: 1–21.

Fischer, M. 1967. Revision der burgenländischen Arten der Gattungen *Synaldis*, *Aphaereta* und *Alysia* (Hymenoptera, Braconidae, Alysiinae). *Wissenschaftliche Arbeiten aus dem Burgenland* **38**: 92–135.

Fischer, M. 1970. Zur Kenntnis der europäischen *Phaenocarpa*-Arten mit besonderer Berücksichtigung der Fauna Niederösterreichs (Hymenoptera, Braconidae, Alysiinae). *Zeitschrift für Angewandte Zoologie* **57**: 409–98.

Fischer, M. 1990. Westpaläarktische *Phaenocarpa*-Arten: Vorläufiger Bestimmungsschlüssel, Deskriptionen und Redeskriptionen (Hymenoptera, Braconidae, Alysiinae). *Annalen Naturhistorischen Museums Wien* 91(B): 105–35.

Fischer, M. 1993. Zur Formenvielfalt der Kieferwespen der Alten Welt: Uber die Gattungen Synaldis Foerster, Trisynaldis Fischer und Kritscherysia Fischer gen. nov. (Hymenoptera, Braconidae, Alysiinae). *Annalen Naturhistorischen Museums Wien* 94/95(B): 451–90.

Both species of *Synaldis* recorded from Britain are included in this key to the Old World species; in German. Most species are known only from Austria, reflecting where Max Fischer works. It is likely that other species of *Synaldis* will be found in Britain and the earlier records will require checking.

Fischer, M. 1993. Einige *Phaenocarpa*-Wespen aus der Alten Welt: Redeskriptionen und Stellung in einem vergleichenden System (Hymenoptera, Braconidae, Alysiinae). *Linzer Biologische Beitrage* 25: 511–63.

Includes key to some species groups and keys to their species, which include a number of the British taxa along with many extralimital ones. Potentially useful for helping to confirm identifications. See also Fischer (1970, 1990).

Godfray, H.C.J. 1984. Intraspecific variation in the leaf-miner parasite *Exotela cyclogaster* Förster (Hymenoptera: Braconidae). *Proceedings and Transactions of the British Entomological and Natural History Society* 17: 47–50.

Concludes that two subspecies of *E. cyclogaster* are simply host-induced variants.

Griffiths, G.C.D. 1964. The Alysiinae (Hym. Braconidae) parasites of the Agromyzidae (Diptera). I. General questions of taxonomy, biology and evolution. *Beiträge zur Entomologie* 14: 823–914.

Contains reasonably straightforward key to genera of Dacnusini and keys to species of *Protodacnusa* and 'some plesiomorph' species of *Chorebus*. Also includes useful discussion of characters.

Griffiths, G.C.D. 1966a. The Alysiinae (Hym. Braconidae) parasites of the Agromyzidae (Diptera). II. The parasites of *Agromyza* Fallén. *Beiträge zur Entomologie* 16: 551–605.

Griffiths, G.C.D. 1966b. The Alysiinae (Hym. Braconidae) parasites of the Agromyzidae (Diptera). III. The parasites of *Paraphytomyza* Enderlein, *Phytagromyza* Hendel and *Phytomyza* Fallén. *Beiträge zur Entomologie* 16: 775–951.

Griffiths, G.C.D. 1967. The Alysiinae (Hym. Braconidae) parasites of the Agromyzidae (Diptera). IV. The parasites of *Hexomyza* Enderlein, *Melanagromyza* Hendel, *Ophiomyia* Braschnikov and *Napomyza* Westwood. *Beiträge zur Entomologie* 17: 653–96.

Griffiths, G.C.D. 1968a. The Alysiinae (Hym. Braconidae) parasites of the Agromyzidae (Diptera). V. The parasites of *Liriomyza* Mik and certain small genera of Phytomyzinae. *Beiträge zur Entomologie* 18: 5–62.

Griffiths, G.C.D. 1968b. The Alysiinae (Hym. Braconidae) parasites of the Agromyzidae (Diptera). VI. The parasites of *Cerodontha* Rondani s. l. *Beiträge zur Entomologie* 18: 63–152.

Griffiths, G.C.D. 1984. The Alysiinae (Hym. Braconidae) parasites of the Agromyzidae (Diptera). VII. Supplement. *Beiträge zur Entomologie* 34: 343–62.

Includes descriptions of a new British *Exotela* and a new British *Dacnusa* and amended key couplets to Griffiths (1966b), and adds a *Chorebus* to the British list.

Königsmann, E. 1960. Revision der palaarktischen Arten der Gattung *Idiasta*. *Beiträge zur Entomologie* 10: 624–54.

Key to Palaearctic species that includes the two found in Britain.

Königsmann, E. 1969. Beitrag zur Revision der Gattung *Orthostigma* (Hymenoptera, Braconidae). *Deutsche Entomologische Zeitschrift* 16: 1–53.

Key to species; in German.

Nixon, G.E.J. 1943–54. A revision of the European Dacnusini (Hym., Braconidae). *Entomologist's Monthly Magazine* 79: 20–34; 159–68; 80: 88–108; 140–151; 193–200; 249–55; 81: 189–204; 217–29; 82: 279–300; 84: 207–24; 85: 289–98; 90: 257–90.

O'Connor, J.P. & Achterberg, C. van 1990. Revision of the Palaearctic genus *Trachyusa* Ruthe (Hymenoptera: Braconidae). *Zoologische Mededelingen* 64: 107–12.

Papp, J. 1968. A survey of the *Phaenocarpa* Förster species of the Carpathian Basin, central Europe (Hymenoptera, Braconidae: Alysiinae). *Beiträge zur Entomologie* 18: 569–603.

Shaw, M.R. 1983. *Aphaereta pallipes* (Say) (Hym: Braconidae) new to Britain and the Palaearctic Region, with remarks on other parasites of Diptera. *Entomologist's Monthly Magazine* 119: 73–4.

Provides notes to help separate a newly discovered North American species (probably introduced) from a closely related British one (*A. minuta*).

Wharton, R.A. 1980. Review of the Nearctic Alysiini (Hymenoptera, Braconidae) with discussion of generic relationships within the tribe. *University of California Publications in Entomology* 88: 1–112.

Provides a key to Nearctic genera of Alysiini that works for quite a large part of the British fauna although there are many more genera represented in North America and there are a few found in the UK not included in Wharton's work. Also provides useful generic diagnoses.

Wharton, R.A. 1985. Characterization of the genus *Aspilota* (Hymenoptera: Braconidae). *Systematic Entomology* 10: 227–37.

Notes on synonymy of *Aspilota* and *Dinotrema*.

Wharton, R.A. 1986. The braconid genus *Alysia* (Hymenoptera): a description of the subgenera and a revision of the subgenus *Alysia*. *Systematic Entomology* 11: 453–504.

Key to subgenera and species groups.

Wharton, R.A. 1988. The braconid genus *Alysia* (Hymenoptera): a revision of the subgenus *Anarcha*. *Contributions of the American Entomological Institute* 25(2): 1–69.

Key to world species of subgenus.

Zaykov, A. 1982. The European species of *Symphya* Förster (Hymenoptera: Braconidae). *Acta Zoologica Academiae Scientiarum Hungaricae* 28: 171–9.

Subfamily Aphidiinae

These small wasps are exclusively parasitoids of aphids. In the past this sub-family has frequently been afforded full family status but although some workers (particularly in the applied field) still adhere to that practice, most sys-tematists now agree that it is just a specialised subfamily of Braconidae. Unlike many groups, considerable work has been carried out on larval as well as adult taxonomy and keys to genera for both first- and final-instar larvae are avail-able, although these are not comprehensive. The taxonomy of the subfamily has been subject to considerable work but there are still numerous areas where species-level identification is problematic. This is partly because of host- and environment-induced variation (Pungerl, 1983), but also because of the exis-tence of numerous cryptic species. These are only now being worked on to any great extent using allozyme, DNA, chromosomal and other data.

Achterberg, C. van 1989. Revision of the subtribe Monoctonina Mackauer sensu stricto (Hymenoptera: Braconidae: Aphidiinae). *Zoologische Mededelingen* 63: 1–22.

Finlayson, T. 1990. The systematics and taxonomy of final-instar larvae of the family Aphidiidae (Hymenoptera). *Memoirs of the Entomological Society of Canada* 152: 1–74.
   Although based on Canadian taxa predominantly, the key to tribes and to many of the genera will probably work for British material. Not all genera are included.

Gärdenfors, U. 1986. Taxonomic and biological revision of Palaearctic *Ephedrus* Haliday (Hymenoptera: Braconidae, Aphidiinae). *Entomologica Scandinavica* (suppl.) 27: 1–95.
   Separate keys to males and females; for males also see Pennacchio et al. (1988). Describes new subgenera affecting British fauna.

Höller, C. 1991. Evidence for the existence of a species closely related to the cereal aphid parasitoid *Aphidius rhopalosiphi* De Stefani-Perez based on host ranges, morphological characters, isoelectric focusing banding patterns, cross-breeding experiments and sex pheromone specificities (Hymenoptera, Braconidae, Aphidiinae). *Systematic Entomology* 16: 15–28.
   Does not formally erect a new species but provides strong evidence for the existence of a cryptic species of potential economic importance, which is likely also to occur in Britain even though it has not been recorded there so far. See also Powell (1982).

Mackauer, M. 1959a. Die mittel-, west- und nordeuropäische Arten der Gattung *Trioxys* Haliday (Hymenoptera: Braconidae, Aphidiinae). *Beiträge zur Entomologie* 9: 144–79.

Mackauer, M. 1959b. Die europäischen Arten der Gattungen *Praon* und *Areopraon* (Hymenoptera: Braconidae, Aphidiinae). *Beiträge zur Entomologie* 9: 810–65.

O'Donnell, D.J. 1989. A morphological and taxonomic study of first-instar larvae of Aphidiinae (Hymenoptera: Braconidae). *Systematic Entomology* 14: 197–219.

Pennacchio, F., Gardenfors, U. & Tremblay, E. 1988. Taxonomic discrimination of males of Palearctic *Ephedrus* Haliday (Hymenoptera, Braconidae, Aphidiinae). *Bolletino del Laboratorio di Entomologia Agraria 'Filippo Silvestri'* **45**: 181–202.

Key makes use of (difficult) male genitalia characters in addition to traditional ones.

Pennacchio, F. & Tremblay, E. 1986. Biosystematic and morphological study of two *Aphidius ervi* Haliday (Hymenoptera, Braconidae) «biotypes» with the description of a new species. *Bolletino del Laboratorio di Entomologia Agraria 'Filippo Silvestri'* **43**: 105–17.

Describes *A. microlophii* and shows how it can be distinguished from *A. ervi*.

Powell, W. 1982. The identification of hymenopterous parasitoids attacking cereal aphids in Britain. *Systematic Entomology* **7**: 465–73.

Provides keys to genera for adults and for mummies of aphidiines associated with cereal aphids (*Sitobion avenae, Metopolophium dirhodum* and *Rhopalosiphum padi*). Also provides a preliminary key to the *Aphidius* species on cereal aphids but notes that the group is still in need of further taxonomic treatment. See also Pungerl (1986) and Höller (1991).

Pungerl, N.B. 1983. Variability in characters commonly used to distinguish *Aphidius* species (Hymenoptera: Aphidiidae). *Systematic Entomology* **8**: 425–30.

Shows that several characters commonly used to distinguish between three common *Aphidius* species are too variable to permit identification.

Pungerl, N.B. 1986. Morphometric and electrophoretic study of *Aphidius* species (Hymenoptera: Aphidiidae) reared from a variety of aphid hosts. *Systematic Entomology* **11**: 327–54.

Provides a key to *Aphidius* females and a separate key for those attacking cereal aphids.

Starý, P. 1960. The generic classification of the family Aphidiidae (Hymenoptera). *Acta Societatis Entomologicae Cechosloveniae.* **57**: 239–52.

As with the following, a good practical key to aphidiine genera.

Starý, P. 1966. *Aphid parasites of Czechoslovakia* 242pp. The Hague.

Starý, P. 1973. A review of *Aphidius*-species (Hymenoptera, Aphidiidae) of Europe. *Annotationes Zoologicae et Botanicae* **84**: 1–85.

Provides a key to species and descriptions; the best work to start with but the genus undoubtedly still needs much further taxonomic work even on the British fauna. See also Powell (1982), Pungerl (1983) and Höller (1991).

Starý, P. 1975. The subgeneric classification of *Lysiphlebus* Förster, 1862 (Hymenoptera, Aphidiidae). *Annotationes Zoologicae et Botanicae* **105**: 1–9.

Provides a key to subgenera and checklist of species in each.

Starý, P., González, D. & Hall, J.C. 1980. *Aphidius eadyi* n. sp. (Hymenoptera: Aphidiidae), a widely distributed parasitoid of the pea aphid, *Acyrthosiphon pisum* (Harris) in the Palearctic. *Entomologia Scandinavica* **11**: 473–80.

Provides key to separate *A. eadyi* from very similar *A. smithi*, but this is a difficult group with numerous sibling species and consequently identification of British specimens is not likely to be without problems. *A. eadyi* is believed to occur in Britain, whether *A. smithi* does is currently uncertain.

Tremblay, E. & Eady, R.D. 1978. *Lysiphlebus confusus* n. sp. per *Lysiphlebus ambiguus* *sensu* Auct. nec Haliday (1834) (Hymenoptera Ichneumonoidea). *Bolletino del Laboratorio di Entomologia Agraria 'Filippo Silvestri'* **35**: 180–4.

Description of a new species affecting a change in name for a common species on the British list.

Subfamily Blacinae

Common subfamily of small parasitoids of beetle larvae. Only three genera, *Blacus*, *Blacometeorus* and *Dyscoletes*, occur in Britain. Males of some species are often found swarming.

Achterberg, C. van 1976. A revision of the tribus Blacini (Hymenoptera, Braconidae, Helconinae). *Tijdschrift voor Entomologie* **118**: 159–322.

Achterberg, C. van 1988. Revision of the subfamily Blacinae Foerster (Hymenoptera, Braconidae). *Zoologische Verhandelingen* **249**: 1–324.

> Achterberg's works largely supersede previous studies. Includes keys to genera and subgenera of the world and separate regional keys to some species of some subgenera of *Blacus*. New synonymies and records affecting the British fauna.

Haeselbarth, E. 1973. Die *Blacus*-Arten Europas und Zentral-Asiens (Hymenoptera, Braconidae). *Veröffentlichungen der Zoologischen Staatssammlung München* **16**: 69–170.

> In German.

Mason, W.R.M. 1976. A revision of *Dyscoletes* Haliday (Hymenoptera: Braconidae). *Canadian Entomologist* **108**: 855–58.

> Key to world species: only one species occurs in Britain.

Shaw, M.R. 1996. British records of two species of *Blacometeorus* Tobias (Hymenoptera: Braconidae, Blacinae). *Entomologist's Gazette* **47**: 267–8.

Subfamily Braconinae

The British braconine fauna is very depauperate compared with that of the tropics. Several genera have been recorded in the literature but are probably absent (e.g. *Iphiaulax* and *Atanycolus*). *Baryproctus*, a predominantly East European genus, had not been caught in Britain for many years and was thought probably to be extinct in the UK, but it has recently been rediscovered (Shaw, 1994). Identification of species of the largest genus, *Bracon*, is extremely problematical and many British specimens may be impossible to identify with accuracy at present.

Achterberg, C. van 1983. Three new Palaearctic genera of Braconinae (Hymenoptera, Braconidae). *Entomologica Scandinavica* **14**: 69–76.

> Although none of the genera treated is known to occur in Britain, it is possible that one or more of them may turn up in the future.

Achterberg, C. van 1988. *Bracon lineifer* spec. nov., a peculiar new species from The Netherlands (Hymenoptera: Braconidae). *Entomologische Berichten, Amsterdam* **49**: 191–4.

> *B. lineifer* has now been found in Britain (Shaw & Bailey, 1991).

Achterberg, C. van 1985c. *Pigeria* gen. nov., a new Palaearctic genus of the Braconidae (Hymenoptera, Braconidae). *Zoologische Mededelingen* **59**: 168–74.

> *Pigeria* is not mentioned in many keys. It was erected to accommodate an aberrant species previously included in *Bracon*.

Beyarslan, M. & Fischer, M. 1990. Bestimmungsschlüssel zur Identifikation der
paläarktischen *Bracon*-Arten des Subgenus *Glabrobracon* Tobias (Hymenoptera,
Braconidae, Braconinae). *Annalen Naturhistorischen Museums Wien* **91B**:
137–45.

> Provides a key to subgenera of *Bracon*, and a key to the species of the subgenus
> *Glabrobracon*, but the latter is probably too simplified to be reliable (Shaw & Huddleston,
> 1991).

Fischer, M. 1967. Über gezüchtete Raupenwespen (Hymenoptera, Braconidae).
Pflanzenschutzberichte **37**: 97–140.

> Key to species of *Habrobracon* (= *Bracon* subgenus *Habrobracon*), but synonymises some
> taxa that have subsequently been regarded as separate. The systematics of this genus is in
> great need of revision using molecular techniques and the possibility exists that there are
> several cryptic species.

Haeselbarth, E. 1967. Zur Kenntnis der palaearktischen Arten der Gattung *Coeloides*
Wesmael (Hymenoptera, Braconidae). *Mitteilungen der Münchner
Entomologischen Gesellschaft* **57**: 20–53.

> Good key to this important genus of parasites of subcortical beetle larvae. In German.

Quicke, D.L.J. 1987. The Old World genera of braconine wasps (Hymenoptera:
Braconidae). *Journal of Natural History* **21**: 43–157.

> Contains key to world genera, which will work (albeit laboriously) for all of the relatively
> few British genera. It is recommended should something unexpected turn up.

Quicke, D.L.J. & Sharkey, M.J. 1989. A key to and notes on the genera of Braconinae
(Hymenoptera: Braconidae) from America North of Mexico with descriptions
of two new genera and three new species. *Canadian Entomologist* **121**:
337–361.

> This key to North American genera includes all British genera, along with a number of
> extralimital ones, except for *Baryproctus*, which will key to the Nearctic endemic genus
> *Lapicida*, and the genus *Pseudovipio* which was misidentified in the UK as *Vipio* and is
> only known from two certain specimens.

Quicke, D.L.J. 1998. Subfamily Braconinae. *In* Wharton, R.A., Marsh, P.M. & Sharkey,
M.J. (Eds) *Identification manual to the New World genera of Braconidae*. Special
Publications of the International Society for Hymenoptera.

> Same comments as for Quicke & Sharkey (1989).

Shaw, M.R. 1994. Some recent British specimens of *Baryproctus barypus* (Marshall)
(Hym., Braconidae, Braconinae). *Entomologist's Monthly Magazine* **130**: 219–21.

Shaw, M.R. & Bailey, M. 1991. Parasitoids (Hymenoptera: Braconidae,
Ichneumonidae, Pteromalidae) and notes on the biology of the fern-boring
sawfly Heptamelus ochroleucus (Stephens) (Hymenoptera: Tenthredinidae) in
the English Lake District. *The Entomologist* **110**: 103–9.

> Records *Bracon lineifer* from Britain for the first time.

Tobias, V.I. 1986. Subfamily Braconinae. *Handbook to the Insects of the European part
of the USSR*. Part 4. *Hymenoptera*. [In Russian.]

> Keys to the genera of Braconinae include several extralimital taxa. Key to the species of
> *Bracon* might be useful although it does not include all British species and covers numer-
> ous ones that do not occur in Britain.

Subfamily Cenocoeliinae

A mostly tropical group with approximately 70 species worldwide, which are parasitoids of wood- or twig-boring beetle larvae. Three species are recorded from Britain. However, one of these is dubiously British (Shaw & Huddleston, 1991). A third, recently recorded, species belongs to *Promachus*.

Achterberg, C. van 1994. Generic revision of the subfamily Cenocoeliinae Szépligeti (Hymenoptera: Braconidae). *Zoologische Verhandelingen* 292: 1–52.

> Key to world genera and description of *Promachus aartseni*, which is now known to occur in Britain, and a new combination.

Shaw, M.R. & Huddleston, T. 1991. Classification and biology of braconid wasps (Hymenoptera: Braconidae). *Handbooks for the Identification of British Insects* 7(11): 126pp.

Subfamily Charmontinae

Recently erected subfamily with one British genus, *Charmon*. In the past, variously included in Helconinae or Orgilinae (usually as *Eubadizon*). Generic limits clarified by van Achterberg (1979), supplanting all previous keys, which often included unrelated taxa in *Eubadizon*.

Achterberg, C. van 1979. A revision of the subfamily Zelinae auct. (Hymenoptera, Braconidae). *Tijdschrift voor Entomologie* 122: 241–479.

> Key to species and a new status affecting the British list. There has been some debate about the validity of some of the species, and van Achterberg does not accept some previously published synonymies.

Subfamily Cheloninae

The commonest British genus, *Chelonus*, is difficult to identify to species level. The other two genera have been subject to recent revisions.

Achterberg, C. van 1982. Revisionary notes on *Chelonus* Jurine and *Anomala* von Block (Hymenoptera, Braconidae, Cheloninae). *Entomologische Berichten, Amsterdam* 42: 185–90.

Achterberg, C. van 1990. Revision of the Western Palaearctic Phanerotomini (Hymenoptera: Braconidae). *Zoologische Verhandelingen* 255: 1–106.

> Provides a key to the West Palaearctic genera of Cheloninae. Synonymises *Bracotritoma* (replacement name for *Tritoma*) with *Phanerotoma*, thus reducing the number of nominal genera on the British list. Provides a good key to the species of *Phanerotoma* in the West Palaearctic. Records *P. ocularis*, from Britain for the first time. Also purports to record *P. dentata* from Britain for the first time, but it was already known.

Hellén, W. 1958. Die *Chelonus*-Arten Finnlands (Hym., Braconidae). *Notulae Entomologicae* 38: 25–32.

> Worth a try.

Huddleston, T. 1984. The Palaearctic species of *Ascogaster* (Hymenoptera,
    Braconidae). *Bulletin of the British Museum (Natural History)* (Entomology) **49**:
    341–92.

Papp, J. 1990. A revision of Thomson's *Microchelonus* species (Hymenoptera;
    Braconidae, Cheloninae). *Acta Zoologica Hungarica* **36**: 295–317.
    Many workers regard *Microchelonus* as a synonym of *Chelonus*, but Tobias (1971), for
    example, provided a key to separate them.

Tobias, V.I. 1971. A review of the Braconidae (Hymenoptera) of the USSR. *Trudy
    Vsesoyuznogo Entomologicheskogo Obshchestva* **54**: 156–268. [In Russian:
    English translation by Amerind, 1975. New Delhi.]

Tobias, V.I. 1986. Subfam. Cheloninae. *Key to Insects of the European part of USSR*, III,
    4: 293–335.
    With the dearth of keys to W. Palaearctic *Chelonus*, this work may enable a start to be
    made, although there are obviously many faunal differences.

Subfamily Doryctinae

In Britain this subfamily has a rather depauperate fauna. No recent generic
keys exist for the Western Palaearctic, but most British specimens will key
correctly to genus in Marsh (1965).

Hedqvist, K.-J. 1976. New species of *Spathius* Nees, 1818 and a key to the species of
    Europe and Canary Islands (Hym., Ichneumonoidea, Braconidae). *Eos* **51** [1975]:
    51–63.
    Supersedes Nixon (1943) but the latter also provides useful information.

Marsh, P.M. 1965. The Nearctic Doryctinae. I. A review of the subfamily with a taxo-
    nomic revision of the tribe Hecabolini (Hymenoptera: Braconidae). *Annals of
    the Entomological Society of America* **58**: 668–99.

Nixon, G.E.J. 1943. A revision of the Spathiinae of the Old World (Hym., Braconidae).
    Transactions of the Royal Entomological Society of London **93**: 172–456.

Shaw, M.R. 1988. *Spathius curvicaudis* Ratzeburg (Hym.: Braconidae) new to Britain
    and parasitising *Agrilus pannonicus* (Piller and Mitterpacher) (Col.:
    Buprestidae). *Entomologist's Record and Journal of Variation* **100**: 215–16.

Shaw, M.R. 1997. The genus *Heterospilus* Haliday in Britain, with description of a
    new species and remarks on related taxa (Hymenoptera: Braconidae:
    Doryctinae). *Zoologische Mededelingen* **71** (5): 33–41.
    Records *Heterospilus* correctly from the UK for the first time, describes a new species and
    discusses associated taxonomic problems.

Subfamily Euphorinae

As treated here, excludes *Meteorus* and *Zele* (Meteorinae) but includes
Cosmophorus as they are treated in most keys to the genera of Euphorinae,
although they are possibly not closely related. Euphorines are parasitoids of

a wide range of insect hosts. Many attack adult insects or nymphs of hemi-metabolous ones.

Achterberg, C. van 1977. A new Holarctic genus, *Spathicopis* gen. nov., belonging to the Euphorinae, Centistini (Hymenoptera, Braconidae). *Entomologische Berichten, Amsterdam* 37: 27–31.

Includes a key to the genera and subgenera of the tribe Centistini. *Spathicopis* itself has not yet been found in the UK but it occurs in The Netherlands so it is quite likely that it will be found in Britain sooner or later.

Achterberg, C. van 1985. The genera and subgenera of Centistini, with description of two new taxa from the Nearctic Region (Hymenoptera: Braconidae: Euphorinae). *Zoologische Mededelingen* 59: 348–62.

Achterberg, C. van 1992. Revision of the European species of the genus *Pygostolus* Haliday (Hymenoptera: Braconidae: Euphorinae), with a key to the Holarctic species. *Zoologische Mededelingen* 66: 349–58.

Achterberg, C. van 1994. The Palaearctic species of the genus *Chrysopophthorus* Goidanich (Hymenoptera: Braconidae: Euphorinae). *Zoologische Mededelingen* 68: 301–07.

This genus was very recently found in Britain (Shaw, 1996); some other species have very wide geographic distributions and could also occur in the UK.

Čapek, M. & Snoflák, J. 1959. Beitrag zur Kenntnis der europäischen Arten der Gattung *Streblocera* Westwood (Hym., Braconidae). *Acta Societatis Entomologicae Cechosloveniae* 56: 343–54.

Foissner, W. & Achterberg, C. van 1997. The valid name for the genus *Loxocephalus* Foerster (Insecta, Hymenoptera: Braconidae), preoccupied by *Loxocephalus* Eberhard, 1862 (Protozoa: Ciliophora). *Zoologische Mededelingen* 71: 31–2.

Haeselbarth, E. 1971. Notizen zur Gattung *Pygostolus* Haliday (Hymenoptera, Braconidae). *Opuscula Zoologica* 112: 1–8.

Provides a key to the three species found in Britain; in German. Illustrates ovipositors of each.

Haeselbarth, E. 1988. Zur Braconidengattung *Townesilitus* Haeselbarth & Loan, 1983. *Entomofauna* 9: 429–60.

Species in this recently described genus were previously treated under *Microctonus*.

Loan, C.C. 1974. *Microctonus apiophaga*, new species, (Hymenoptera: Braconidae, Euphorinae) a parasite of adult *Apion* weevils in Britain (Coleoptera: Curculionidae). *Proceedings of the Entomological Society of Washington* 76: 186–9.

Loan, C.C. 1974. The European species of *Leiophron* Nees and *Peristenus* Foerster (Hymenoptera: Braconidae, Euphorinae). *Transactions of the Royal Entomological Society of London* 126: 207–38.

Supersedes Richards (1967). Includes rather poor figures of male genitalia.

New, T.R. 1970. The life histories of two species of *Leiophron* Nees (Hymenoptera, Braconidae) parasitic on Psocoptera in southern England. *Entomologist's Gazette* 21: 39–48.

O'Connor, J.P., O'Connor, M.A. & Achterberg, C. van 1991. Some records of Irish Braconidae (Hymenoptera) including seven species new to Ireland. *Irish Naturalist's Journal* **23**: 451–4.
Records *Peristenus grandiceps* from the UK for the first time.

Richards, O.W. 1960. On some British species of *Perilitus* Nees (Hymenoptera: Braconidae). *Proceedings of the Royal Entomological Society of London* (B) **29**: 140–4.

Richards, O.W. 1967. Some British species of *Leiophron* Nees (Hymenoptera: Braconidae, Euphorinae), with the description of two new species. *Transactions of the Royal Entomological Society of London* **119**: 171–86.
Provides a partial key to British and European species and descriptions of some new species. See Loan (1974) for a more recent work.

Shaw, M.R. 1989. *Cosmophorus cembrae* Ruschka new to Britain (Hymenoptera: Braconidae: Euphorinae). *Entomologist's Gazette* **40**: 241–3.
Cosmophorus is a rare genus of parasitoids of adult scolytids.

Shaw, M.R. 1996. *Chrysopophthorus hungaricus* (Zilahi-Kiss) (Hymenoptera: Braconidae, Euphorinae) new to Britain, a parasitoid of adult Chrysopidae (Neuroptera). *Entomologist's Gazette* **47**: 185–7.

Shaw, S.R. 1985. A phylogenetic study of the subfamilies Meteorinae and Euphorinae (Hymenoptera: Braconidae). *Entomography* **3**: 277–370.
Provides a key to the world genera of Euphorinae, illustrated with scanning electron micrographs, and diagnoses for each genus.

Subfamily Gnamptodontinae

Small subfamily of nepticulid parasitoids with one British genus.

Achterberg, C. van 1983. Revisionary notes on the subfamily Gnaptodontinae, with description of eleven new species (Hymenoptera, Braconidae). *Tijdschrift voor Entomologie* **126**: 25–57.
Keys to genera and species. New combination recorded for one British species.

Subfamily Helconinae

Achterberg, C. van 1987. Revision of the European Helconini (Hymenoptera: Braconidae: Helconinae). *Zoologische Mededelingen* **61**: 263–85.
Deals with the small, relatively uncommon genera *Helcon*, *Helconidea* and *Wroughtonia*, which all have British representatives.

Achterberg, C. van 1990. Revision of the genera *Foersteria* Szépligeti and *Polydegmon* Foerster (Hymenoptera: Braconidae) with the description of a new genus. *Zoologische Verhandelingen* **257**: 1–32.

Hellén, W. 1958. Zur Kenntnis der Braconiden (Hym.) Finnlands. II. Subfamilie Helconinae (part.). *Fauna Fennica* **4**: 3–37.

Snoflák, J. 1952. La monographie de *Triaspis* Hal. (*Sigalphus* Latr.) (Hym. Bracon.) de la Tchécoslovaquie. *Acta Entomologica Musei Nationalis Pragae* 28 [1950]: 285–395.

Subfamily Histeromerinae

Small, specialised subfamily that attack beetle larvae within wood. In the past usually included as a tribe in Doryctinae or Braconinae. In Britain only one species has been found so far (usually collected by coleopterists in dead wood), but a second Holarctic species is known from the Netherlands and so may also occur in the UK

Achterberg, C. van 1992. Revision of the genus *Histeromerus* Wesmael (Hymenoptera: Braconidae). *Zoologische Mededelingen* 66: 189.

Subfamily Homolobinae

Achterberg, C. van 1979. A revision of the subfamily Zelinae auct. (Hymenoptera, Braconidae). *Tijdschrift voor Entomologie* 122: 241–479.
Deals with the Zelinae of Authors, and so includes Homolobinae, Charmontinae and the genus *Zele*, which belongs to the Meteorinae.

Subfamily Ichneutinae

A small subfamily with no recent keys to the species occurring in Western Europe. Three species, belonging to two genera in different tribes have been recorded from Britain.

Sharkey, M.J. & Wharton, R.A. 1994. A revision of the genera of the world Ichneutinae (Hymenoptera: Braconidae). *Journal of Natural History* 28: 873–912.
Key to world genera; keys to species of some extralimital taxa.

Subfamily Macrocentrinae

Achterberg, C. van & Haeselbarth, E. 1983. Revisionary notes on the European species of *Macrocentrus* Curtis *sensu stricto* (Hymenoptera, Braconidae). *Entomofauna* 4: 37–59.
Provides a provisional key to the European species that is largely superseded by Achterberg (1993).

Achterberg, C. van 1993. Revision of the subfamily Macrocentrinae Foerster (Hymenoptera: Braconidae) from the Palaearctic region. *Zoologische Verhandelingen* 286: 1–110.

Eady, R.D. & Clarke, J.A.J. 1964. A revision of the genus *Macrocentrus* Curtis (Hym., Braconidae) in Europe, with descriptions of four new species. *Entomologist's Gazette* 15: 97–127.

Subfamily Meteorinae

Sometimes classified within the Euphorinae to which they are clearly closely related. In Britain there are only two genera, *Meteorus* and *Zele*.

Achterberg, C. van 1979. A revision of the subfamily Zelinae auct. (Hymenoptera, Braconidae). *Tijdschrift voor Entomologie* **122**: 241–479.
Key to species of *Zele*, but see also Achterberg (1984). Synonymises *Zemiotes* with *Zele*, thus effecting changes to the British list.

Achterberg, C. van 1984. Addition to the revision of the genus *Zele* Curtis (Hymenoptera: Braconidae). *Entomologische Berichten, Amsterdam* **44**: 110–12.
Provides a key to the species and subspecies of the *Zele albiditarus* complex.

Huddleston, T. 1980. A revision of the western Palaearctic species of the genus *Meteorus* (Hymenoptera, Braconidae). *Bulletin of the British Museum (Natural History)* (Entomology) **41**: 1–58.
Includes a good key and many taxonomic changes, and descriptions of two new British species.

Shaw, M.R. 1988. *Meteorus brevicauda* Thomson, (Hymenoptera: Braconidae) reared from larvae of *Zeugophora subspinosa* (Fabricius) (Coleoptera: Chrysomelidae). *Entomologist's Gazette* **39**: 205–6.
Confirms the previously uncertain occurrence of this species in the UK.

Subfamily Microgastrinae

One of the most frequently encountered subfamilies; its members are exclusively parasitoids of lepidopteran larvae. The systematics of the group has been controversial. Mason (1981) proposed major changes to the groups generic structure, but this has not been wholly accepted (Walker *et al.* 1990). Many people have been confused by the change in name of a lot of familiar species that were traditionally placed in *Apanteles*, many such as *A. glomeratus*, the common parasitoid of the large white butterfly, to the genus *Cotesia*. The British fauna has been subject to extensive studies by Nixon and more recently by Papp.

Anon. 1980. Exhibit by Dr M. R. Shaw. *Proceedings and Transactions of the British Entomological and Natural History Society* **13**: 15.
First record of *Apanteles pedius* from Britain.

Mason, W.R.M. 1981. The polyphyletic nature of *Apanteles* Foerster (Hymenoptera: Braconidae): a phylogeny and reclassification of Microgastrinae. *Memoirs of the Entomological Society of Canada* **115**: 1–147.
Important paper that splits *Apanteles* into a number of genera, some but not all of which have become accepted by current authorities on the subfamily.

Nixon, G.E.J. 1965. A reclassification of the tribe Microgasterini (Hymenoptera: Braconidae). *Bulletin of the British Museum (Natural History)* (Entomology) (suppl.) **2**: 1–284.

Provides keys to genera (but see notes above and Mason, 1981) and European species of *Hypomicrogaster*. Nixon included Miracinae as a tribe within the Microgastrinae.

Nixon, G.E.J. 1968. A revision of the genus *Microgaster* Latreille (Hymenoptera: Braconidae). *Bulletin of the British Museum (Natural History)* (Entomology) 22: 33–72.

Nixon, G.E.J. 1970. A revision of the N.W. European species of *Microplitis* Förster (Hymenoptera: Braconidae). *Bulletin of the British Museum (Natural History)* (Entomology) 25: 1–30.

Nixon, G.E.J. 1972. A revision of the north-western European species of the *laevigatus*-group of *Apanteles* Förster (Hymenoptera: Braconidae). *Bulletin of Entomological Research* 61: 701–43.

Nixon, G.E.J. 1973. A revision of the north-western European species of the *vitripennis, pallipes, octonarius, triangulator, fraternus, formosus, parasitellae, metacarpalis* and *circumscriptus*-groups of *Apanteles* Förster (Hymenoptera: Braconidae). *Bulletin of Entomological Research* 63: 169–230.

Nixon, G.E.J. 1974. A revision of the north-western European species of the *glomeratus*-group of *Apanteles* Förster (Hymenoptera: Braconidae). *Bulletin of Entomological Research* 64: 453–524.

Nixon, G.E.J. 1976. A revision of the north-western European species of the *merula, lacteus, vipio, ultor, ater, butalidis, popularis, carbonarius* and *validus*-groups of *Apanteles* Förster (Hymenoptera: Braconidae). *Bulletin of Entomological Research* 65: 687–732.

Papp, J. 1976. Key to the European *Microgaster* Latr. species, with a new species and taxonomical remarks (Hymenoptera: Braconidae, Microgasterinae). *Acta Zoologica Academiae Scientiarum Hungaricae* 22: 97–117.

Papp, J. 1976. A survey of the European species of *Apanteles* Först. (Hymenoptera, Braconidae: Microgasterinae). I. The species-groups. *Annales Historico-Naturales Musei Nationalis Hungarici* 68: 251–74.

Papp, J. 1978. A survey of the European species of *Apanteles* Först. (Hymenoptera, Braconidae: Microgasterinae). II. The *laevigatus*-group, 1. *Annales Historico-Naturales Musei Nationalis Hungarici* 70: 265–301.

Papp, J. 1979. A survey of the European species of *Apanteles* Först. (Hymenoptera, Braconidae: Microgasterinae). III. The *laevigatus*-group, 2. *Annales Historico-Naturales Musei Nationalis Hungarici* 71: 235–50.

Papp, J. 1980. A survey of the European species of *Apanteles* Först. (Hymenoptera, Braconidae: Microgasterinae). IV. The *lineipes-, obscurus-* and *ater*-group. *Annales Historico-Naturales Musei Nationalis Hungarici* 72: 241–72.

Papp, J. 1981. A survey of the European species of *Apanteles* Först. (Hymenoptera, Braconidae: Microgasterinae). V. The *lacteus-, longipalpis-, ultor-, butalidis-* and *vipio*-group. *Annales Historico-Naturales Musei Nationalis Hungarici* 73: 263–91.

Papp, J. 1982. A survey of the European species of *Apanteles* Först. (Hymenoptera, Braconidae: Microgasterinae). VI. The *laspeyresiella-, merula-, facatus-* and *validus*-group. *Annales Historico-Naturales Musei Nationalis Hungarici* 74: 225–67.

Papp, J. 1983. A survey of the European species of *Apanteles* Först. (Hymenoptera, Braconidae: Microgasterinae). VII. The *carbonarius-*, *circumscriptus-*, *fraternus-*, *pallipes-*, *parasitellae-*, *vitripennis-*, *liparidis-*, *octonarius-* and *thompsoni-*group. *Annales Historico-Naturales Musei Nationalis Hungarici* 75: 247–83.

Papp, J. 1984. A survey of the European species of *Apanteles* Först. (Hymenoptera, Braconidae: Microgasterinae). VIII. The *metacarpalis-*, *formosus-*, *popularis-* and *suevus-*group. *Annales Historico-Naturales Musei Nationalis Hungarici* 76: 265–95.

Papp, J. 1984. Palaearctic species of *Microgaster* Latreille (=*Microplitis* Förster) with description of seven new species (Hymenoptera, Braconidae, Microgastrinae). *Entomologische Abhandlungen und Berichte aus dem Staatlichen Museum für Tierkunde in Dresden* 47: 95–140.

Papp, J. 1986. A survey of the European species of *Apanteles* Först. (Hymenoptera, Braconidae: Microgasterinae). IX. The *glomeratus-*group, 1. *Annales Historico-Naturales Musei Nationalis Hungarici* 78: 225–47.

Papp, J. 1987. A survey of the European species of *Apanteles* Först. (Hymenoptera, Braconidae: Microgasterinae). X. The *glomeratus-*group 2 and the *cultellatus-*group. *Annales Historico-Naturales Musei Nationalis Hungarici* 79: 207–58.

Papp, J. 1988. A survey of the European species of *Apanteles* Först. (Hymenoptera, Braconidae: Microgasterinae). XI. 'Homologization' of the species-groups of *Apanteles* s.l. with Mason's generic taxa. Checklist of genera. Parasitoid/host list 1. *Annales Historico-Naturales Musei Nationalis Hungarici* 80: 145–75.

Shaw, M.R. 1992. A new species of *Hygroplitis* Thomson in England (Hymenoptera: Braconidae, Microgastrinae). *Entomologist's Gazette* 43: 283–8.
Describes a new species, *H. pseudorussatus,* from Britain.

Tobias, V.I. 1986. Subfamily Microgasterinae. *Handbook to the Insects of the European part of the USSR*. 3. Hymenoptera. Part 4: 344–459. Leningrad. (In Russian.)

Walker, A.K., Kitching, I.J. & Austin, A.D. 1990. A reassessment of the phylogenetic relationships within the Microgastrinae (Hymenoptera: Braconidae). *Cladistics* 6: 291–306.

Subfamily Microtypinae

This subfamily has only recently been recorded from Britain (Shaw, 1992); there may be more British species than the one recorded to date.

Čapek, M. & Achterberg, C. van 1992. A revision of the genus *Microtypus* Ratzeburg (Hymenoptera: Braconidae). *Zoologische Mededelingen* 66: 323–38.

Achterberg, C. van 1992. Revision of the genera of the subfamily Microtypinae (Hymenoptera: Braconidae). *Zoologische Mededelingen* 66: 369–80.

Shaw, M.R. 1992. *Microtypus wesmaelii* Ratzeburg (Hymenoptera: Braconidae, Microtypinae), a species and subfamily new to Britain. *Entomologist's Gazette* 43: 289–91.

Subfamily Miracinae

This small subfamily is known from Britain from only a single species to date.

Papp, J. 1984. Two new species of *Mirax* Haliday in the Palaearctic Region (Hymenoptera: Braconidae, Adeliinae). *Folia Entomologica Hungarica* **45**: 167–71.

Richards, O.W. 1957. A note on the genus *Mirax* Hal. (Hym., Braconidae, Microgasterinae). *The Entomologist* **90**: 120–2.

Subfamily Neoneurinae

Huddleston, T. 1976. A revision of *Elasmosoma* Ruthe (Hymenoptera, Braconidae) with two new species from Mongolia. *Annales Historico-Naturales Musei Nationalis Hungarici* **68**: 215–25.

The genus *Elasmosoma* has not been found in Britain, but this paper provides a key to the genera of Neoneurinae. There are no recent keys that treat the three species of *Neoneurus* recorded from Britain.

Subfamily Opiinae

An extremely difficult subfamily taxonomically, the main works on the Palaearctic fauna being those by Fischer. Generic limits are in need of redefinition and many species will need to be reclassified. The plethora of subgenera, especially within *Opius*, has added more confusion than help to the situation. Quicke *et al.* (1997) includes many new combinations affecting the British list, but is not a key work. Two papers that are expected to be published in 1998 by van Achterberg & van Zuijlen should greatly simplify matters.

Achterberg, C. van 1975. About the identity of *Biosteres* (*Biosteres*) *impressus* (Wesmael) (Braconidae, Opiinae). *Entomologische Berichten, Amsterdam* **35**: 175–6.

Achterberg, C. van & Zuijlen, J.A.W. van, In press. Revision of the subfamily Opiinae from Northwest Europe (Hymenoptera: Braconidae). Part I. The Palaearctic genera of the subfamily Opiinae (Hymenoptera: Braconidae), with additional notes on the genus *Xynobius* Foerster.

Achterberg, C. van & Zuijlen, J.A.W. van, In press. Revision of the subfamily Opiinae from Northwest Europe (Hymenoptera: Braconidae). Part II. Revision of the Northwest European species of the genera *Apodesmia* Foerster, *Opius* Wesmael, *Psyttalia* Walker, *Phaedrotoma* Foerster, and *Utetes* Foerster (Braconidae: Opiinae).

Fischer, M. 1972. Hymenoptera Braconidae (Opiinae I). *Das Tierreich* **92**: 1–620.
Major work on the subfamily, but already considerably out of date.

Fischer, M. 1984. Aufteilung des Formenkreises um das Subgenus *Cryptonastes* Foerster des Genus *Opius* Wesmael sowie Erganzunngen zum Subgenus *Tolbia* Cameron (Hymenoptera, Braconidae, Opiinae). *Zeitschrift der Arbeitsgemeinschaft Österreich Entomologen* **36**: 33–40.
Redefinition and description of some subgenera and keys to Palaearctic species.

Fischer, M. 1986. Neue Bestimmungsschlüssel für paläarktische Opiinae, neue Subgenera, Redeskriptionen und eine neue Art. (Hymenoptera, Braconidae). *Annalen Naturhistorischen Museums Wien* **88/89** (B): 607–62.
Keys to the thirteen subgenera belonging to Section B of *Opius*, and keys to species of most of them; also a diagnosis of Section B.

Godfray, H.C.J. 1986. Four species of Opius (Hym.; Braconidae) new to Britain. *Entomologist's Monthly Magazine* **122**: 127.

Godfray, H.C.J. 1988. *Biosteres spinaciae* (Thomson) [Hymenoptera, Braconidae, Opiinae], a new parasitic wasp to Britain. *Entomologist's Monthly Magazine* **124**: 251–2.
Records *Biosteres spinaciae* new to the British fauna.

Quicke, D.L.J., van Achterberg, C. & Godfray, H.C.J. 1997. Comparative morphology of the venom gland and reservoir in opiine and alysiine braconid wasps (Insecta, Hymenoptera, Braconidae). *Zoologica Scripta* **25**: 1–28.
Many new combinations affecting the British list.

Subfamily Orgilinae

Only one genus, *Orgilus*, occurs in Britain.

Achterberg, C. van 1987. Revisionary notes on the subfamily Orgilinae (Hymenoptera, Braconidae). *Zoologische Verhandelingen* **242**: 1–111.
Key to world genera and species.

Taeger, A. 1989. *Die Orgilus-Arten der Paläarktis (Hymenoptera, Braconidae).* Akademie der Landwirtschaftswissenschaften der Deutschen Demokratischen Republik, Berlin.
Includes key to all Palaearctic species and descriptions of new species, several of which occur in Britain.

Subfamily Rogadinae s.l.

Included here are Exothecinae, Hormiinae, Pambolinae, Rhysipolinae and Rhyssalinae as well as Rogadinae *sensu stricto*. The Rogadinae *sensu stricto* are represented in Britain by two genera, *Aleoides* and *Rogas*, but in older works these are all referred to as *Rogas*. Two other generic names encountered, *Heterogamus* and *Petalodes*, are referred to *Aleiodes* (van Achterberg, 1991). Identification of *Aleiodes* species is very difficult. A revision of *Aleiodes* is currently in preparation by van Achterberg & Shaw; this takes into account much new biological data. Older keys are of doubtful use.

Achterberg, C. van 1983. Revisionary notes on the Palaearctic genera and species of the tribe Exothecini Foerster (Hymenoptera, Braconidae). *Zoologische Mededelingen* 57: 339–55.

Keys to genera and to European species of *Shawiana* and *Xenarcha*; notes on *Colastes*.

Achterberg, C. van 1985. The *Aleiodes* dispar-group of the Palaearctic region (Hymenoptera: Braconidae: Rogadinae). *Zoologische Mededelingen* 59: 178–87.

Achterberg, C. van 1991. Revision of the genera of the Afrotropical and W. Palaearctic Rogadinae Foerster (Hymenoptera: Braconidae). *Zoologische Verhandelingen* 273: 1–102.

Key includes, in addition to many extralimital genera, the three occurring in Britain, and also relevant generic synonymies.

Godfray, H.C.J. & McGavin, G.C. 1985. *Colastes pubicornis* (Thomson) (Hym., Braconidae, Exothecini) new to Britain, with a first host record. *Entomologist's Monthly Magazine* 121: 109–10.

Papp, J. 1975. Three new European species of *Colastes* Hal. with taxonomic remarks (Hymenoptera: Braconidae, Exothecinae). *Acta Zoologica Academiae Scientiarum Hungaricae* 21: 411–23.

Some of the species treated by Papp as *Colastes* actually belong to *Shawianus* (see van Achterberg, 1983).

Shaw, M.R. 1978. *Aleiodes pallidator* (Thunberg) (= *unicolor* Wesmael) (Hym., Braconidae) new to Britain. *Entomologist's Monthly Magazine* 113 [1977]: 81.

Subfamily Sigalphinae

Only one distinctive species, *Acampsis alternipes*, occurs in Britain (Shaw & Huddleston, 1991).

Shaw, M.R. & Huddleston, T. 1991. Classification and biology of braconid wasps (Hymenoptera: Braconidae). *Handbooks for the Identification of British Insects* 7(11): 126pp.

Superfamily Cynipoidea: Gall-wasps, etc.

This superfamily, with 190 British species in 3 families, is perhaps best known because of the gall-wasps that induce galls on oak and other plants. However, the gall-causers comprise only one of the families (Cynipidae); the other cynipoids are internal parasitoids of Diptera, Hymenoptera, Neuroptera, Aphididae or Psyllidae. The higher classification of the Cynipoidea has recently been reviewed (Ronquist, 1995) and now only three families are recognised in Europe: the Cynipidae, Ibaliidae and Figitidae, all three occurring in Britain.

Cynipoids are not easy to identify; they are generally small (most are under 5 mm long), robust, brown or black insects. They can be recognised by the distinctive triangular radial cell in the forewing, and the absence of a costal cell and pterostigma.

The literature on the Cynipoidea is mostly in the form of small papers; the only comprehensive works are those of Weld (1952), which is difficult to obtain, and Dalla Torre & Kieffer (1910), which is very out of date. Fortunately, the British Cynipoidea are completely covered by three hand-books (Eady & Quinlan, 1963; Quinlan, 1978 and Fergusson, 1986) and there is a general account of the group in Gauld & Bolton (1988).

Family Cynipidae

There are just under 100 British species of Cynipidae; all have phytophagous larvae that either form galls (gall-wasps) or are inquilines (i.e. they use galls of other insects). The female gall-wasp oviposits into the host-plant but the galls are produced by action of the living cynipid larva on the plant. The larva stimulates plant growth and produces a form of a gall specific to that particular cynipid. The cynipid feeds and develops within the gall and eventually the adult wasp emerges. There is a great structural diversity of galls; although many are inconspicuous, the more obvious examples are a subject of general interest.

Keys to both gall wasps and their galls are given in the handbook by Eady & Quinlan (1963). Galls are easier to identify than the gall-causers and there are several popular books on plant galls; Darlington (1975) is unfortunately out of print but Redfern & Askew (1992) and Stubbs (1986) are very useful illustrated guides. Askew (1984) provides a more advanced review of the biology of gall wasps and there is a journal devoted to plant galls: *Cecidology*, the journal of the British Plant Gall Society.

In Britain the family Cynipidae is represented by four tribes, the Aylacini, Rhoditini, Cynipini and the Synergini. The Aylacini has recently been revised by Nieves-Aldrey (1994). Most species have a simple life cycle and they form simple galls, mostly on stems, on a range of about 40 plant species (Askew, 1984). The Rhoditini is represented in Britain by *Diplolepis;* the species of this genus have a specialised 'plough-blade'-like hypopygium and they form mostly stem or leaf galls on Rosaceae, mainly on rose. The oak gall-wasps (Cynipini) have complicated heterogonous reproductive cycles, which involve an agamic, female only, generation alternating with a sexual generation. The two generations often have dissimilar galls and gall sites and sometimes even different species of host plant. The life cycles of few species have been properly investigated (see Askew, 1984). The inquilines (Synergini) inhabit the galls of other Cynipidae, they oviposit into developing galls and exploit the food supply of the gall-causer. Some species develop inside the gall without coming into contact with the host, whereas other species may kill the host as a result of competition.

Family Ibaliidae

These are the largest British cynipoids, length 8–18 mm. They have characteristic transverse ridges on the mesoscutum and the fore wing has a radial cell at least nine times as long as broad. There is one British genus with two species, *Ibalia leucospoides* and *I. rufipes;* they are both endoparasitoids of woodwasps (Hymenoptera, Symphyta, Siricoidea) in coniferous trees. The host larva, or egg, lies deep in the wood and in order to reach it *Ibalia* has a remarkable ovipositor, which is very long (about 1.5 times the body length) and coiled in a full circle inside a laterally flattened and blade-like metasoma. In use the ovipositor is uncoiled and fed down the siricid's oviposition hole. The life cycle of *Ibalia* is very long, often three or four years. It develops inside its host, pupates and then finally the adult *Ibalia* uses its powerful mandibles to chew its way out of the wood and emerge. The handbook by Fergusson (1986) includes the Ibaliidae and has notes, a key to the British species and references to their biology.

Family Figitidae

These are small parasitoid cynipoids, usually between 1 and 4 mm long, and the family is characterised by subtle venation and other characters (see Ronquist, 1995; Fergusson, 1986). Many species are difficult to identify and they have not been the subject of much amateur interest. This family now includes two groups, the Charipinae and Eucoilinae, that were previously considered to be families. There are just under 100 British species of Figitidae.

Apart from the overwintering generation, most of these small parasitoids have a short life cycle and some are multivoltine. As far as we know, the British Figitidae are all endoparasitoids. Oviposition is into the haemocoel (some exceptions) of an early larval-instar (usually) but the host is not killed until after it has formed a puparium (or cocoon or aphid mummy). Then the parasitoid larva consumes the last of the host larva and finally pupates, still within the host's puparium (cocoon or aphid mummy). Finally the adult chews a hole in the puparium and emerges. Hymenoptera, Hemiptera and Neuroptera are attacked, but about 70% of these parasitoids have dipterous hosts: some are parasitoids on aphidophagous Diptera, some are parasitoids of economically important species. Many are parasitoids on Diptera associated with dung, carrion, rotting plant material, etc., and consequently many Figitidae are attracted to the odours from these habitats.

The handbook by Quinlan (1978) covers the 'Eucoilidae' (now known as the Eucoilinae); the remaining Figitidae (including the 'Charipidae', now the

Charipinae) are included in the handbook by Fergusson (1986). The latter work also has notes on the biology and a good bibliography.

The subfamilies Eucoilinae, Charipinae, Aspiceratinae, Anacharitinae and Figitinae are represented in Britain. Species of the Eucoilinae are easily recognised by a unique modification of the scutellum into a teardrop-shaped plate. The Charipinae contains very small species that have a smooth and shiny thorax; they are all the hyperparasites of either aphids or Psyllidae. The commonest genera, *Alloxysta* and *Phaenoglyphis*, are hyperparasitoids of aphids via Aphidiinae (Braconidae) or Aphelinidae (Chalcidoidea) primary parasites. The egg is laid into the haemocoel of the host parasitoid, which is within the haemocoel of the host aphid. The *Alloxysta* larva develops and consumes the primary parasitoid and pupates within the mummified skin of the aphid. The cynipoid may be parasitised by other hyperparasitoids: tertiary parasitism. Two rare British species (*Dilyta subclavata* and *Apocharips xanthocephala*) are hyperparasitoids of Psyllidae via encyrtid (Chalcidoidea) primary parasitoids. The Aspiceratinae is represented by three British genera; they have a distinctive short and saddle-shaped second metasomal tergite. They are parasitoids of Diptera, especially Syrphidae; *Callaspidia defonscolombei* is known to oviposit into the brain of its host, thus avoiding early exposure to the host's immune defence reaction. Anacharitines are parasitoids of neuropterous larvae. The female anacharitine oviposits into the haemocoel of the host and the parasitoid larva develops internally until the host spins a cocoon, then the anacharitine larva kills its host and pupates within the hosts cocoon and finally emerges as an adult. The Figitinae is a difficult group to define and the species have a biology similar to that of the Aspiceratinae or Eucoilinae.

References

Askew, R.R. 1984. The biology of gall wasps. *In* Ananthakrishnan, T.N. (Ed.) *Biology of gall insects*, pp. 223–71. Arnold, London.
A good review of the complexities of gall wasp and gall biology; useful references.
Cecidology. The Journal of the British Plant Gall Society.
Articles on galls of all types, general interest, local recording etc. Good value and not too formal in style. The society holds field meetings and provides various services. Contact addresses are provided on the back of the latest issue of the journal.
Dalla Torre, K.W. & Kieffer, J.J. 1910. Cynipidae. *Das Tierreich* 24: 1–891.
Badly out-of-date German tome, but still the only comprehensive work on the superfamily.
Darlington, A. 1975. *The pocket encyclopaedia of plant galls in colour*. Blandford Press, London.
Good illustrations and colour photographs. Unfortunately this book is out of print, but it can be found in many public libraries.

Eady, R.D. & Quinlan, J. 1963. Hymenoptera Cynipoidea. Key to families and subfamilies and Cynipinae (including galls). *Handbooks for the Identification of British Insects* 8(1a): 81pp.

Keys to genera, species and galls of the Cynipidae. The section on Aylacini has been superseded by Nieves-Aldrey (1994).

Fergusson, N.D.M. 1986. Charipidae, Ibaliidae & Figitidae (Hymenoptera: Cynipoidea). *Handbooks for the Identification of British Insects* 8(1c): 55pp.

Includes keys to genera and species, notes on collecting, biology, and a good bibliography. Although the higher classification has changed the basic groupings are still extant and the keys to genera and species remain current.

Nieves-Aldrey, J.L. 1994. Revision of West-European genera of the tribe Aylacini Ashmead (Hymenoptera, Cynipidae). *Journal of Hymenoptera Research*, 3: 175–206.

A recent revision of this tribe, with keys.

Quinlan, J. 1978. Hymenoptera Cynipoidea Eucoilidae. *Handbooks for the Identification of British Insects* 8(1b): 58pp.

Keys to genera and species of Eucoilinae.

Redfern, M. & Askew, R.R. 1992. *Plant galls.* Naturalists' Handbooks No. 17: 99pp. Richmond Publishing, Slough.

Excellent pocket-sized popular book on galls, well illustrated with much information, useful addresses, further reading, techniques, etc. Not expensive and a very good book with which to start an interest in galls.

Ronquist, F. 1995. Phylogeny and early evolution of the Cynipoidea (Hymenoptera). *Systematic Entomology* 20: 309–35.

New family structure for the Cynipoidea, no key.

Stubbs, F.B. 1986. *Provisional keys to British plant galls.* 95pp. British Plant Gall Society, Leicester.

Good, pocket-sized popular book on galls, illustrated with many sketches that are more accurate than first appearances may suggest. Excellent value.

Weld, L.H. 1952. *Cynipoidea (Hym.) 1905–1950.* 351pp. Privately published, Ann Arbor.

Key to cynipoid genera of the world, but badly out of date and hard to obtain.

Superfamily Proctotrupoidea

The classification of this superfamily is uncertain, but of the three families recorded from Britain, the Heloridae appears to be a relict group containing a single, morphologically isolated genus, the Proctotrupidae is moderately speciose and the neglected Diapriidae is a very large family, species of which are commonly encountered in moist habitats.

The biology of Proctotrupoidea is very diverse, most families parasitising saprophagous, mycophagous or even carnivorous insects that develop in leaf litter, fungi, soil, rotting matter and similar situations, including larval or pupal Coleoptera or Diptera, rarely Neuroptera or other Hymenoptera. Most appear to develop as koinobionts.

Identification of British Proctotrupoidea to family is covered by Gauld & Bolton (1988) and the world treatment of Masner (1993) is also useful. The world catalogue by Johnson (1992) should be used to supplement the current British checklist (Fitton, 1978). News and views are available in *Proctos*, a newsletter for those interested in Proctotrupoidea *sensu lato* (Masner & Denis, 1995).

General references

Clausen, C.P. 1940. *Entomophagous insects*. McGraw Hill, New York and London.
   Biology and figures of sub-imaginal stages of diapriids, helorids and proctotrupids.
Fitton, M.G. *et al*. 1978. A check list of British Insects. *Handbooks for the Identification of British Insects* 11(4): 159pp.
   Checklist of British species of Proctotrupoidea.
Gauld, I.D. & Bolton, B. (Eds) 1988. *The Hymenoptera*. British Museum (Natural History), London, and Oxford University Press, Oxford. [Reprinted 1996, with minor alterations and additions.]
   General account of British proctotrupoids with key to families.
Gauld, I.D. & Hanson, P.E. 1995. The evolution, classification and identification of the Hymenoptera. *In* Hanson, P. E. & Gauld, I. D. (Eds) *The Hymenoptera of Costa Rica*, pp. 138–56. The Natural History Museum, London, and Oxford University Press, Oxford.
   Classification of proctotrupoids.
Johnson, N. 1992. Catalog of World Proctotrupoidea excluding Platygastridae. *Memoirs of the American Entomological Institute* 51: 1–825.
   Comprehensive, annotated, world taxonomic catalogue, with notes on which publications contain identification keys.
Masner, L. 1993. Superfamily Proctotrupoidea. *In* Goulet H. & Huber, J.T. (Eds) *Hymenoptera of the world: an identification guide to families*, pp. 537–57. Agriculture Canada.
   Classification and identification of proctotrupoid families.
Masner, L. & Denis, J. (Eds) 1975 onwards. *Proctos*.
   Newsletter for those interested in proctotrupoid (*sensu lato*) families. See also the Web version of *Proctos* at: http://iris.biosci.ohio-state.edu/newsletters/proctos/homepg.html.
Masner, L. & Dessart, P. 1967. La reclassification des categories taxonomiques supérieures des Ceraphronoidea (Hymenoptera). *Bulletin d'Institut Royal des Sciences Naturelles de Belgique* 43(22): 1–33.
   Classification of proctotrupoid (*sensu lato*) families. In French.

Family Diapriidae

The Diapriidae is a large and neglected family, with about 300 species in 41 genera recorded from Britain. Adults are commonly found in damp shaded habitats such as forests, marshes, in or near water, and in soil and litter.

Despite the facts that the Diapriidae is one of the larger families of parasitic Hymenoptera, that many species are extremely common (Chambers, 1971), and that there are keys to most British species, little is known about their life history. Most species are pupal (or more rarely larval-pupal) endoparasitoids of Diptera, but a few also attack Formicidae, Dryinidae and Coleoptera.

The group is divided into four subfamilies, Ambositrinae, Belytinae, Ismarinae and Diapriinae, of which only the last three occur in Britain and are treated below in more detail. Identification of the British fauna to sub-family is covered by Nixon (1980).

Subfamily Belytinae

Occurring in moist habitats and particularly common in forests, belytines have mostly been reared from mycetophilid and sciarid flies. *Scorpioteleia longiventris* and *Cinetus lanceolatus* have been observed ovipositing into the fungus of the genus *Suillus*. When ovipositing, most of the gaster is inserted into the spore duct and the terminal segments are extruded to form a semi-hyaline tube. At rest these segments are telescoped into the gaster (Huggert, 1979). One species, *Synacra paupera,* is important as a control agent of the greenhouse pest *Bradysia paupera* (Diptera, Sciaridae) (Hellqvist, 1994; Notton, 1997).

The last comprehensive British treatment of Nixon (1957) with keys to British genera and species is still useful for identification although numerous small additions and revisions have been made since then which need to be taken into consideration. Macek (1989–95) provided numerous small works on the European fauna, which cover British species, and both Chambers (1974, 1985) and Notton (1994, 1997) added new genera and species to the British fauna with notes on their recognition. Works on adjacent areas that may also be useful include those of Hellén (1964), who keyed the Finnish genera and species, and Kozlov (1987), who keyed genera and species of the European part of the former USSR.

Subfamily Diapriinae

There are numerous genera and species of Diapriinae occurring in a wide range of habitats. Among the British species there are some that live in the intertidal zone, some in deep soil, and at least one that is a myrmecophile. British diapriines are endoparasitoids of dipterous pupae within the puparia (Notton, 1991), rarely of Coleoptera pupae (Staphylinidae). Oviposition may be into the host pupa inside the puparium or into the larva (Cros, 1935). Most host records for diapriines are from cyclorrhaphan families, especially

Calliphoridae, Chloropidae, Lonchaeidae, Muscidae, Phoridae, Platy-pezidae, Sepsidae, Syrphidae and Tachinidae. There are isolated records of diapriines from other dipteran families including some from orthorrhaphan and nematocerous Diptera (Cros, 1935; Nixon, 1980). Many diapriines are gregarious endoparasitoids with 30–50 individuals emerging from a puparium (Clausen, 1940). Some species are pseudohyperparasitoids (*Trichopria* and *Tetramopria* spp., which attack tachinids) and others may be true hyperparasitoids. A non-British species of *Coptera* is a primary parasitoid of tephritid puparia, but it will develop as a facultative hyperparasitoid within larvae of *Opius* (Braconidae) or *Tetrastichus* (Eulophidae) developing in the primary host (Pemberton & Willard, 1918). There are three larval instars, the first usually having a strongly sclerotised head capsule and large mandibles; the later instars are hymenopteriform and largely featureless except for three pairs of spiracles and weakly to moderately developed mandibles (Pemberton & Willard, 1918; Wright *et al.*, 1946; Simmonds, 1952). A number of species of diapriine are associated with ants. Several have been found in ants' nests but little is known about what they are really doing. The only clearly ant-associated species in Britain is the rare *Tetramopria cincticollis*, which is a symphilic myrmecophile (Notton, 1994). Other diapriids living in unusual habitats include the wingless *Platymischus dilatatus*, which is frequently found in large numbers on the seashore where it parasitises *Orygma luctuosa* (Sepsidae) in rotting seaweed. *Viennopria lacustris* is an amphibious species found in lentic freshwater habitats in Britain and Europe and capable of swimming underwater. Some species of *Basalys* are economically important as they attack dipterous pests such as the carrot fly *Psila rosae* (Psilidae) and the frit fly *Oscinella frit* (Chloropidae) (Wright *et al.*, 1946; Simmonds, 1952).

Nixon (1980) covers the identification of the British species, but should be used in conjunction with the papers of Notton (1992, 1993, 1994a,b), which give recent additions to the British fauna and keys and notes for their recognition, and also Notton (1995), which gives updated nomenclature for *Diapria* genus-group species. Some works on adjacent areas may be useful, including those of Hellén (1963) who keys the Finnish genera and species, and Kozlov (1987), who keys the genera and species of the western part of the former USSR.

Subfamily Ismarinae

Although uncommon, four species of the single genus *Ismarus* are widespread in Britain. Ismarines are hyperparasitoids; they develop within dryinid larvae attacking cicadellids. The European *I. flavicornis* and *I. halidayi* develop in *Anteon* species parasitising idiocerine and macropsine nymphs (Chambers, 1955), *I. dorsiger* attacks *Aphelopus* species on typhlocybine adults (Jervis, 1979) and *I. rugulosus* parasitises either *Chelogynus* or *Prenanteon* species developing on adult deltocephalines (Waloff, 1975). Although details of the biology of these species are unknown, they apparently oviposit into the dryinid larva before it has left its host.

Nixon (1957) covers the identification of the British species but is also useful for north-west Europe. Masner's (1976) work on the Nearctic and Neotropical species adds more information on some British species.

References

Chambers, V.H. 1955. Some hosts of *Anteon* spp. (Hym., Dryinidae) and a hyperparasite *Ismarus* (Hym., Belytidae). *Entomologist's Monthly Magazine* 91: 114–15.
Biology.

Chambers, V.H. 1971. Large populations of Belytinae (Hym., Diapriidae). *Entomologist's Monthly Magazine* 106: 149–54.
Biology.

Chambers, V.H. 1974. Taxonomic notes on the Belytinae with a new species of *Pantoclis* Förster. *Journal of Entomology* (B) 42(2): 127–31.
Additions to the British fauna with keys for their recognition.

Chambers, V.H. 1985. A new genus of Belytinae. *Entomologist's Monthly Magazine* 121: 207–9.
Panbelista longiscapa described from Britain with a key for its recognition.

Cros, A. 1935. Biologie du *Trichopria stratiomyiae* Kieffer (Hymén. Proctotrypidae). *Bulletin de la Société d'Histoire Naturelle de l'Afrique du Nord* 26: 131–6.
Biology. In French.

Hellén, W. 1964. Die Ismarinen und Belytinen Finnlands (Hymenoptera: Proctotrupoidea). *Fauna Fennica* 18: 1–68.
Keys to genera and species of Belytinae and Ismarinae in Finland. In German.

Hellén, W. 1963. Die Diapriinen Finnlands (Hymenoptera: Proctotrupoidea). *Fauna Fennica* 14: 1–35.
Keys to genera and species of Diapriinae in Finland. In German.

Hellqvist, S. 1994. Biology of *Synacra* sp. (Hym., Diapriidae), a parasitoid of *Bradysia paupera* (Dipt., Sciaridae) in Swedish greenhouses. *Journal of Applied Entomology* 117: 491–7.
Biology.

Huggert, L. 1979. *Cryptoserphus* and belytine wasps (Hym., Proctotrupoidea) parasitising fungus- and soil-inhabiting Diptera. *Notulae Entomologicae* 59: 139–44.
Biology.

Jervis, M.A. 1979. Parasitism of *Aphelopus* species (Hymenoptera: Dryinidae) by *Ismarus dorsiger* (Curtis) (Hymenoptera: Diapriidae). *Entomologist's Gazette* 30: 127–9.
Biology.

Kozlov, M.A. 1987. Superfamily Proctotrupoidea (Proctotrupoids). Diapriidae. *In* Medvedev, G.S. (Ed.) *Keys to the insects of the European part of the USSR* III, Part 2, pp. 1000–110. Amerind Publishing, New Delhi, India.
A synthesis of keys to genera and species covering the European part of the former USSR.

Macek, J. 1989. Studies on the Diapriidae Part 1. *Annales Zoologici* **42**(17): 353–62.
Key to genera of European Pantolytini.

Macek, J. 1990. Revision of European Psilommina 1. *Psilomma* and *Acanosema* complex. *Acta Entomologica Musei Nationalis Pragae* **43**: 335–60.
Key to genera of European Psilommina, keys to species of *Psilomma, Acanopsilus, Psilommacra, Acanosema* and *Cardiopsilus*.

Macek, J. 1993a. Revision of Holarctic *Polypeza. Folia Heyrovskyana* **1**(2): 19–23.
Key includes the British species of *Polypeza*.

Macek, J. 1993b. Revision of European *Pantolyta* Förster. *Folia Heyrovskyana* **1**(5): 41–5.
Key to European *Pantolyta*.

Macek, J. 1995a. Revision of the European species of *Belyta* Jurine. *Acta Musei Nationalis Pragae ser. B, Historia naturalis* **51**(1–4): 1–22.
Key to European species of *Belyta*.

Macek, J. 1995b. Revision of West Palaearctic *Lyteba* Th. *Folia Heyrovskyana* **3**(3): 29–39.
Key to European species of *Lyteba*.

Macek, J. 1995c. Revision of genus *Opazon* Haliday in Europe. *Folia Heyrovskyana* **3**(7–8): 80–104.
Key to European species of *Opazon*.

Macek, J. 1995d. A taxonomic revision of European Psilommina Part 2. The *Synacra* complex. *European Journal of Entomology* **92**: 469–82.
Key to European species of *Synacra* including *Sundholmiella* and *Paratelopsilus*.

Nixon, G.E.J. 1957. Hymenoptera, Proctotrupoidea, Diapriidae subfamily Belytinae. *Handbooks for the Identification of British Insects* **8**(3dii): 107pp.
Concise keys to all the then known British genera and species of Belytinae and Ismarinae; now dated.

Nixon, G.E.J. 1980. Diapriidae (Diapriinae) Hymenoptera, Proctotrupoidea. *Handbooks for the Identification of British Insects* **8**(3di): 55pp.
Concise keys to all the then known British genera and species of Diapriinae.

Notton, D.G. 1991. Some Diptera host records for species of *Basalys* and *Trichopria. Entomologist's Monthly Magazine* **127**: 123–6.
Biology.

Notton, D.G. 1992. *Aneuropria* Kieffer, 1905 and *Viennopria* Jansson, 1953 new to Britain. *Entomologist's Gazette* **43**: 59–63.
Additions to the British fauna with keys for their recognition.

Notton, D.G. 1993. New species of *Trichopria* and *Diapria* from the British Isles. *Entomologist's Monthly Magazine* **129**: 139–49.
Additions to the British fauna with keys for their recognition.

Notton, D.G. 1994a. A British *Trichopria* new to science and *Sundholmiella giraudi* Kieffer new to Britain. *Entomologist's Monthly Magazine* **130**: 201–3.
Additions to the British fauna with notes for their recognition.

Notton, D.G. 1994b. A description of the male of *Entomacis penelope* Nixon. *Entomologist's Gazette* **45**: 57–8.

Contains an improvement to Nixon's (1980) key to *Entomacis.*

Notton, D.G. 1994c. New Eastern Palaearctic Myrmecophile *Lepidopria* and *Tetramopria* (Hymenoptera, Proctotrupoidea, Diapriidae, Diapriini). *Insecta Koreana* **11**: 64–74.

Key to Palaearctic *Tetramopria* species.

Notton, D.G. 1995. A catalogue of type material of British *Diapria* genus group (Hymenoptera, Proctotrupoidea, Diapriidae). *Beitrage zur Entomologie* **45**(2): 269–98.

Updated nomenclature for British *Diapria, Tetramopria, Trichopria* and *Viennopria.* In English with German abstract.

Notton, D.G. 1997. *Synacra paupera* Macek (Hymenoptera, Diapriidae) new to Britain: a parasitoid of the greenhouse pest *Bradysia paupera* Tuomikoski (Diptera, Sciaridae). *Entomologist's Monthly Magazine* **133**: 257–9.

Synacra paupera added to the British fauna with notes on its recognition and biology.

Pemberton, C.E. & Willard, H.F. 1918. A contribution to the biology of fruit-fly parasites in Hawaii. *Journal of Agricultural Research* **15**: 419–65.

Biology and figures of larva of *Coptera silvestrii.*

Simmonds, F.J. 1952. Parasites of the frit fly *Oscinella frit* (L.) in eastern North America. *Bulletin of Entomological Research* **43**: 503–42.

Biology and figures of larva of *Basalys.*

Waloff, N. 1975. The parasitoids of the nymphal and adult stages of leafhoppers (Auchenorrhyncha: Homoptera) of acidic grassland. *Transactions of the Royal Entomological Society of London* **126**: 637–86.

Biology.

Wright, D.W., Geering, Q.A. & Ashby, D.G. 1946. The insect parasitoids of the carrot fly *Psila rosae* Fab. *Bulletin of Entomological Research* **37**: 507–29.

Biology and figures of larva of *Basalys.*

Family Heloridae

There are three species in one genus recorded from Britain: they are fairly uncommon insects, but may be encountered in numbers when rearing chrysopids. Helorids appear to develop exclusively as solitary endoparasitoids of the larvae of *Chrysopa* and related genera of Chrysopidae, adults emerging from the host cocoon. The helorid oviposits into the host larva and the egg is free-floating in the haemocoel. Eclosion occurs about two days after oviposition and the larva remains in its first-instar until the host has spun a cocoon. This may be for as little as three days in summer, or up to about 8 months in the late summer generation, which overwinters as a first-instar larva in a diapausing host larva. The first-instar larva is polypodeiform, with paired ventral fleshy lobes on the abdominal segments; the head is strongly sclerotised (Clancy, 1946). When the host has spun its cocoon the helorid larva usually completes development rapidly, passing through two further instars. The third-instar larva partly emerges from the skin of the host and pupates. The pupal stage lasts 8–12 days and several generations may be produced in a season.

British helorids may be identified by using either Townes (1977) or Pschorn-Walcher (1955). Neither is ideal for British workers: Pschorn-Walcher's works are in German, while Townes' work requires the use of a longer key since it treats the world species. Using both in conjunction is awkward because of differences in nomenclature between the two authors.

Clancy, D.W. 1946. The insect parasites of the Chrysopidae (Neuroptera). *University of California Publications in Entomology* 7: 403–96.
  Biology and larva of *Helorus anomalipes*.
Pschorn-Walcher, H. 1955. Revision der Heloridae (Hymenopt., Proctotrupoidea). *Mitteilungen der Schweizerischen Entomologischen Gesellschaft* 28: 233–50.
  Revision and identification of European *Helorus*, in German.
Pschorn-Walcher, H. 1971. Heloridae et Proctotrupidae. *Insecta Helvetica* 4, Hymenoptera. Fotorotar, Zurich, Switzerland. 64pp.
  Identification of European *Helorus*, in German.
Townes, H. 1977. A revision of the Heloridae (Hymenoptera). *Contributions of the American Entomological Institute* 15(2): 1–12.
  World revision with keys including all British species.

Family Proctotrupidae

The Proctotrupidae is a moderate-sized family with nine genera and 40 species recorded from Britain. In damp woodlands some species can be exceedingly common. Proctotrupids parasitise larvae found under bark, in leaf litter, in fungi and in similar damp secluded places. The majority of host records involve coleopterous larvae, but some species parasitise mycetophilids. An old record of a species of *Phaneroserphus* reared from a lithobiid centipede (Newman, 1867) has not been confirmed. Townes & Townes (1981) summarise the host data. Proctotrupids are solitary or gregarious endoparasitoids (see, for example, Critchley, 1973) with up to 52 individuals emerging from a single host. Oviposition is into the host larva. Species that parasitise mycetophilids have been observed on macrofungi inserting virtually the whole of the gaster into spore ducts or between lamellae in an attempt to reach mycetophilid larvae (Huggert, 1979). The first-instar larva is polypodeiform, with paired fleshy protuberances ventrally on the abdominal segments; its cephalic capsule is strongly sclerotised. This larva apparently remains almost quiescent until the host is about to pupate, whereupon the proctotrupid larva develops rapidly. The host larva becomes sluggish and dies. The parasitoid larvae may or may not consume the entire body contents of the host, but they all emerge in a characteristic fashion (with individuals of the gregarious species often in neat rows and facing the same direction) from the ventral surface of the host, usually through intersegmental membranes (Clausen, 1940; Askew, 1971, fig. 82). They pupate with the terminal segments of the gaster still embedded in the host (Basden, 1959). At least some proctotrupids apparently overwinter as first-instar larvae in their diapausing hosts. Luff (1976) described the rather elaborate courting pattern of a species of *Phaenoserphus*, during which the male antennae are entwined around those of the female. There are a number of accounts of the sub-imaginal stages of several species but one of the most complete is by Eastham (1929). A few proctotrupids are important because they attack pests. For example, *Paracodrus apterogynus* attacks wireworms, the larvae of click-beetles (Elateridae), in Europe.

Identification of genera and species may be accomplished by using the world revision of Townes & Townes (1981) although these authors use idiosyncratic nomenclature; some may find Pschorn-Walcher's (1971) key to the Swiss species easier to use as it includes only the European species. Nixon's (1938, 1942) papers specifically covered the identification of British species, but included some misidentifications and are slightly dated, although nevertheless they remain useful works. Otherwise, also a little dated but with European coverage, the key of Kozlov (1987) covers the species of the western part of the former USSR.

Askew, R.R. 1971. *Parasitic insects*. Heinemann, London.
  General biology and figure of proctotrupid pupae.
Basden, E.B. 1959. *Phaenoserphus viator* (Hal.) (Hym., Proctotrupidae), a parasite of carabid larvae in Scotland. *Entomologist's Monthly Magazine* **95**: 35–6.
  Biology.
Critchley, B.R. 1973. Parasitism of the larvae of some Carabidae (Coleoptera). *Journal of Entomology* (A) **48**: 37–42.
  Biology.
Eastham, L.E.S. 1929. The post-embryonic development of *Phaenoserphus viator* Hal. (Proctotrypoidea), a parasite of the larva of *Pterostichus niger* (Carabidae), with notes on the anatomy of the larva. *Parasitology* **21**: 1–21.
  Biology and sub-imaginal stages.
Huggert, L. 1979. *Cryptoserphus* and Belytinae wasps (Hymenoptera, Proctotrupoidea) parasitising fungus- and soil-inhabiting Diptera. *Notulae Entomologicae* **59**: 139–44.
  Biology.

Kozlov, M.A. 1987. Superfamily Proctotrupoidea (Proctotrupoids). Proctotrupidae. *In* Medvedev, G.S. (Ed.) *Keys to the insects of the European part of the USSR* III, Part 2, pp. 991–1000. Amerind Publishing, New Delhi, India.
Keys to genera and species for the European part of the former USSR.

Luff, M.L. 1976. Notes on the biology of the developmental stages of *Nebria brevicollis* (F.) (Col., Carabidae) and their parasites, *Phaenoserphus* spp. (Hym., Proctotrupidae). *Entomologist's Monthly Magazine* 111: 249–55.
Biology.

Newman, E. 1867. A *Proctotrupes* parasitic on a myriapod. *The Entomologist* 3: 342–4.
Biology.

Nixon, G.E.J. 1938. A preliminary revision of the British Proctotrupinae (Hym., Proctotrupoidea). *Transactions of the Royal Entomological Society of London* 87: 431–65.
Keys to British genera and species of proctotrupid.

Nixon, G.E.J. 1942. Notes on the males of *Cryptoserphus* together with the description of a new species. *The Entomologist* 75: 195–7.
Keys to British species of *Cryptoserphus*.

Pschorn-Walcher, H. 1971. Heloridae et Proctotrupidae. *Insecta Helvetica* 4, Hymenoptera. Fotorotar, Zurich, Switzerland.
Keys to genera and species of Swiss proctotrupids, in German.

Townes, H. & Townes, M. 1981. A revision of the Serphidae (Hymenoptera). *Memoirs of the American Entomological Institute* 32: 1–541.
World revision with keys to genera and species, extensive bibliography and summary of host data; however, nomenclature idiosyncratic.

Superfamily Platygastroidea

Although Königsmann (1978) suggested that the Platygastridae and Scelionidae have a close affinity with the Diapriidae, this has not been widely accepted and it seems that the Platygastridae and Scelionidae belong in a separate superfamily (Naumann & Masner, 1985; Masner, 1993 and references therein) linked largely on the basis of similarities of the ovipositor; however, these authors do not provide convincing evidence for the monophyly of the Scelionidae, and this family may be paraphyletic with respect to the Platygastridae. Most platygastroids are solitary idiobiont or koinobiont endoparasitoids attacking the eggs of insects and arachnids and also other egg-like insect structures, such as whitefly 'pupae'. They are found in a wide variety of habitats with a high proportion of aquatic species.

Identification of British Platygastroidea to family is covered by Gauld & Bolton (1988) and the world treatment of Masner (1993) is also useful. News and views are available in *Proctos*, a newsletter for those interested in Proctotrupoidea *sensu lato* (Masner & Denis, 1975).

General references

Clausen, C.P. 1940. *Entomophagous insects.* New York.
Biology and figures of sub-imaginal stages of platygastrids and scelionids.

Fitton, M.G. *et al.* 1978. A check list of British Insects. *Handbooks for the Identification of British Insects* 11(4): 159pp.
Checklist of British species of Platygastroidea.

Gauld, I.D. & Bolton, B. (Eds) 1988. *The Hymenoptera.* British Museum (Natural History), London, and Oxford University Press, Oxford. [Reprinted 1996, with minor alterations and additions.]
Key to families.

Königsmann, E. 1978. Das phylogenetische System der Hymenoptera. Teil 3: Terebrantes (Unterordnung Apocrita). *Deutsches Entomologische Zeitschrift* 25: 1–55.
Classification of Platygastroidea.

Masner, L. 1993. Superfamily Platygastroidea *In* Goulet, H. & Huber, J.T. (Eds) *Hymenoptera of the world: an identification guide to families,* pp. 558–65. Agriculture Canada.
Classification and identification of world platygastroid families.

Masner, L. 1995. The proctotrupoid families. *In* Hanson, P.E. & Gauld, I.D. (Eds) *The Hymenoptera of Costa Rica,* pp. 209–46. The Natural History Museum, London, and Oxford University Press, Oxford.
Classification of Platygastroidea.

Masner, L. & Denis, J. (Eds) 1975 onwards. *Proctos.*
Newsletter for those interested in proctotrupoid sensu lato families (including platygastroids). See also Web site of *Proctos* at:
http://iris.biosci.ohio-state.edu/newsletters/proctos/homepg.html.

Naumann, I.D. & Masner, L. 1985. Parasitic wasps of the proctotrupoid complex: a new family from Australia and a key to world families (Hymenoptera: Proctotrupoidea s. l.). *Australian Journal of Zoology* 33: 761–83.
Classification and identification of platygastroid families.

Family Platygastridae

The Platygastridae is a large and taxonomically poorly known family with many species: two subfamilies, Platygastrinae and Sceliotrachelinae, occur in Britain. Fitton *et al.* (1978) list 157 species in 15 genera as British, but this list should be regarded with caution as most of the species were described by Francis Walker in 1835, their taxonomic status has not recently been assessed and many more await discovery. The Platygastridae remains taxonomically the most poorly studied family of Hymenoptera in Britain.

Platygastrids are endoparasitoids; most species attack Diptera, particularly the gall-forming Cecidomyiidae. *Inostemma lycon* is known to parasitise the broom-pod cecidomyiid, *Contarinia pulchripes* (Parnell, 1963). Some (e.g. species of *Allotropa*) are important parasitoids of mealy-bugs (Pseudococcidae) and a species of *Amitus* recently discovered in Britain (Polaszek, 1997) is known to parasitise whiteflies (Aleyrodidae). *Fidiobia* species are parasitic in the eggs of curculionid and chrysomelid coleopterans and a species of *Tetrabaeus* has been reared from the nest of a sphecid (Krombein, 1964; Muesebeck, 1979). Muesebeck also noted that occasional platygastrine species have been reared from hymenopterous galls, but it is noteworthy that some cecidomyiids live as inquilines in hyme-nopterous galls (Skuhravá *et al.*, 1984). The biologies of few platygastrids have been investigated and virtually all observations relate to a few species of *Platygaster* and *Inostemma* that parasitise cecidomyiids. Oviposition is into the egg of the host, or into the newly emerged larva, but development is not completed until the host larva has become fully grown. The platygastrid generally pupates within the larval skin of the host (Parnell, 1963). Although the first-instar larvae of some platygastrids are more or less hymenopteri-form, those of many species are quite extraordinary in appearance and superficially resemble cyclopoid copepods, with an inflated 'cephalothorax' carrying large curved mandibles, and a slender abdomen that often termi-nates in a bifurcate 'tail' (Clausen, 1940). When fully fed these larvae may be extremely bloated and ovoid in shape (Clancey, 1944). The exact number of larval instars appears to vary from one to three, depending upon species. The majority of developmental time is passed as an embryo or larva within the trophamnion; once the larva emerges from the trophamnion it rapidly consumes its host. Some gregarious species are known to spin cocoons within the host puparium. A number of species are of economic importance as parasitoids of cecidomyid midges, which are pests of cereals and other crops, and also as parasitoids of whiteflies (Aleyrodidae).

Very little work has been done on the British Platygastridae since the works of Walker (1835 *et seqq.*) and Haliday (1833) and it is probably the least well known family of British Hymenoptera and the most difficult to identify to species. The most recent generic keys that cover the British fauna are those of Masner & Huggert (1989), which key the genera of Platygastrinae and Sceliotrachelinae formerly assigned to Inostemmatinae, and Kozlov (1987) on genera of Platygastridae occurring in the European part of the former USSR. There are no modern keys to the majority of British species. Some British species may be keyed using Vlug (1985); however, these keys do not fully cover the fauna and appear to have been based on a small amount of type material and so should be used with great caution. *Euxestonotus* may be keyed using Buhl (1995b). The key by Gerling (1990) will allow easy recognition of reared material of the genus *Amitus*. Otherwise Kozlov's (1987) identification keys for the European part of the former USSR may be of some use. There are a number of recent additions to the British fauna (Day, 1971; Buhl, 1995a,c); given the poor state of knowledge, there are likely to be many more. The comprehensive world catalogue of Vlug (1995) should be used to supplement the current British checklist (Fitton, 1978) and Masner & Huggert (1989) should be consulted for the latest subfamily placements of genera formerly belonging to Inostemmatinae. Currently reliable identifications are not possible for the majority of specimens, particularly from the huge genus *Platygaster*, much basic taxonomic work is necessary before identification keys can be written.

Buhl, P.N. 1995a. On two European species of *Platygaster* Latreille, 1809. *Frustula Entomologica* 18(31): 147–52.
    Platygaster nottoni Buhl described from Britain.
Buhl, P.N. 1995b. Taxonomic studies on *Euxestonotus* Fouts (Hym., Platygastridae). *Entomologist's Monthly Magazine* 131: 115–21.
    Key to world species of Euxestonotus and E. hasselbalchi described from Britain.
Buhl, P.N. 1995c. Some species of Platygastridae new to the British list. *Entomologist's Monthly Magazine* 131: 122.
    New additions to the British fauna but unfortunately without keys.
Clancey, D.W. 1944. Biology of *Allotropa burrelli*, a gregarious parasite of *Pseudococcus comstocki*. *Journal of Agricultural Research* 69: 159–63.
    Biology and larval morphology.
Day, M.C. 1971. A new species of *Platygaster* Latreille (Hym., Proctotrupoidea, Platygasteridae) reproducing by thelytokous parthenogenesis. *Entomologist's Gazette* 22: 37–42.
    Platygaster virgo described from Britain with comparative notes and details of biology.
Gerling, D. 1990. Natural Enemies of Whiteflies: Predators and Parasitoids. *In* Gerling, D. (Ed.) *Whiteflies: their bionomics, pest status and management*, pp. 147–85. Intercept, Andover, Hants.
    Key to genera of whitefly (Aleyrodidae) parasitoids, allowing for easy recognition of Amitus among material known to have been reared from whiteflies.
Haliday, A.H. 1833. An essay on the Classification of the parasitic Hymenoptera of Britain, which correspond to the Ichneumones minuti of Linnaeus. *Entomological Magazine* 1: 259–76.
    Descriptions of some species from Britain, for the enthusiast. Partly in Latin.
Kozlov, M.A. 1987. Superfamily Proctotrupoidea (Proctotrupoids). Platygastridae. *In* Medvedev, G.S. (Ed.) *Keys to the insects of the European part of the USSR* III, Part 2, pp. 1180–212. Amerind Publishing, New Delhi, India.

A comprehensive synthesis of previous works covering the European part of the former USSR with keys to genera and species.

Krombein, K. V. 1964. Natural history of Plummers Island, Maryland, XVIII. The hibiscus wasp, an abundant rarity and its associates (Hymenoptera: Sphecidae). *Proceedings of the Biological Society of Washington* **77**: 73–112.
Biology.

Marchal, P. 1906. Recherches sur la biologie et le développement des Hyménoptères parasites. I. La polyembryonie specifique ou germinogonie. *Archives de Zoologie Expérimentale et Générale* **2**: 257–335.
Biology.

Masner, L. & Huggert, L. 1989. World review and keys to genera of the subfamily Inostemmatinae with reassignment of the taxa to the Platygastrinae and Sceliotrachelinae (Hym., Platygastridae). *Memoirs of the Entomological Society of Canada* **147**: 1–214.
Keys to world genera of Inostemmatinae illustrated with line drawings and scanning electron micrographs.

Muesebeck, C.F.W. 1979. Proctotrupoidea and Ceraphronoidea. *In* Krombein, K.V., Hurd, P.D., Smith, D.R. & Burks, B.D. (Eds) *Catalog of Hymenoptera in America North of Mexico.* **1**: 1121–95.

Myers, J.G. 1927. Natural enemies of the pear leaf-curling midge, *Perissia pyri* Bouché (Dipt., Cecidom.). *Bulletin of Entomological Research* **18**: 129–138.
Biology.

Parnell, J.R. 1963. Three gall midges (Diptera: Cecidomyiidae) and their parasites found in the pods of the broom (*Sarothamnus scoparius* (L.) Wimmer). *Transactions of the Royal Entomological Society of London* **115**: 261–75.
Biology.

Polaszek, A. 1997. *Amitus* Haldeman (Hym., Platygastridae): A genus of whitefly parasitoids new to Britain. *Entomologist's Monthly Magazine* **133**: 77–9.
Amitus, a genus new to Britain, with a note of caution on species identification and details of biology.

Skuhravá, M., Skuhravy, V. & Brewer, J.W. 1984. Biology of gall midges. *In* Ananthakrishnan, T.N. (Ed.) *The biology of gall insects*, pp. 169–222. London.
Biology.

Vlug, H.J. 1985. The types of Platygastridae (Hymenoptera, Scelionidae) described by Haliday and Walker and preserved in the National Museum of Ireland and in the British Museum (Natural History). 2. Keys to species, redescriptions, synonymy. *Tijdschrift voor Entomologie* **127**(1984): 179–224.
Contains keys to 97 British species of Platygastridae but should be used with caution as it is not a comprehensive treatment of the British fauna and was apparently based largely on an examination of types without a consideration of the full range of variation exhibited by non-type material.

Vlug, H.J. 1995. Catalogue of the Platygastridae (Platygastroidea) of the World. *Hymenopterorum Catalogus (nova editio)* **19**: 1–168.
Comprehensive, world, taxonomic catalogue.

Walker, F. 1835. On the species of *Platygaster* &c. *Entomologist's Monthly Magazine* 97: 240–4.

> Descriptions of a major part of the British fauna, only for the most determined enthusiast. In Latin.

Family Scelionidae

The Scelionidae is a large and taxonomically poorly known family with numerous species in Britain. Three subfamilies, Scelioninae, Teleasinae and Telenominae, are recognised (Masner, 1976), and all are represented in Britain. Fitton *et al.* (1978) list 102 species in 14 genera as occurring in Britain but this is probably an underestimate as certain genera such as *Trimorus*, *Trissolcus* and *Telenomus* may contain many, very similar, species.

Scelionids are endoparasitoids in the eggs of insects and some other arthropods, and development is always completed within a single host egg. The overwhelming majority of species are primary parasitoids, but one European species may develop as a facultative hyperparasitoid (Viktorov, 1966). Most species are solitary and strongly avoid superparasitism. A few species of telenomine that attack large eggs are known to be gregarious. In these cases usually between five and ten individuals complete development in a single host egg. The majority of scelionids are solitary parasitoids, but most attack hosts that are clumped, although a few species do search out and oviposit into solitary eggs.

Scelionids as a group attack a wide variety of arthropod eggs in various situations. However, most species are fairly host-specific; many are restricted to a single host species, and most others attack species in one family. Only occasional records exist of scelionids ovipositing into the eggs of species of several families, and no one species is known to attack hosts of different orders (Johnson, 1984). Species of the subfamily Telenominae are mostly associated with lepidopteran (*Telenomus*) and heteropteran (*Trissolcus*) eggs, and a number of species are important as biocontrol agents (reviews: Orr, 1988; Bin & Johnson, 1982). The Scelioninae is the most primitive subfamily morphologically, and shows the most diverse host relationships. Some genera (e.g. *Scelio*) attack orthopteran eggs, others (e.g. *Gryon*) heteropteran eggs and several (e.g. *Idris, Baeus*) spiders' eggs. The subfamily Teleasinae is the least well known even though there are many British species (e.g. *Trimorus*). As far as is known teleasines parasitise eggs of Coleoptera. Phoresy is probably a more common phenomenon in the Scelionidae than in any other family of Parasitica (Clausen, 1976). Several *Telenomus* species have often been found among the long pubescence of adult lasiocampids and other large moths. The European species *Mantibaria seefelderiana* is phoretic on the preying mantis (*Mantis religiosa*) (Couturier, 1941). These phoretic scelionids ride on the body of the adult whose eggs they parasitise, and alight while the insect is ovipositing in order to be able to parasitise freshly laid eggs (Clausen, 1940). Female scelionids oviposit into the eggs in almost any stage of development, although some species apparently can only successfully parasitise relatively freshly laid eggs. Some are known to introduce a venom that curtails further host development (Strand *et al.*, 1986).

British Scelionidae are poorly known and identification is often difficult. The genera may be keyed by using Masner (1980). Kozlov's (1987) generic key is less modern. There are no modern keys to the majority of British species. A few genera are keyed to species; *Platytelenomus* (now part of *Telenomus*) is keyed by Fergusson (1983) and *Idris* is keyed by Huggert (1979). Given the absence of keys covering the British species, keys for adjacent regions of Europe may be helpful for *Gryon* (Kozlov & Kononova, 1989; Mineo, many small papers, see Johnson, 1992), for Telenominae (Kozlov & Kononova, 1983) and for Scelioninae (Kozlov & Kononova, 1990). Given the poor state of knowledge, there are likely to be numerous additions to the fauna in future. The comprehensive world catalogue of Johnson (1992) should be used to supplement the current British checklist (Fitton, 1978). Currently reliable identifications are not possible for the majority of specimens, particularly from the larger genera, and much basic taxonomic work is necessary before identification keys can be written.

Askew, R.R. 1971. *Parasitic insects*. Heinemann, London.
    General biology and figure of egg marking in *Trissolcus basalis*.
Bin, F. & Johnson, N.F. 1982. Potential of Telenominae in biocontrol with egg parasitoids (Hym., Scelionidae). *Colloques de l'INRA* 9: 275–87.
    Review of biocontrol using telenomines.
Clausen C.P. 1976. Phoresy among entomophagous insects. *Annual Review of Entomology* 21: 343–68.
    Biology.
Couturier, A. 1941. Nouvelles observations sur *Rielia manticida* Kief. Hyménoptère (Proctotr. Scelion.) parasite de la mante religieuse, II. Coportement de l'insecte parfait. *Revue de Zoologie Agricole et Appliquée* 40: 49–62.
    Biology.
Fedde, G.F. 1977. A laboratory study of egg parasitization capabilities of *Telenomus alsophilae*. *Environmental Entomology* 6: 773–6.
    Biology.
Fergusson, N.D.M. 1983. A review of the genus *Platytelenomus* Dodd (Hym., Proctotrupoidea). *Entomologist's Monthly Magazine* 119: 199–206.

Key to world *Platytelonomus*; addition of *Platytelonomus* to the British list
(*Platytelonomus* is now considered a synonym of *Telenomus*).

Huggert, L. 1979. Revision of the west Palaearctic species of the genus *Idris* Forster
s.l. (Hymenoptera, Proctotrupoidea: Scelionidae). *Entomologica Scandinavica.*
(suppl.) **12**: 1–60.
Key to females of west Palaearctic species of *Idris*.

Johnson, N.F. 1984. Systematics of Nearctic *Telenomus*. Classification and revisions
of the *podisi* and *phymatae* species groups (Hymenoptera: Scelionidae). *Bulletin
of the Ohio Biological Survey (New Series)* **6**(3): 1–113.
Biology.

Johnson, N.F. 1992. Catalog of World Proctotrupoidea excluding Platygastridae.
Memoirs of the American Entomological Institute **51**: 1–825.
Comprehensive, annotated, world, taxonomic catalogue, with notes on which publica-
tions contain identification keys.

Kozlov, M.A. 1987. Superfamily Proctotrupoidea (Proctotrupoids). Platygastridae. *In*
Medvedev, G.S. (Ed.) *Keys to the insects of the European part of the USSR* III, Part
2, pp. 1110–79. Amerind Publishing, New Delhi, India.
A synthesis of keys to genera and species of Scelionidae of the European part of the
former USSR.

Kozlov, M.A. & Kononova, S.V. 1983. Telenominae of the USSR (Hymenoptera,
Scelionidae, Telenominae). *Opredeliteli po Faune SSSR* **136**: 1–329.
Key to genera and species of Scelionidae, Telenominae of the former USSR. In Russian.

Kozlov, M.A. & Kononova, S.V. 1989. New species of the genus *Gryon* Haliday
(Hymenoptera, Scelionidae) of the USSR and neighbour countries. *Trudy
Zoologicheskogo Instituta Akademii Nauk SSSR* **188**: 78–100.
Key to Palaearctic species of *Gryon*. In Russian.

Kozlov, M.A. & Kononova, S.V. 1990. Stselionidy fauny SSSR: (Hymenoptera,
Scelionidae, Scelioninae). [Scelioninae of the fauna of the USSR: (Hymenoptera,
Scelionidae, Scelioninae).] *Opredeliteli po Faune SSSR* **161**: 1–344.
Key to genera and species of Scelionidae, Scelioninae of the former USSR. In Russian.

Masner, L. 1976. Revisionary notes and keys to World genera of Scelionidae
(Hymenoptera: Proctotrupoidea). *Memoirs of the Entomological Society of
Canada* **97**: 1–87.
Key to world genera covering those occurring in Britain.

Masner, L. 1980. Key to genera of Scelionidae of the Holarctic region with descriptions of new genera and species. *Memoirs of the Entomological Society of Canada*
113: 1–54.
Comprehensive key to Holarctic genera covering those occurring in Britain, illustrated
with scanning electron micrographs.

Mineo, G. 1978a. Studi Morfo-biologici comparativi sugli studi preimmaginali degli
scelionidi (Hym. Proctotrupoidea). II. Su alcune specie del genere *Gryon* Haliday
and *Telenomus heydeni* Mayr. *Bollettino dell'Istituto di Entomologia Agraria e
dell'Osservatorio di Fitopatologia di Palermo* **10**: 81–94.
Biology and morphology of the subimaginal stages of *Gryon monspeliense* and
Telenomus heydeni. In Italian.

Mineo, G. 1978b. Studi Morfo-biologici comparativi sugli studi preimmaginali degli scelionidi (Hym. Proctotrupoidea). III. Nota su *Mantibaria manticida* (Kieff.). *Bollettino dell'Istituto di Entomologia Agraria e dell'Osservatorio di Fitopatologia di Palermo* **10**: 95–103.

Biology and morphology of the subimaginal stages of *Mantibaria seefelderiana*. In Italian.

Mineo, G. & Sinacori, A. 1978. Studi Morfo-biologici comparativi sugli studi preimmaginali degli scelionidi (Hym. Proctotrupoidea). *Bollettino dell'Istituto di Entomologia Agraria e dell'Osservatorio di Fitopatologia di Palermo* **10**: 105–12.

Biology and morphology of the subimaginal stages of *Telenomus lopicida*. In Italian.

Orr, D.B. 1988. Scelionid wasps as biocontrol agents: a review. *Florida Entomologist* **71**: 506–28.

Review of biocontrol using scelionids.

Schneider, F. 1940. Schadinsekten und ihre Bekämpfung in ostindischen Gambirkulteren. *Mitteilungen der Schweizerischen Entomologischen Gesellschaft* **18**: 77–207.

Biology. In German.

Strand, M.R., Meola, S.M. & Vinson, S.B. 1986. Correlating pathological symptoms in *Heliothis virescens* eggs with development of the parasitoid *Telenomus heliothidis*. *Journal of Insect Physiology* **32**: 389–402.

Biology.

Viktorov, G.A. 1966. [*Telenomus sokolovi* Mayr (Hymenoptera, Scelionidae) as a secondary parasite of the eggs of *Eurygaster integriceps* Put.] *Doklady Akademia Nauk SSSR (Biol.)* **169**: 741–4.

Biology. In Russian.

Waage, J.K. 1982. Sib-mating and sex ratio strategies in scelionid wasps. *Ecological Entomology* **7**: 103–12.

Biology.

Superfamily Ceraphronoidea

The Ceraphronoidea was previously often included within the Proctotrupoidea, though currently most authors now accept it as a distinct superfamily (see, for example, Masner & Dessart, 1967; Königsmann, 1978). Two extant families are recognised, but some fossil families may also belong in this group. Ceraphronoids are usually solitary endoparasitoids of Diptera, or sometimes of Thysanoptera, Neuroptera or Homoptera, or hyperparasitic via aphidiine or microgasterine braconids, chalcidoids or cynipoids.

Identification of British Ceraphronoidea to family is covered by Dessart & Cancemi (1986). Gauld & Bolton (1988), Fergusson (1978) and the world treatment of Masner (1993) are also useful. The most recent British checklist is that of Fitton (1978). News and views are available in *Proctos*, a newsletter for those interested in Proctotrupoidea *sensu lato* (Masner & Denis, 1975).

General references

Clausen, C.P. 1940. *Entomophagous insects.* McGraw Hill, New York and London.
   Biology and figures of subimaginal stages of ceraphronoids.
Dessart, P. & Cancemi, P. 1986. Identification Key to ceraphronoid genera with com-
   ments and new species. *Frustula Entomologica, Nouvelles Series* VII–VIII
   (XX–XXI): 307–72.
   Illustrated key to world families and genera of Ceraphronoidea.
Fitton, M.G. *et al.* 1978. A check list of British Insects. *Handbooks for the
   Identification of British Insects* 11(4): 159pp.
   Checklist of British species.
Gauld, I.D. & Bolton, B. (Eds) 1988. *The Hymenoptera.* British Museum (Natural
   History), London, and Oxford University Press, Oxford. [Reprinted 1996, with
   minor alterations and additions.]
   Key to British families of Ceraphronoidea.
Königsmann, E. 1978. Das phylogenetische System der Hymenoptera. Teil 3:
   Terebrantes (Unterordnung Apocrita). *Deutsches Entomologische Zeitschrift* 25:
   1–55.
   Systematics of families of Ceraphronoidea; in German.
Masner, L. 1993. Superfamily Ceraphronoidea. *In* Goulet, H. & Huber, J.T. (Eds.)
   Hymenoptera of the world: an identification guide to families, pp. 566–9.
   Agriculture Canada.
   Classification and identification of ceraphronoid families.
Masner, L. & Denis, J. (Eds.) 1975 onwards. *Proctos.*
   **Newsletter for those interested in proctotrupoid (*sensu lato*) families (including platy-
   gastroids). See also Web site of *Proctos* at:**
   http://iris.biosci.ohio-state.edu/newsletters/proctos/homepg.html.
Masner, L. & Dessart, P. 1967. La reclassification des catégories taxonomiques
   supérieures des Ceraphronoidea (Hymenoptera). *Bulletin d'Institut Royal des
   Sciences Naturelles de Belgique* 43(22): 1–33.
   Classification of Ceraphronoidea; in French.

Family Ceraphronidae

The Ceraphronidae is a moderately large family that is represented in Britain
by 26 species in three genera. Very few observations have been published
about the biology of ceraphronids, but references suggest they develop as
endoparasitoids (Parnell, 1963; Dessart, 1964). Many are known to attack the
larvae of nematocerous dipterans, especially those of predatory cecidomyiids. For example, *Aphanogmus venustus* is known to be an endoparasitoid of
the larvae of *Lestodiplosis* (Cecidomyiidae). The ceraphronid larva pupates
within the skin of the mature host larva (Parnell, 1963). In Europe,
Aphanogmus strobilorum is known to parasitise a phytophagous cecidomyiid

(Bakke, 1955) and another species of the same genus is an endoparasitoid of the pre-adult stages of thrips (Dessart & Bournier, 1971). Some tropical species are apparently larval/pupal endoparasitoids of Lepidoptera or hyperparasitoids through microgasterine braconids; other species are known to parasitise chrysopid and coniopterygid lacewings (Priesner, 1936; Chiu *et al.*, 1981).

British ceraphronids may be identified to genus by using Dessart & Cancemi (1986). There are no modern keys to the majority of species although there are many small papers on European species by Dessart (1963 onwards), which are helpful, and Parr (1960) provided comparative descriptions of some British *Aphanogmus*. The keys by Alekseev (1987) to species occurring in the European part of the former USSR may also be useful.

Alekseev, V. N. 1987. Superfamily Ceraphronoidea. Ceraphronidae. *In* Medvedev, G.S. (Ed.) *Keys to the insects of the European part of the USSR* III, Part 2, pp. 1240–57. Amerind Publishing, New Delhi, India.
Keys to genera and species of Ceraphronidae occurring in the European part of the former USSR.

Bakke, A. 1963. Insects reared from spruce cones in Northern Norway, 1951. *Norsk Entomologisk Tidsskrift* 9: 152–212.
Biology.

Chiu, S.-C., Chou, L.Y. & Chou, K.C. 1981. A preliminary survey on the natural enemies of *Kerria lacca* (Kerr) in Taiwan. *Journal of Agricultural Research of China* 30: 420–5.
Biology.

Dessart, P. 1963. Contribution à l'étude des hyménoptères Proctotrupoidea (II). Révision de *Aphanogmus* (Ceraphronidae) decrits par C. G. Thomson. (III) Révision du genre *Allomicrops* Kieffer, 1914, et description de *Ceraphron masneri* sp. n. (Ceraphronidae). *Bulletin et Annales de la Société Royale Belge d'Entomologie* 99: 387–416; 513–39.
Descriptions of some European *Aphanogmus* but without keys. In French.

Dessart, P. 1964. Contribution à l'étude des hyménoptères Proctotrupoidea (IV). Trois Ceraphronidae parasites de la cecidomyie du colza: *Dasyneura brassicae* (Winnertz) en France. *Bulletin et Annales de la Société Royale Belge d'Entomologie* 100: 109–30.
Descriptions of some European *Aphanogmus* and *Ceraphron* but without keys. In French.

Dessart, P. 1965. Contribution à l'étude des hyménoptères Proctotrupoidea (VI). Les Ceraphronidae et quelques Megaspilinae (Ceraphronidae) du Musée Civique d'Histoire Naturelles de Gênes. *Bulletin et Annales de la Société Royale Belge d'Entomologie* 101: 105–92.
Descriptions of some European *Ceraphron;* key to *Ceraphron* with reduced wings. In French.

Dessart, P. 1996. Notule hymenopterologiques n°s 10–21 (Ceraphronoidea; Chalcidoidea Pteromalidae). *Bulletin et Annales de la Société Royale Belge d'Entomologie* **132**: 277–99.
Nomenclature of Ceraphronidae. In French.

Dessart, P. & Bournier, A. 1971. *Thrips tabaci* Lindman (Thysanoptera), hôte inattendu d'*Aphanogmus fumipennis* (Thomson) (Hym. Ceraphronidae). *Bulletin et Annales de la Société Royale Belge d'Entomologie* **107**: 116–17.
Biology.

Parnell, J.R. 1963. Three gall midges (Diptera: Cecidomyiidae) and their parasites found in the pods of broom (*Sarothamnus scoparius* (L.) Wimmer). *Transactions of the Royal Entomological Society of London* **116**: 255–73.
Biology.

Parr, M.J. 1960. Three new species of *Aphanogmus* (Hymenoptera: Ceraphrontidae) from Britain, with a redescription of *A. fumipennis* Thoms., 1858, a species new to Britain. *Transactions of the Society for British Entomology* **14**: 115–30.
Additions of *Aphanogmus* to the British fauna with comparative notes.

Priesner, H. 1936. *Aphanogmus steinitzi* spec. nov., ein Coniopterygiden-Parasit (Hymenoptera-Proctotrupoidea). *Bulletin de la Société Royale Entomologique d'Egypte* **20**: 248–51.
Biology.

Family Megaspilidae

The Megaspilidae is a moderately large family well represented in Britain. It is divided into two subfamilies, the Lagynodinae and Megaspilinae. The former subfamily, in which the females have reduced wings and a strongly modified thorax, contains only two British species in one genus. The Megaspilinae is represented in Britain by 59 species in four genera.

Megaspilids are ectoparasitoids of a wide range of primary hosts including Homoptera (coccids), Mecoptera (boreids), Neuroptera (hemerobiids, chrysopids and coniopterygids) and Diptera (cecidomyiids, syrphids, chloropids, chamaemyids and muscids). Many megaspilids are hyperparasitic and attack a variety of aphidiine braconids, chalcidoids and cynipoids (Cooper & Dessart, 1975). These are sometimes reared from agriculturally important hosts. One genus in particular, *Dendrocerus*, is economically important because most species are hyperparasitoids of aphids, via aphelinid and aphidiine braconid primary parasitoids. Haviland (1920) studied the biology of *D. carpenteri*, one of the commonest hyperparasitic species. She observed that this species was a pseudohyperparasitoid as it only oviposited on aphidiine larvae or pupae inside a completely consumed aphid. The megaspilid developed as an ectoparasitoid and passed through four larval instars. The first three larval instars were observed to be hymenopteriform and the final-instar larva was similar but possessed a short anal cornus. Clausen (1940) figured the larva of a *Dendrocerus* species.

British megaspilids may be identified to genus by using Dessart & Cancemi (1986); Fergusson (1980) also provides a generic key, which is less up-to-date but easier to use as it covers only British genera. There are some modern keys covering British species; Lagynodinae are keyed by Dessart (1987; see also 1966a), *Dendrocerus* are keyed by Fergusson (1978) and European megaspilids attacking syrphid flies are keyed by Dessart (1974). A number of small papers on European species by Dessart (1965 onwards) and Alekseev's (1987) keys to species occurring in the European part of the former USSR may also be helpful.

Alekseev, V.N. 1987. Superfamily Ceraphronoidea. Megaspilidae. *In* Medvedev, G.S. (Ed.) *Keys to the insects of the European part of the USSR* III, Part 2, pp. 1216–57. Amerind Publishing, New Delhi, India.
  Keys to genera and species occurring in the European part of the former USSR.

Cooper, K. W. & Dessart, P. 1975. Adult, larva and biology of *Conostigmus quadratogenalis* Dessart & Cooper, sp. n. (Hym. Ceraphronoidea), parasite of *Boreus* (Mecoptera) in California. *Bulletin et Annales de la Société Royale Belge d'Entomologie* **111**: 37–53.
  Biology.

Dessart, P. 1965a. Contribution à l'étude des hyménoptères Proctotrupoidea (VI). Les Ceraphronidae et quelques Megaspilinae (Ceraphronidae) du Musée Civique d'Histoire Naturelles de Gênes. *Bulletin et Annales de la Société Royale Belge d'Entomologie* **101**: 105–92.
  Descriptions of some European *Dendrocerus* and *Trichosteresis* but without a key. In French.

Dessart, P. 1965b. Contribution à l'étude des hyménoptères Proctotrupoidea (IX). Révision du genre *Macrostigma* Rondani, 1877 (Ceraphronidae, Megaspilinae). *Redia* **49**: 157–63.
  Description of *Dendrocerus aphidum*. In French.

Dessart, P. 1966a. Contribution à l'étude des hyménoptères Proctotrupoidea (X). Révision des genres *Lagynodes* Forster, 1840, et *Plastomicrops* Kieffer, 1906 (Ceraphronidae). *Bulletin de l'Institut Royal des Sciences Naturelles de Belgique* **42**(18): 1–85.
  Key to female *Lagynodes*. In French.

Dessart, P. 1966b. Contribution à l'étude des hyménoptères Proctotrupoidea (XII). A propos des Ceraphronidae Megaspilinae males a antennes rameuses. *Bulletin de l'Institut Royal des Sciences Naturelles de Belgique* **42**(32): 1–16.
  On *Dendrocerus* males. In French.

Dessart, P. 1974. Les mégaspilides européens (Hym. Ceraphronoidea) parasites des diptères syrphides avec une révision du genre *Trichosteresis*. *Annales de la Société Entomologique de France* **10**: 395–448.
  Description of some European *Dendrocerus, Trichosteresis* and *Conostigmus*; key to European species reared from Syrphidae. In French.

Dessart, P. 1985. Les *Dendrocerus* a notaulices incompletes (Hymenoptera Ceraphronoidea Megaspilidae). *Bulletin et Annales de la Société Royale Belge d'Entomologie* 121(9) 1985: 409–58.

   Identification key to males of *Dendrocerus* with incomplete notauli. In French.

Dessart, P. 1987. Revision des Lagynodinae (Hymenoptera Ceraphronoidea Megaspilidae). *Bulletin de l'Institut Royal des Sciences Naturelles de Belgique, Entomologie* 57: 5–30.

   Key to world species of Lagynodinae. In French.

Dessart, P. 1996a. Hymenoptera Ceraphronoidea nouveaux ou peu connus (n°. 2). *Bulletin et Annales de la Société Royale Belge d'Entomologie* 132: 45–62.

   Conostigmus linearis added to the British fauna. In French.

Dessart, P. 1996b. Notules hymenopterologiques n°s 10–21 (Ceraphronoidea; Chalcidoidea Pteromalidae). *Bulletin et Annales de la Société Royale Belge d'Entomologie* 132: 277–99.

   Nomenclature of Megaspilidae. In French.

Fergusson, N.D.M. 1980. A revision of the British species of *Dendrocerus* Ratzeburg (Hymenoptera: Ceraphronoidea) with a review of their biology aphid hyperparasites. *Bulletin of the British Museum (Natural History)* (Entomology) 41: 255–314.

   Key to British subfamilies of Ceraphronoidea, genera of Megaspilidae and species of *Dendrocerus*.

Haviland, M.D. 1920. On the bionomics and development of *Lygocerus testaceimanus*, Kieffer, and *Lygocerus cameroni* Kieffer, (Proctotrypoidea-Ceraphronidae) parasites of *Aphidius* (Braconidae). *Quarterly Journal of Microscopical Science* 65: 101–27.

   Biology.

Superfamily Chalcidoidea

This superfamily of parasitic wasps is one of the largest and most poorly known group of insects. In the British Isles approximately 1800 species have been found to date, but undoubtedly several hundreds are yet to be discovered. Chalcids probably have a greater range of biological diversity than any other similarly sized group of insects. Most species are parasitoids, but several families contain species that are phytophagous, e.g. Eurytomidae, Pteromalidae, and Torymidae. In Britain, phytophagous species feed on the endosperm of seeds of legumes (Fabaceae), pines (Pinaceae) and Rosaceae, or form galls in grass stems (Poaceae). Some chalcids have larvae that are predaceous on eggs or larvae of the host, e.g. some Pteromalidae, Euryomidae and Encyrtidae. Parasitoid biology is extremely varied in the Chalcidoidea. There are solitary and gregarious species; ectoparasitoids and endoparasitoids; primary, secondary and tertiary parasitoids; polyembryonic species; and species with planidial larvae. Some species are extremely polyphagous whereas others may be very host-specific. All stages of hosts are attacked, from the egg (which may be parasitised by species of Mymaridae, Trichogrammatidae, Eulophidae, Encyrtidae and Aphelinidae) to the pupa (attacked by several groups of Pteromalidae in particular) or adult (some Encyrtidae and Pteromalidae). Chalcids attack insects in 339 families representing 15 different orders, including all endopterygote orders, many exopterygotes and also some arachnids (including pseudoscorpions, ticks and mites). Homoptera (especially Coccoidea and Aphidoidea) are attacked as eggs, nymphs or adults by a variety of chalcids, but especially species of the Encyrtidae and Aphelinidae. The eggs of Psocoptera are parasitised by mymarids and the eggs of Thysanoptera by trichogrammatids. Nymphal Thysanoptera are parasitised by some Eulophidae, and immature Acari are attacked by encyrtids. A few species of Eulophidae, Encyrtidae and Pteromalidae are associated with spider egg sacs. This superfamily is the most successful group used in applied biological control.

Because of their relatively small size chalcids receive very little attention from amateur entomologists and for this reason virtually no species have common names. In general, identification is very difficult. This is due to a variety of reasons. Their small size makes them very difficult to study except with good-quality optical equipment. Further to this, the classification of the Chalcidoidea is far from ideal with an overabundance of families, at present 21 families being recognised worldwide with 17 occurring in Britain. As a result of this their identification, even to family level, can be very difficult. Keys to the currently recognised families of Chalcidoidea based on the British fauna can be found in Graham (1969; see references for Pteromalidae) and Gauld & Bolton (1988; see general introduction to Hymenoptera). Other keys, based on the same family arrangement, can be found in Goulet & Huber (1993) (see general introduction to Hymenoptera), Peck *et al.* (1966), Boucek (1988) and Gibson *et al.* (1987).

General references

Askew, R.R. 1961. On the biology of the inhabitants of oak galls of Cynipidae (Hym.) in Britain. *Transactions of the Society for British Entomology* 14(11): 237–68.
    Includes a key to many of the British species of Chalcidoidea associated with cynipid galls on oak.
Askew, R.R. & Shaw, M.R. 1974. An account of the Chalcidoidea (Hymenoptera) parasitising leaf-mining insects of deciduous trees in Britain. *Biological Journal of the Linnean Society* 6(4): 289–335.
    Provides no means by which to identify parasitoids, but a good summary information relating to chalcidoid parasitoids of the most common leaf-miners of deciduous trees in Britain.

Askew, R.R. & Ruse, J.M. 1974. The biology of some Cecidomyiidae (Diptera) galling the leaves of birch (*Betula*) with special reference to their chalcidoid (Hymenoptera) parasites. *Transactions of the Royal Entomological Society of London* **126**: 129–167.
Provides no means by which to identify parasitoids, although there are short diagnoses for many species and includes a good summary information relating to the biology of chalcidoid parasitoids of some British Cecidomyiidae.

Boucek, Z. 1988. *Australasian Chalcidoidea (Hymenoptera). A biosystematic revision of genera of fourteen families, with a reclassification of species.* CAB International, Wallingford, Oxon.
Provides a good key to the families of Chalcidoidea and also includes hundreds of excellent line drawings of Australasian representatives of all families included.

Ferrière, C. & Kerrich, G.J. 1958. Hymenoptera 2. Chalcidoidea. Section (a) Agaontidae, Leucospidae, Chalcididae, Eucharitidae, Perilampidae, Cleonymidae and Thysanidae. *Handbooks for the Identification of British Insects* **8**(2a): 40pp.
Includes a key to the families of Chalcidoidea based on an out-of-date classification; also includes keys to genera and species of the some smaller families and subfamilies.

Gibson, G.A.P., Huber, J.T. & Woolley, J.B. (Eds) 1997. *Annotated keys to the genera of Nearctic Chalcidoidea (Hymenoptera).* National Research Council Canada, NRC Research Press.
Provides a good key to the families of Chalcidoidea and also includes hundreds of excellent line drawings and electron micrographs of North American taxa, many of which occur in the British Isles.

Nikol'skaya, M. 1952. Chalcids of the fauna of the USSR (Chalcidoidea). *Opredeliteli po Faune SSSR, Izdavaemie Zoologicheskim Institutom Akademii Nauk SSR* **44**: 575pp. Akademiya Nauk SSSR, Moscow and Leningrad. (In Russian; English translation: 1963: Israeli Program for Scientific Translations, Jerusalem, 593pp.)
Can be useful for identification of British species of chalcidoids provided it is treated with a degree of caution since most keys are not based on reliably determined material.

Noyes, J.S. 1982. Collecting and preserving chalcid wasps (Hymenoptera: Chalcidoidea). *Journal of Natural History* **16**: 315–34.
A summary of preferred methods of collecting and preservation.

Noyes, J.S. 1985. Chalcidoids and biological control. *Chalcid Forum* **5**: 5–10.
A review of all classical biological programmes up to 1970 with an analysis of the importance of the various groups of natural enemies.

Parker, H.L. 1924. Recherches sur les formes postembryonaires de chalcidiens. *Annales de la Société Entomologique de France* **93**: 261–379.
Reviews the larval and pupal forms of a wide variety of chalcidoids with line drawings of the immature stages of most taxa included.

Peck, O., Boucek, Z. & Hoffer, A. 1964. Keys to the Chalcidoidea of Czechoslovakia (Insecta: Hymenoptera). *Memoirs of the Entomological Society of Canada* **34**: 1–120.
Still one of the best and easiest to use keys to families; also includes and keys to genera of chalcidoids occurring in the former Czechoslovakia, many of which are found in the British Isles.

Prinsloo, G.L. 1980. An illustrated guide to the families of African Chalcidoidea
(Insecta: Hymenoptera). *Department of Agriculture and Fisheries Science
Bulletin, Republic of South Africa* **395**: 1–66.
    Contains a good illustrated key to the families of chalcidoids occurring in Africa; even
    though of African chalcidoids the illustrations may prove helpful in recognising taxa that
    occur in the British Isles.

Family Chalcididae

Only eight species of this family are known to occur in Britain. Chalcidids are
predominantly solitary, primary endoparasitoids of Lepidoptera and Diptera,
although a few species attack Hymenoptera, Coleoptera or Neuroptera. The
British species have been recorded as endoparasitoids of Diptera, Coleoptera
and Symphyta. Most species oviposit into more or less fully grown hosts, such as
mature larvae or young pupae, although *Chalcis myrifex* oviposits into submerged *Stratiomys* larvae (Diptera: Stratiomyiidae). Females may lay up to
about 200 eggs, which are elongately oval and may sometimes have a very short
petiole. The first-instar larva may be caudate or hymenopteriform, with or
without spiracles, but with well-developed cuticular spines. Subsequent instars
are more or less hymenopteriform. Pupation takes place in the host pupa. Most
chalcidids overwinter as adult females, or as mature larvae in the hosts.

Boucek, Z. 1952. The first revision of the European species of the family Chalcididae
(Hymenoptera). *Sborník Entomologického Oddeleni Národního Musea v Praze*
27(suppl. 1): 1–108.
    A very good, well-illustrated revision of the European species, including all currently
    recorded British species.
Ferrière, C. & Kerrich, G.J. 1958. Hymenoptera 2. Chalcidoidea. Section (a)
    Agaontidae, Leucospidae, Chalcididae, Eucharitidae, Perilampidae,
    Cleonymidae and Thysanidae. *Handbooks for the Identification of British Insects*
    8(2a): 40pp.
    Keys the British species of Chalcididae with some omissions and one misidentification
    (*Inreiva subaenea*, a misidentification of *Psiliocharis subarmata* (Förster)).

Family Eurytomidae

About 100 species of Eurytomidae have been recorded in Britain. The family
contains species that exhibit a wide range of biologies. There are two groups
of phytophagous eurytomids in Britain: those that develop on endosperm in
seeds, and those that feed in plant stems, especially stems of grasses. The
seed-feeding group is represented by *Systole*, species of which feed in the
seeds of Umbelliferae, and *Bruchophagus*, many species of which develop in
the seeds of legumes (Fabaceae). The stem-mining group is represented in

Britain by *Tetramesa*, which mainly develop in the central cavity of grass stems, feeding above the nodes. In some species this may result in stunting of the flower head. The majority of species are entomophagous for at least part of their larval development, although several are known to complete their development by feeding on plant tissue. Most of these entomophagous species are ectoparasitoids of insect larvae feeding within plant tissue. Hosts attacked include Coleoptera, gall-forming Hymenoptera (mostly Cynipinae), Diptera (especially Tephritidae) and Lepidoptera. Some species can develop in a variety of ways. Another species, *Eurytoma brunniventris*, may parasitise a cynipid gall-maker, its *Synergus* inquiline or other chalcid parasitoids, or even feed on the tissue of the gall. The eurytomid egg is very characteristic, with a short process at the micropylar end and a long flattened filamentous process at the opposite end. Many externally deposited eggs are clothed with short spines, the form of which may be useful for separating sibling species (Claridge & Askew, 1960). The comparative morphology of the final-instar larvae of eight species of eurytomid was studied by Roskam (1982). Most species overwinter as mature larvae.

The identification of the genera occurring in the British Isles can be done relatively easily by using Claridge (1961b) or Peck *et al.* (1964; see general introduction to Chalcidoidea). Species identification is more difficult, especially for larger genera such as *Bruchophagus*, *Eurytoma* and *Tetramesa*, but this can be assisted by using some of the papers listed below.

Claridge, M.F. 1959a. A contribution to the biology and taxonomy of the British
    species of the genus *Eudecatoma* Ashmead (Hym., Eurytomidae). *Transactions
    for the Society of British Entomology* 13: 149–68.
    **Includes a key to the British species of *Sycophila* (=*Eudecatoma*), with description of all
    species, but very few illustrations.**

Claridge, M.F. 1959b. Notes on the genus *Systole* Walker, including a previously unde-
    scribed species (Hym., Eurytomidae). *Entomologist's Monthly Magazine* 95:
    38–43.
    Distinguishes the two recorded British species of *Systole*.

Claridge, M.F. 1961a. A contribution to the biology and taxonomy of some Palaearctic
    species of *Tetramesa* Walker (=*Isosoma* Walk.; =*Harmolita* Motsch.)
    (Hymenoptera: Eurytomidae) with particular reference to the British fauna.
    Transactions of the Entomological Society of London 113: 175–216.
    **This paper includes a key to the British species of *Tetramesa* with species diagnoses and
    illustrated descriptions of the galls formed for many of the species included.**

Claridge, M.F. 1961b. An advance towards a natural classification of eurytomid
    genera (Hym., Chalcidoidea) with particular reference to British forms.
    Transactions of the Society for British Entomology 14: 167–85.
    **Provides a key to all genera of Eurytomidae occurring in the British Isles with diagnoses
    of each genus.**

Claridge, M.F. & Askew, R.R. 1960. Sibling species in the *Eurytoma rosae* group (Hym., Eurytomidae). *Entomophaga* 5(2): 141–153.

Distinguishes the species of the *Eurytoma rosae* group using adult morphology and characters of the deposited egg.

Ferrière, C. 1950. Notes sur les *Eurytoma* (Hym., Chalcidoidea). I. Les types de Thomson et de Mayr. *Mitteilungen der Schweizerischen Entomologischen Gesellschaft* 23: 377–410.

Includes a key to all genera of Euryomidae occurring in Britain with keys to most European species of *Eurytoma* and *Bruchophagus* which are based on the examination of type or other reliably identified material.

Roskam, J.C. 1982. Larval characters of some eurytomid species (Hymenoptera, Chalcidoidea). *Proceedings, Koninklijke Nederlandse Akademie van Wetenschappen* 85: 293–305.

Compares the morphology of the final instars of eight species of larvae of the genera *Eurytoma* and *Tetramesa*.

Szelényi, G. 1976a. Mongolian eurytomids (Hymenoptera: Chalcidoidea). II. *Acta Zoologica Academiae Scientiarum Hungaricae* 22(1/2): 173–87.

This paper includes a simple key to most of the described Palaearctic species of *Eurytoma*, although it should be treated with caution because it is doubtfully based on reliably identified material.

Szelényi, G. 1976b. Mongolian eurytomids (Hymenoptera: Chalcidoidea). III. *Acta Zoologica Academiae Scientiarum Hungaricae* 22(3/4): 397–405.

Includes a simple key to most of the described Palaearctic species of *Bruchophagus*, although is should be treated with caution because it is doubtfully based on reliably identified material.

Family Torymidae

Over 100 species of Torymidae have been recorded in Britain with more than two-thirds of these belonging to the genus *Torymus*. Torymids are generally entomophagous, although a number of species are phytophagous either directly or as inquilines in galls, i.e. the larvae feed in turn on gall-maker and gall tissue. Several torymids are associated with aculeate Hymenoptera, e.g. *Diomorus armatus*, parasitic on stem-nesting sphecids, and *Monodonto-merus obscurus*, a common ectoparasitoid of various solitary bees. The majority of entomophagous species are ectoparasitoids of the inhabitants of plant galls. The phytophagous species are specialists that feed on the endo-sperm in developing seeds. The majority of these belong to the genus *Megastigmus*, the larvae of which develop in seeds of various Cupressaceae, Pinaceae and arborescent Rosaceae. Most species are restricted to a single plant genus and several achieve pest status, e.g. *Megastigmus spermatrophus* on *Pseudotsuga menziesii* and *Megastigmus suspectus* on *Abies alba*. Several species of *Torymus* are known to feed in seeds of various arborescent Rosaceae such as *Sorbus*, *Pyrus*, *Malus* and *Crataegus*.

Torymids lay kidney-shaped to very elongate ovoid eggs in, on, or near to the food source. The eggs of *Monodontomerus* bear numerous minute recurved spines. The larvae are hymenopteriform, and their cuticle bears setae; these setae are most conspicuous in parasitic species, but are inconspicuous in phytophagous species. Overwintering is generally as a fully fed larva, and pupation generally occurs the following year. The female pupa has the ovipositor externally visible and bent over its dorsum.

In general, the identification of torymids can be difficult, although keys to genera are given by Peck *et al.* (1964; see general introduction to Chalcidoidea), Hoffmeyer (1930–31) and Grissell (1995). Of the larger genera, the species of *Megastigmus* are perhaps the easiest to identify since a key to all British species is provided by Boucek (1970b); species of *Torymus* can be identified using Graham & Gijswijt (1998).

Boucek, Z. 1970a. On some new or otherwise interesting Torymidae, Ormyridae, Eurytomidae and Pteromalidae (Hymenoptera), mainly from the Mediterranean subregion. *Bollettino del Laboratorio di Entomologia Agraria 'Filippo Silvestri' di Portici* 27: 27–54.
   Includes a key to the known European species of *Gylphomerus*.
Boucek, Z. 1970b. On some British *Megastigmus* (Hym. Torymidae), with a revised key to the West European species. *Entomologist's Gazette* 21: 265–75.
   Provides a key to all recorded British species of *Megastigmus*, with some notes on hosts.
Graham, M.W.R. de V. & Gijswijt, M.J. 1998. Revision of the European species of *Torymus* Dalman (s.lat.) (Hymenoptera: Torymidae). *Zoologische Verhandelingen* 317: 1–202.
   Includes a key and host records for all British species.
Grissell, E.E. 1995. Toryminae (Hymenoptera: Chalcidoidea: Torymidae): a redefinition, generic classification and annotated world catalogue of species. *Memoirs on Entomology, International* 2: 474pp.
   Results of a phylogenetic analysis of the taxa of Toryminae with keys to subfamilies of Torymidae and tribes and genera of Toryminae with a complete catalogue and host-parasitoid list of world species of Toryminae.
Hoffmeyer, E.B. 1930. Beiträge zur Kenntnis der dänischen Callimomiden, mit Bestimmungstabellen der europäischen Arten. (Hym. Chalc.) (Slutning) *Entomologiske Meddelelser* 17: 232–60.
   Key to genera of Torymidae based on outdated classification; keys to most British species of *Torymus* (as *Callimome*), *Monodontomerus*, *Gylphomerus* and *Pseudotorymus*.
Hoffmeyer, E.B. 1931. Beiträge zur Kenntnis der dänischen Callimomiden, mit Bestimmungstabellen der europäischen Arten. (Hym., Chalc.) (Callimomidenstudien 5.). *Entomologiske Meddelelser* 17: 261–85.
   Key to species of most British species of *Megastigmus* with a review of hosts of Torymidae.

Family Ormyridae

A small family with only three recorded British species. Many species of ormyrid are parasitoids of various gall-forming insects. They are known to attack cecidogenic cynipids, chalcidoids and Diptera, and a few may also parasitise phytophagous eurytomids in seeds.

British species can be identified by using Hoffmeyer (1931) or Nikol'skaya (1952; see general introduction to Chalcidoidea).

Doganlar, M. 1991. Systematic positions of some taxa in Ormyridae and descriptions of a new species in *Ormyrus* from Turkey and a new genus in the family (Hymenoptera, Chalcidoidea). *Türkiye Entomoloji Dergisi* 15(1): 1–13.
Includes a key to genera of Ormyridae.
Hoffmeyer, E.B. 1931. Beiträge zur Kenntnis der dänischen callimomiden, mit Bestimmungstabellen der europäischen Arten. (Hym., Chalc.) (Callimomidenstudien 5.). *Entomologiske Meddelelser* 17: 261–85.
Includes a key to all recorded British species of *Ormyrus*.
Nieves Aldrey, J.L. 1984d. First data on the representatives of the family Ormyridae in Spain, with the description of a new species (Hymenoptera, Chalcidoidea). *Graellsia* 40: 119–28. [In Spanish.]
Contains diagnostic notes and summary of hosts of species occurring in British Isles.

Family Eucharitidae

The single species of this family, *Eucharis adscendens*, has been recorded in the British Isles only from South Wales prior to 1846 and Norfolk in 1907. Boucek (1956) has made some observations on its biology as a parasitoid of *Formica rufa* (Hymenoptera: Formicidae). The adult female lays large numbers of eggs in the closed flower of *Falcaria vulgaris* (Apiaceae) near the nest of its host. The first-instar larva is of the planidial type and has been described in detail by Boucek. The first-instar larva attaches itself to a passing worker ant and is then carried into the nest. Inside the nest it actively searches out ant larvae, on which it then feeds.

The adult of *Eucharis adscendens* has been figured by Ferrière & Kerrich (1958; see general introduction to Chalcidoidea).

Identification: Ferrière & Kerrich (1958) (British species, with whole insect figure: see introduction to Chalcidoidea).

Boucek, Z. 1956. A contribution to the biology of *Eucharis adscendens* (F.) (Hymenoptera). *Acta Societatis Zoologicae Bohemoslovenicae* 20: 97–99.
Notes on the biology of *Eucharis adscendens*.

Family Perilampidae

A family containing only about ten species in Britain. Many species are hyperparasitic, developing on tachinid and ichneumonoid primary parasitoids of Lepidoptera and Symphyta. Some of these may be facultative hyperparasitoids and capable of developing equally well on the primary or parasitoid host. Primary hosts include Diprionidae, Tenthredinidae, Chrysopidae and Nitidulidae. Species are frequently encountered feeding on flowers, although some feed on aphid honeydew or on the exudate of damaged leaves. Eggs are attached to or imbedded into foliage in the vicinity of the primary host. The first-instar larva is planidial and has sclerotised bands on most segments, which aid its locomotion. Species can be endoparasitic (subfamily Perilampinae) or ectoparasitic (subfamily Chrysolampinae). The first-instar larva of endoparasitic species normally enters the host by penetrating the skin. If, in turn, it encounters a primary parasitoid larva it will also enter this. Further development usually does not occur until the host pupates. When this occurs, the larva exits from the host and feeds externally on the host pupa. Pupation takes place in the host cocoon or puparium. Some ectoparasitic species are known to develop gregariously on beetle larvae, although only one will mature (Askew, 1980).

The family is divided into two subfamilies, the Perilampinae (in Britain represented only by the genus *Perilampus*) and the Chrysolampinae (in Britain represented only by the genus *Chrysolampus*). The latter subfamily has been included within the family Pteromalidae, but is now normally included with the Perilampidae. The British species of *Perilampus* and *Chrysolampus* can be identified by using Ferrière & Kerrich (1958; see general introduction to Chalcidoidea), Steffan (1952) or Boucek (1956).

Askew, R.R. 1980. The biology and larval morphology of *Chrysolampus thenae* (Walker) (Hymenoptera, Pteromalidae). *Entomologist's Monthly Magazine* 115: 155–9.
    An interesting account of the larva and mode of parasitism of *Chrysolampus thenae*.
Boucek, Z. 1956. Notes on the Czechoslovak Perilampidae. *Acta Faunistica Entomologica Musei Nationalis Pragae* 1: 83–98.
    Text in Czech but key to species in English and includes all British species of *Perilampus* and *Chrysolampus*.
Steffan, J.R. 1952. Les espèces françaises du genre *Perilampus* Latr. (Hym. Perilampidae). *Bulletin de la Société Entomologique de France* 52: 68–74.
    Provides a key to *Perilampus* which includes most British species.

Family Pteromalidae

The largest family of Chalcidoidea in the British Isles with nearly 630 recorded species. As might be expected of such a large group, the life histories of taxa in this family are extremely varied, the group embracing many of the life-styles exhibited by insect parasitoids. There are solitary and gregarious species, ectoparasitoids and endoparasitoids, koinobionts and idiobionts, primary and secondary parasitoids and even predators. Many pteromalids develop as solitary or gregarious ectoparasitoids of larvae and pupae of Diptera, Coleoptera, Hymenoptera, Lepidoptera and Siphonaptera. Large numbers of species attack hosts concealed in plant tissue, such as wood-borers, stem- and leaf-miners, gall-formers, etc. Some species associated with galls develop as inquilines, feeding on the gall tissue, or as parasitoids feeding externally on larvae, pupae or even adults of the gall-former (Askew, 1961). Other pteromalids are strictly ectoparasitoids but develop within the puparium of dipteran hosts. Numerous other pteromalids are endoparasitoids, e.g. *Pteromalus puparum*, a common endoparasitoid of various butterfly pupae, especially those of Papilionidae, Pieridae and Nymphalidae. Species of *Tomicobia* are unusual in developing as endoparasitoids of adult Coleoptera. Several species develop as predators rather than parasitoids. The larvae of some species of *Systasis* feed on a succession of small cecidomyiid larvae whereas species of *Panstenon* are predators of the eggs of delphacids concealed in grass sheaths, and eunotines feed on the eggs of Coccoidea. A few pteromalids are obligate hyperparasitoids, attacking aphelinids and aphidiine braconids parasitising aphids. Pupation may take place inside the dead host, externally in the vicinity of the host, or sometimes inside a host cocoon. Overwintering is usually as mature larvae, but some species overwinter as pupae or as adults, often in the foliage of conifers, in leaf litter in tree boles, in tussocks, haystacks, birds' nests, etc.

  The family is probably polyphyletic and contains several groups which may not be directly related: it currently includes 32 separate subfamilies, of which 14 are represented in the British Isles. There are several keys to facilitate the identification of genera of pteromalids occurring in Britain, the simplest of which are those found in Peck *et al.* (1964; see general introduction to Chalcidoidea) and Boucek & Rasplus (1991). However, by far the most comprehensive treatment is that of Graham (1969) which, although very difficult to use, provides keys to almost all of the British species. Ferrière & Kerrich (1958; see general introduction to Chalcidoidea) also includes keys to the genera and species of some of the smaller subfamilies.

Askew, R.R. 1961. A study of the biology of species of the genus *Mesopolobus* Westwood (Hymenoptera: Pteromalidae) associated with cynipid galls on oak. *Transactions of the Royal Entomological Society* **113**: 155–73.

Detailed notes on the parasitic development of pteromalids associated with oak galls in Britain as well as keys to the species.

Askew, R.R. 1972. A revision of the British species of *Halticoptera* (Hymenoptera: Pteromalidae) allied to *H. circulus* (Walker). *Journal of Entomology* (B) **41**: 45–52.

Includes an updated key to the species of *Halticoptera* most similar to *H. circulus*.

Askew, R.R. 1980. The European species of *Coelopisthia* (Hymenoptera: Pteromalidae). *Systematic Entomology* **5**: 1–6.

Provides a key to the European species of *Coelopisthia*.

Askew, R.R. & Kennaugh, J.H. 1992. A review of the British species of *Semiotellus* Westwood (Hym., Pteromalidae). *Entomologist's Monthly Magazine* **128**: 215–18.

A revised key to the four British species of *Semiotellus*.

Boucek, Z. 1958. Eine Cleonyminen-Studie; Bestimmungsstabelle der Gattungen mit Beschreibungen und Notizen, eingeschlossen einige Eupelmidae (Hym. Chalcidoidea). *Sborník Entomologického Oddeleni Národního Musea v Praze* **32**: 353–86.

A revisionary study of the genera and species of Cleonyminae.

Boucek, Z. 1961. Beiträge zur Kenntnis der Pteromaliden-fauna von Mitteleuropa, mit Beschreibungen neuer Arten und Gattungen (Hymenoptera). *Sborník Entomologického Oddeleni Národního Musea v Praze* **34**: 55–95.

Notes, with good drawings, on many British taxa.

Boucek, Z. & Rasplus, J.-Y. 1991. *Illustrated key to West-Palaearctic genera of Pteromalidae (Hymenoptera: Chalcidoidea)*. Institut National de la Recherche Agronomique, Paris.

A well-illustrated, relatively easy to use key to the genera of pteromalids known to occur in the west Palaearctic.

Graham, M.W.R. de V. 1969. The Pteromalidae of north-western Europe (Hymenoptera: Chalcidoidea). *Bulletin of the British Museum (Natural History) (Entomology)* Supplement **16**: 908pp.

The most comprehensive account of pteromalids occurring in Europe; includes all the then known British species; keys to genera and species notoriously difficult to use and require much practice to get the best out of them.

Kerrich, G.J. & Graham, M.W.R. de V. 1957. Systematic notes on British and Swedish Cleonymidae, with description of a new genus (Hym., Chalcidoidea). *Transactions of the Society for British Entomology* **12**: 265–311.

Keys to genera and species of British Cleonyminae now superseded by Graham (1969).

Parnell, J.R. 1964. Investigations on the biology and larval morphology of the insects associated with the galls of *Asphondylia sarothamni* H. Loew (Diptera: Cecidomyiidae) on broom (*Sarothamnus scoparius* (L.) Wimmer). *Transactions of the Royal Entomological Society of London* **116**: 255–73.

Includes notes on the immature stages of some pteromalids associated with cecidomyiid galls.

Varley, G.C. 1937. Description of the eggs and larvae of four species of chalcidoid Hymenoptera parasitic on the knapweed gall-fly. *Proceedings of the Royal Entomological Society of London* (B) **6**: 122–130.

A well-known work that includes notes on the morphology of some pteromalids associated with cecidomyiid galls.

Family Eupelmidae

A family with only about 20 recorded species in the British Isles. The vast majority of species of Eupelmidae are parasitic or facultatively hyperparasitic on the immature stages of other insects, with hosts recorded in the orders Lepidoptera, Homoptera, Hymenoptera, Coleoptera, Neuroptera and Orthoptera. A small number of species are predators on the eggs or larvae of various insects, or on the eggs of spiders. A few are solitary, primary endoparasitoids of eggs of Lepidoptera, Orthoptera and Hemiptera. Most eupelmids are solitary, but some species are gregarious. Most are ectoparasitoids, including some that develop gregariously on dipterous pupae within puparia. A few species are solitary endoparasitoids of Coccoidea. Overwintering is usually as a mature larva or pupa. Species of the subfamily Eupelminae are of interest because of a unique adaptation of the mesothorax for jumping, which probably rivals that of fleas (Siphonaptera). The adaptation and mechanism of jumping is quite complex and has been described in some detail by Gibson (1986). Because of these modifications for jumping, eupelmines often die in a contorted state with the head and gaster reflexed upwards and often nearly meeting over the thorax. The middle legs are often held in front of the head after death.

Only two of the three recognised subfamilies are found in Britain. The genera known to occur in Britain can be keyed out most easily by using Peck *et al.* (1964; see general introduction to Chalcidoidea), Kalina (1984) or Gibson (1989; Calosotinae only) and Gibson (1995, Eupelminae only). Of the five genera known to occur in Britain, two are represented by one species only. Of the remaining three genera, the species of *Calosota* can be separated using Graham (1969), *Eupelmus* (=*Macroneura*) with Kalina (1981b and 1988) and Nikol'skaya (1952; see general introduction to Chalcidoidea) and *Anastatus* with Kalina (1981a).

Gibson, G.A.P. 1986. Mesothoracic skeletomusculature and mechanics of flight and jumping in Eupelminae (Hymenoptera, Chalcidoidea: Eupelmidae). *Canadian Entomologist* 118(7): 691–728.

Description of thoracic structure and the jumping mechanism of female Eupelminae.

Gibson, G.A.P. 1989. Phylogeny and classification of Eupelmidae, with a revision of the World genera of Calosotinae and Metapelmatinae. *Memoirs of the Entomological Society of Canada* **149**: 121pp.
A well-illustrated revision of the world genera of two subfamilies with a catalogue of all included taxa.

Gibson, G.A.P. 1995. Parasitic wasps of the subfamily Eupelminae: classification and revision of world genera (Hymenoptera: Chalcidoidea: Eupelmidae). *Memoirs on Entomology, International* **5**: v+421pp.
A revision of the world genera of Eupelminae with a key to genera and some notes on biology.

Graham, M.W.R. de V. 1969. Some Eupelmidae (Hymenoptera: Chalcidoidea) new to Britain, with notes on new synonymy in this family. *Proceedings of the Royal Entomological Society of London* (B) **38**: 89–94.
Includes a key to the two species of *Calosota* occurring in Britain.

Kalina, V. 1981a. The Palaearctic species of the genus *Anastatus* Motschulsky, 1860 (Hymenoptera, Chalcidoidea, Eupelmidae) with descriptions of new species. *Silvaecultura Tropica et Subtropica, Prague* **8**: 3–25.
Publication difficult to obtain, but the paper includes a key to all Palaearctic species of *Anastatus*, two of which are known to occur in Britain.

Kalina, V. 1981b. The Palaearctic species of the genus *Macroneura* Walker, 1837 (Hymenoptera, Chalcidoidea, Eupelmidae), with descriptions of new species. *Sbornik Vedeckych Lesnickeho Ustavu Vysoke Skoly Zemedelske v Praze* **24**: 83–111.
Publication difficult to obtain, but paper includes a key to Palaearctic species of *Eupelmus* subg. *Macroneura*.

Kalina, V. 1984. New genera and species of Palearctic [sic] Eupelmidae (Hymenoptera, Chalcidoidea). *Silvaecultura Tropica et Subtropica, Prague* **10**: 1–29.
Publication difficult to obtain, but this paper includes a key to Palaearctic genera of eupelmids.

Kalina, V. 1988. Descriptions of new Palaearctic species of the genus *Eupelmus* Dalman with a key to species (Hymenoptera, Chalcidoidea, Eupelmidae). *Silvaecultura Tropica et Subtropica, Prague* **12**: 3–33.
Publication difficult to obtain, but paper includes a key to Palaearctic species of *Eupelmus* s.s.

Family Encyrtidae

The third largest family of Chalcidoidea in Britain with 234 recorded species. About half of the included species are associated with scale insects (Homoptera: Coccoidea), as endoparasitoids of immatures or adults, or more rarely as egg predators, or as hyperparasitoids via other Encyrtidae, or Aphelinidae, Pteromalidae, Braconidae, Dryinidae, etc. Almost all species of one subfamily (Tetracneminae) are parasitoids of Pseudococcidae; species of the other subfamily (Encyrtinae) are known to be parasitoids of a wider variety of coccoids (occasionally also of Pseudococcidae) and other insects, mites, ticks and spiders. Species belonging to the genera *Copidosoma* and *Ageniaspis* are polyembryonic parasitoids of larvae of Lepidoptera, and often cause the host larvae to become grotesquely deformed and twisted when they are killed as prepupae. Some species of *Copidosoma* can produce several thousand individuals from a single deposited egg and in some species there are two morphs of larvae: a smaller one, which develops normally, and a larger 'guard' morph, which emerges from the embryonic envelope first but fails to ecdyse and eventually disintegrates. The morphology of the egg and first-instar larva of encyrtids has been summarised by Maple (1947). The egg is characteristically dumb-bell-shaped and is laid inside the host. In many cases the stalk of the egg may remain protruding through the body wall of the host thus enabling the larva, when it hatches, to utilise atmospheric oxygen. Later-instar larvae and the pupae may be enclosed in a sheath that has anastomosed with the tracheal system of the host, e.g. *Encyrtus*. In these taxa the host may not be killed until after the adult parasitoid emerges. Overwintering is generally as a mature larva or pupa within the body of the host.

Worldwide, the Encyrtidae is one of the most important chalcidoid families for the biological control of insect pests. Several species are currently being used for the control of mealybug pests in British greenhouses, e.g. *Leptomastix dactylopii* and *Leptomastidea bifasciata*.

The identification of British taxa is not easy, although virtually all genera can be keyed relatively easily by using Peck *et al.* (1964; see general introduction to Chalcidoidea). Perhaps the best key to genera is that provided by Trjapitzin (1973, 1989) but unfortunately this is in Russian with no easily available English translation. A shortened version of this key, which has been translated into English, is provided later by Trjapitzin (1978). Most British species can be identified by using Trjapitzin (1978 and 1989), but unfortunately these are often unreliable since they are largely based on original descriptions only and not on the examination of reliably identified material. Individual keys to the species of certain genera are listed below.

Claridge, M.F. 1958. The British and Scandinavian species of the genus *Cheiloneurus* Westwood (Encyrtidae). *Entomologist's Monthly Magazine* **94**: 156–61.
  A revision and key to British species of *Cheiloneurus*.

Claridge, M.F. 1964. The Palaearctic species of *Bothriothorax* (Hym., Encyrtidae) allied to *B. paradoxus* (Dalman). *Entomophaga* **9**: 21–5.
  Describes *B. intermedia* and adds it to the British list, emending Graham's (1958) key to the genus.

Graham, M.W.R. de V. 1958. Notes on some genera and species of Encyrtidae (Hym., Chalcidoidea), with special reference to Dalman's types. *Entomologisk Tidskrift* **79**: 147–75.
Includes a key to the European species of *Bothriothorax*.

Graham, M.W.R. de V. 1969. Synonymic and descriptive notes on European Encyrtidae (Hym; Chalcidoidea). *Polskie Pismo Entomologiczne* **39**: 211–319.
Includes a key to the European species of *Encyrtus*.

Graham, M.W.R. de V. 1991. Revision of western European species of *Ericydnus* Haliday (Hym., Encyrtidae) including one species new to science. *Entomologist's Monthly Magazine* **127**: 177–89.
Includes and keys all British species of *Ericydnus*.

Hoffer, A. 1952. Monograph of the Czechoslovak species of the genus *Echthroplexiella* Mercet. *Acta Entomologica Musei Nationalis Pragae* **28**(399): 57–69.
Provides diagnoses of many species with a key to the European species then known.

Hoffer, A. 1957. Czechoslovak species of the genus *Metallon* Walker. Seventh preliminary paper for the monographic investigation of the Czechoslovak Encyrtidae (Hym., Chalcidoidea). *Acta Societatis Entomologicae Czechoslovenicae* **54**: 41–53.
Provides a key to all species of *Ectroma* (under the name *Metallon*) occurring in Britain.

Hoffer, A. 1960. The Czechoslovak species of the genus *Paralitomastix* Mercet (Hym., Chalcidoidea). *Acta Societatis Entomologicae Czechoslovenicae* **57**(2): 136–42.
Distinguishes between *Paralitomastix varicornis* and *P. subalbicornis*.

Hoffer, A. 1970. Zweiter Beitrag zur Taxonomie der palaearktischen Arten der Gattung *Aphidencyrtus* Ashm. (Hym., Chalc., Encyrtidae). *Studia Entomologica Forestalia* **1**(5): 65–80.
A key to Palaearctic species of *Syrphophagus* (= *Aphidencyrtus*) includes all recorded British species that are hyperparasitic on aphids and psyllids.

Hoffer, A. 1976. Uber die Gattung *Mayrencyrtus* Hincks, 1944 (Hym., Chalc., Encyrtidae). *Studia Entomologica Forestalia* **2**: 95–100.
The key to European species includes both recorded British species.

Jensen, P.B. 1989. Revision of the genus *Aschitus* (Hymenoptera: Encyrtidae) in Europe and Soviet Asia. *Entomologica Scandinavica* **19**(3): 293–323.
Includes all British species of the genus, some of which have been placed previously in *Microterys*, with notes on their hosts.

Jensen, P.B. & Sharkov, A.V. 1989. Revision of the genus *Trichomasthus* (Hymenoptera: Encyrtidae) in Europe and Soviet Asia. *Entomologica Scandinavica* **20**: 23–54.
Includes all recorded British species of *Trichomasthus* with notes on their hosts.

Maple, J.D. 1947. The eggs and first-instar larvae of Encyrtidae and their morphological adaptations for respiration. *University of California Publications in Entomology* **8**(2): 25–117.
An excellent review of egg and early larval morphology in the Encyrtidae.

Springate, N.D. & Noyes, J.S. 1990. A review of British species of *Anagyrus* Howard
(Hymenoptera: Encyrtidae) with new records and descriptions of other
Chalcidoidea. *Entomologist's Gazette* 41: 213–30.
Includes a key to all British species of *Anagyrus* with a summary of their known hosts.

Sugonjaev, E.S. 1964. Palaearctic species of the genus *Blastothrix* Mayr
(Hymenoptera, Chalcidoidea) with remarks on their biology and economic
importance. Part 1. *Entomologicheskoe Obozrenie* 43(2): 368–90. (In Russian;
English translation: *Entomological Review, Washington* 43: 189–98).
Includes a key to all known Palaearctic species with a review of the hosts of some of the
species.

Sugonjaev, E.S. 1965. Palaearctic species of the genus *Blastothrix* Mayr
(Hymenoptera, Chalcidoidea) with remarks on their biology and economic
importance. Part II. *Entomologicheskoe Obozrenie* 44: 395–410. (In Russian;
English translation: *Entomological Review, Washington* 44: 225–33).
Reviews the hosts of the species not covered in Part I.

Szelényi, G. 1972. Neue Encyrtiden aus Ungarn (Hymenoptera, Chalcidoidea).
Annales Historico-Naturales Musei Nationalis Hungarici 64: 347–53.
Provides a key to all known Palaearctic species of *Helegonatopus*.

Tachikawa, T. 1981. Hosts of encyrtid genera in the World (Hymenoptera:
Chalcidoidea). *Memoirs of the College of Agriculture, Ehime University* 25(2):
85–110.
Reviews the hosts of Encyrtidae from all parts of the world.

Thorpe, W.H. 1936. On a new type of respiratory interrelation between an insect
(chalcid parasite) and its host (Coccidae). *Parasitology* 28(4): 517–40.
An account of the relationship between an encyrtid parasitoid of a scale insects that
pupates inside its host without killing it.

Timberlake, P.H. 1919. Revision of the parasitic chalcidoid flies of the genera
Homalotylus Mayr and *Isodromus* Howard, with descriptions of two closely
related genera. *Proceedings of the United States National Museum* 56: 133–94.
Includes a key to the species of *Isodromus*.

Trjapitzin, V.A. 1972. Encyrtidae (Hymenoptera) collected by Soviet-Mongolian zoological expeditions in 1967–1969. I. *Nasek. Mongolyy* 1: 613–44.
Provides a key to all known Palaearctic species of *Cheiloneurus*.

Trjapitzin, V.A. 1973a. The classification of the family Encyrtidae (Hymenoptera,
Chalcidoidea). Part 1. Survey of the systems of classification. The subfamily
Tetracneminae Howard, 1892. *Entomologicheskoe Obozrenie* 52(1): 163–175. (In
Russian; English translation (1975): *Entomological Review, Washington* 52:
118–25.)
A good review of the different classifications proposed for the family Encyrtidae with a
proposed new classification.

Trjapitzin, V.A. 1973b. Classification of the parasitic Hymenoptera of the family
Encyrtidae (Chalcidoidea). Part II. Subfamily Encyrtinae Walker, 1837.
Entomologicheskoe Obozrenie 52(2): 416–429. (In Russian; English translation:
(1975): *Entomological Review, Washington* 52: 287–95.)
Second part of above review of encyrtid classification.

Trjapitzin, V.A. 1975. Contribution to the knowledge of parasitic Hymenoptera of the genus *Metaphycus* Mercet, 1917 (Hymenoptera, Chalcidoidea, Encyrtidae) of the Czechoslovakian fauna. *Studia Entomologica Forestalia* 2(1): 5–17.
A usable key to the Palaearctic species of *Metaphycus*, which is to be used with caution since it is not based on authentically determined material.

Trjapitzin, V.A. 1978. Hymenoptera II. Chalcidoidea 7. Encyrtidae. *In* Medvedev, G.S. (Ed.) *Opredeliteli Nasekomykh Evropeyskoy Chasti SSR* 3: 236–328. (In Russian. English translation by United States Department of Agriculture and published 1987, Amerind Publishing Co. Pvt. Ltd., New Delhi.)
Provides keys to genera and species of Encyrtidae occurring in the European part of the former USSR and can be useful for separating species of smaller genera such as *Prionomitus*.

Trjapitzin, V.A. 1989. [Parasitic Hymenoptera of the Fam. Encyrtidae of Palaearctic.] *Opredeliteli po Faune SSSR Izdavaemie Zoologicheskim Institutom AN SSSR* 158: 1–489. Leningrad, Nauka.
The best work to be published to date on Palaearctic Encyrtidae although some of the keys are to be treated with caution because they are not based on authentically determined material. Full of useful information on classification of family, hosts and distribution. (In Russian.)

Family Signiphoridae

To date only two species of this family have been recorded in Britain. The two British species have been reared from scale insects (Homoptera; Coccoidea) as hyperparasitoids, although species elsewhere have been recorded from whiteflies (Homoptera: Aleyrodidae), psyllids (Homoptera: Psyllidae) and the puparia of Diptera. Pupation takes place inside the host remains or, in the scale parasitoids, outside the body of the host but under the scale covering. Overwintering usually takes place as a mature larva or pupa within the host remains.

Both British species can be identified by using Ferrière & Kerrich (1958; see general introduction to Chalcidoidea).

Woolley, J.B. 1988. Phylogeny and classification of the Signiphoridae (Hymenoptera: Chalcidoidea). *Systematic Entomology* 13: 465–501.
The classification of signiphorids with a review of their hosts and distribution.

Family Aphelinidae

A moderately large family with 45 species having been recorded from the British Isles. The majority of aphelinids are parasitoids of sternorrhynchous Homoptera, most being parasitic on scale-insects with a few attacking

Aphidoidea, Aleyrodoidea or Psylloidea. Species can be ectoparasitoids, endoparasitoids, egg parasitoids or egg predators. Some species are very unusual in that males and females may have different ontogenies. In these species the females develop as primary endoparasitoids of homopterous hosts (usually coccoids) whereas the males develop as obligate hyperparasitoids of coccoids or psylloids via their eulophid, aphelinid or encyrtid primary parasitoids, sometimes in a primary host completely different from that of the female. In some species the males may even develop as obligate primary parasitoids in a host completely different from that of the female. Larvae of ectoparasitic species (e.g. *Aphytis* spp.) have a functional tracheal system and spiracles, but those that are endoparasitoids have larvae with neither spiracles nor a functioning tracheal system. Pupation may take place inside or outside the host. Some species pupate inside the living host within a pupation chamber, that becomes filled with air, which may be derived from the host's tracheal system, as in the Encyrtidae. Parasitoids of scales and aphids emerge by cutting a hole through the integument of the host mummy, but if the scale has a delicate covering they push their way out from beneath it. Overwintering is normally as a mature larva or pupa.

Many species of aphelinid are useful in the biological control of insect pests. In Britain, *Encarsia formosa* is used to control *Trialeurodes vaporariorum*, a serious pest of numerous horticultural plants in greenhouses, and *Aphelinus mali* successfully controls *Eriosoma lanigerum* on apple in some areas. *Encarsia citrina* also contributes to the control of many potential diaspidid pests in greenhouses.

The genera of British aphelinids are best identified by using the key to world genera provided by Hayat (1983). Ferrière (1965) and Peck *et al.* (1964; see general introduction to Chalcidoidea) can also be used for this purpose, but some of the generic concepts in these publications are out of date. Most British species can be identified also by using Ferrière (1965), but this includes many misidentifications so it is best used in conjunction with Graham (1976) where most of these errors are corrected.

Ferrière, C. 1965. *Hymenoptera Aphelinidae d'Europe et du bassin Mediterranean.* Masson et Cie, Paris.
   Keys to genera and most species of aphelinid that have been recorded in Britain, but should be used together with Graham (1976) to avoid misidentifications.

Graham, M.W.R. de V. 1976. The British species of *Aphelinus* with notes and descriptions of other European Aphelinidae (Hymenoptera). *Systematic Entomology* 1(2): 123–46.
   Includes corrections of many of the misidentifications appearing in Ferrière (1965) and also includes a very useful key to the British species of *Aphelinus*.

Hayat, M. 1983. The genera of Aphelinidae (Hymenoptera) of the World. *Systematic Entomology* 8: 63–102.
    A key to all known world genera of aphelinids with an up-to-date generic classification.
Rosen, D. & DeBach, P. 1979. Species of *Aphytis* of the World (Hymenoptera: Aphelinidae). *Series Entomologica* 17: 801pp.
    Full of useful information on the taxonomy and biology of species of this very important genus, but the key to species should not be used by the faint-hearted.
Viggiani, G. 1984. Bionomics of the Aphelinidae. *Annual Review of Entomology* 29: 257–76.
    A useful review of the different biologies of Aphelinidae.
Walter, G.H. 1983. 'Divergent male ontogenies' in male Aphelinidae (Hymenoptera: Chalcidoidea): a simplified classification and a suggested evolutionary sequence. *Biological Journal of the Linnean Society* 19: 63–82.
    A very good review of the interesting phenomenon of differing host preferences of males and females of some species of aphelinid.
Williams, T. & Polaszek, A. 1996. A re-examination of the host relations in the Aphelinidae (Hymenoptera: Chalcidoidea). *Biological Journal of the Linnean Society* 57: 35–45.
    An updated view on the classification for heteronomous parasitoids proposed by Walter (1983).

Family Elasmidae

A small family with only three recorded British species. Species are usually gregarious primary ectoparasitoids of concealed or otherwise protected lepidopterous larvae, or hyperparasitoids of such hosts, via cocooned braconids and ichneumonids. Overwintering is probably as a pupa.

All British species can be identified by using Graham (1995).

Graham, M.W.R. de V. 1995. European *Elasmus* (Hymenoptera: Chalcidoidea, Elasmidae) with a key and descriptions of five new species. *Entomologist's Monthly Magazine* 131: 1–23.
    Keys the females of all European species of *Elasmus* as well as including reference to several useful works for other areas of the world.

Family Tetracampidae

A small family with only seven species recorded from Britain. Most species appear to be associated with insects that mine in plants. British species of *Foersterella* are endoparasitoids of the eggs of *Cassida* spp. (Coleoptera, Cassididae) and *Dipriocampe* spp. are endoparasitoids of the eggs of diprionids. One species, *Dipriocampe diprioni*, has been introduced into Canada from Europe in an attempt to control various diprionid pests. It has not become established.

The British genera and species can be identified by using Boucek (1958), although Peck *et al.* (1964: see general introduction to Chalcidoidea) can also be used to identify genera.

Boucek, Z. 1958. Revision der europäischen Tetracampidae (Hym. Chalcidoidea) mit einem Katalog der Arten der Welt. *Sborník Entomologického Oddeleni Národního Musea v Praze* **32**: 41–90.
Key to European genera and species, together with a catalogue of world species.

Boucek, Z. & Askew, R.R. 1968. World Tetracampidae. In Delucchi, V. & Remaudière, G. (Eds) *Index of Entomophagous Insects* **4**: 19pp. Le François, Paris.
A catalogue of world genera and species, including information on distribution and hosts.

Family Eulophidae

One of the largest families of chalcidoids in the British Isles with about 520 recorded species. The majority of Eulophidae are primary parasitoids of concealed larvae, especially those inhabiting leaf mines. The best-known species attack Lepidoptera, but many species parasitise larvae of other insects living in similar concealed situations (such as Agromyzidae, heterarthrine Tenthredinidae and Curculionidae). Other eulophids attack various gall-forming species of insects or eriophyid mites. Various other species collectively exhibit a great range of life-styles. *Eulophus* species are gregarious ectoparasitoids of exposed, leaf-feeding lepidopterous larvae. When fully grown these parasitoid larvae often pupate on leaves around their dead host. A number of other eulophids develop as endoparasitoids in insect eggs, including eggs of dytiscid beetles. One species, *Tetrastichus mandanis*, starts its life as an endoparasitoid in the egg of a delphacid, but emerges from the egg during its second instar, and becomes a predator that actively searches out other homopteran eggs. Species may be ectoparasitoids (Eulophinae and Euderinae), or endoparasitoids (Entedontinae and many Tetrastichinae). Many species can behave as facultative hyperparasitoids, and hyperparasitism is usual or even obligatory in some species. The larvae of some leaf-miner parasitoids (e.g. *Diglyphus*, *Chrysocharis*) construct a circle of little faecal pillars about themselves and pupate within this circle, which serves to prevent the collapse of the host mine as the plant tissue dries out. The larvae of *Euplectrus bicolor* pupate near or beneath the host and are enclosed in flimsy cocoons (Swezey, 1924). Overwintering is normally as a prepupa or pupa. A few species (e.g. *Necremnus* spp.) may overwinter as adults. The mating behaviour of some eulophids is very complex; it can be used to separate closely related species and may help in understanding phylogenetic relationships. In *Melittobia acasta* an unmated female may remain with her developing (male) progeny and will mate with one of her emerging sons. If the males are inside a dipterous puparium the female may even gnaw her way in to gain access. Several species of eulophid are important in biocontrol programmes throughout the world; for example, *Chrysocharis laricinellae* is at least partly responsible for the control of *Coleophora laricella* (Lepidoptera: Coleophoridae), a pest of larch in North America (Peck, 1963).

There are several works available that can be useful in identifying the British genera and species of Eulophidae. Most genera occurring in the British Isles are covered by Peck *et al.* (1964; see general introduction to Chalcidoidea), although the classification in this work, particularly that of the Tetrastichinae, is somewhat out of date. The four recognised subfamilies and very nearly all genera and species of Eulophinae and Euderinae are keyed by Askew (1968); Graham (1987, 1991) deals with all taxa of the Tetrastichinae. The British genera and species of Entedoninae are keyed by Graham (1959, 1963) with updates and more modern treatments of some genera of this subfamily by Askew (1979), Bryan (1980) and Hansson (1983, 1985, 1986, 1990, 1994).

Askew, R.R. 1964. On the biology and taxonomy of some European species of the genus *Elachertus* Spinola (Hymenoptera, Eulophidae). *Bulletin of Entomological Research* **55**: 53–8.
Includes a useful key to help in the separation of species of *Hyssopus* (as *Elachertus*).

Askew, R.R. 1968. Hymenoptera 2. Chalcidoidea Section (b). *Handbooks for the Identification of British Insects* **8**(2b): 39pp.
Keys to the four subfamilies of Eulophidae with keys to British genera and species of Eulophinae and Euderinae.

Askew, R.R. 1979. Taxonomy of some European *Chrysonotomyia* Ashmead (Hymenoptera: Eulophidae) with description of *C. longiventris* n. sp. and notes on distribution. *Entomologica Scandinavica* **10**(1): 27–31.
Key to some British species of *Neochrysocharis* and all *Closterocerus* (as *Chrysonotomyia*; see Hansson, 1994).

Askew, R.R. 1984. Variation in *Cirrospilus vittatus* (Hym., Eulophidae) and the description of a new species from Britain. *Entomologist's Monthly Magazine* **120**: 63–8.
Includes notes in colour variation within one British species and a key to similar species occurring in Britain.

Askew, R.R. 1991. Review of species of *Entedon* Dalman having a complete frontal fork with redefinition of the species-group of *cioni* Thomson (Hymenoptera: Eulophidae). *Entomologica Scandinavica* **22**: 219–29.
Reviews the species of *Entedon* with a complete frontal fork, with a key to species and review of their distribution and hosts.

Askew, R.R. 1992. Additions to the British list of *Entedon* Dalman (Hym., Eulophidae) with descriptions of three new species. *Entomologist's Monthly Magazine* **128**: 119–28.
Diagnostic notes on six species of *Entedon* newly recorded from the British Isles.

Assem, J. van den, Bosch, H.A.J. in & Prooy, E. 1982. *Melittobia* courtship behaviour: a comparative study of the evolution of a display. *Netherlands Journal of Zoology* **32**(4): 427–71.

Use of courtship behaviour in taxonomy of *Melittobia*.

Assem, J. van den & Maeta, Y. 1980. On a fourth species of *Melittobia* from Japan. *Kontyû* **48**(4): 477–481.

Use of courtship behaviour in taxonomy of *Melittobia*.

Bosch, H.A.J. & Assem, J. van den 1986. The taxonomic position of *Aceratoneuromyia granularis* Domenichini (Hymenoptera: Eulophidae) as judged by characteristics of its courtship behaviour. *Systematic Entomology* **11**(1): 19–23.

Use of courtship behaviour in taxonomy of *Melittobia*.

Boucek, Z. & Askew, R.R. 1968. Palaearctic Eulophidae sine Tetrastichinae. *Index of Entomophagous Insects* **3**: 260pp. (Ed.: Delucchi, V.; Remaudière, G.) Le François, Paris.

An excellent, comprehensive catalogue of Palaearctic Eulophidae with notes on distribution and hosts.

Bryan, G. 1980. The British species of *Achrysocharoides* (Hymenoptera, Eulophidae). *Systematic Entomology* **5**(3): 245–62.

A revision of the British species of *Achrysocharoides* with a good key to species.

Cameron, E. 1939. The holly leaf-miner (*Phytomyza ilicis*, Curt.) and its parasites. *Bulletin of Entomological Research* **30**: 173–208.

A classic treatment of the parasitoids of holly-leaf miner, which will enable the beginner to identify all the parasitoids reared from this host. A very good place to start the study of leaf-miner parasitoids.

Dahms, E.C. 1984. Revision of the genus *Melittobia* (Chalcidoidea: Eulophidae) with the description of seven new species. *Memoirs of the Queensland Museum* **21**: 271–336.

Includes notes on the use of courtship behaviour in the taxonomy of *Melittobia*.

Graham, M.W.R. de V. 1959. Keys to the British genera and species of Elachertinae, Eulophinae, Entedontinae and Euderinae (Hym., Chalcidoidea). *Transactions of the Society for British Entomology* **13**(10): 169–204.

Gives a key to the four subfamilies of Eulophidae, keys to genera and species of Eulophinae, and keys to the genera of Entedoninae with keys to species of the smaller genera. Uses an out-of-date classification for the Entedoninae, but a good place to start.

Graham, M.W.R. de V. 1963. Additions and corrections to the British list of Eulophidae (Hym., Chalcidoidea). *Transactions of the Society for British Entomology* **15**(9): 167–275.

Several additions to the taxa included in Graham (1959) with some keys revised and keys to species of several genera, including some of the larger genera of Entedoninae, e.g. *Chrysocharis* and *Omphale*. Classification a little out of date, but again a good place to start.

Graham, M.W.R. de V. 1987. A reclassification of the European Tetrastichinae (Hymenoptera: Eulophidae), with a revision of certain genera. *Bulletin of the British Museum (Natural History)* (Entomology) **55**(1): 1–392.

An indispensable reclassification of the Tetrastichinae with keys to all genera and species of *Aprostocetus* and some other smaller genera. Also includes a host–parasitoid checklist for taxa included.

Graham, M.W.R. de V. 1991. A reclassification of the European Tetrastichinae (Hymenoptera: Eulophidae): revision of the remaining genera. *Memoirs of the American Entomological Institute* **49**: 322pp.

A revised key to European genera with keys to the species of the 20 genera not included in the earlier work (Graham, 1987), e.g. *Baryscapus, Oomyzus, Tetrastichus* and *Tamarixia*. Also includes a host–parasitoid checklist for taxa included.

Hansson, C. 1983. Taxonomic notes on the genus *Achrysocharoides* Girault, 1913 (Hymenoptera: Eulophidae), with a redescription and a description of a new species. *Entomologica Scandinavica* **14**: 281–91.

An update on the key to species of genus provided by Bryan (1980) with discussion of some of the characters used in separating species.

Hansson, C. 1985. Taxonomy and biology of the Palaearctic species of *Chrysocharis* Forster, 1856 (Hymenoptera: Eulophidae). *Entomologica Scandinavica* (suppl.) **26**: 1–130.

A good revision with a key to species that includes all British species of *Chrysocharis*.

Hansson, C. 1986. Revision of the Asiatic, European and north American species of *Derostenus* Westwood (Hymenoptera: Eulophidae). *Entomologica Scandinavica* **17**: 313–22.

Provides a key to the species of *Derostenus*, including taxonomic notes on both species recorded from Britain.

Hansson, C. 1990. A taxonomic study on the Palaearctic species of *Chrysonotomyia* Ashmead and *Neochrysocharis* Kurdjumov (Hymenoptera: Eulophidae). *Entomologica Scandinavica* **20**: 29–52.

A redefinition of *Chrysonotomyia* and *Neochrysocharis* with keys to the Palaearctic species of both genera.

Hansson, C. 1994. Re-evaluation of the genus *Closterocerus* Westwood (Hymenoptera: Eulophidae) with a revision of the Nearctic species. *Entomologica Scandinavica* **25**: 1–25.

Reviews the taxonomic status of *Closterocerus* with the result that most species previously included in *Chrysonotomyia* are transferred to *Closterocerus*.

Schauff, M.E. 1991. The Holarctic genera of Entedoninae (Hymenoptera: Eulophidae). *Contributions of the American Entomological Institute* **26**(4): 109pp.

Provides a key to Holarctic genera that includes all British genera.

Family Trichogrammatidae

A moderately large family of chalcidoids, but with only about 30 species having been recorded as British. Virtually all trichogrammatids are primary, solitary or gregarious endoparasitoids of the eggs of other insects, notably those of Lepidoptera, Hemiptera, Coleoptera, Thysanoptera, Hymenoptera, Diptera and Neuroptera. One species of *Trichogramma* is known to occasionally develop as a facultative hyperparasitoid using a species of *Telenomus* in a lepidopteran egg as host (Strand & Vinson, 1984). Elsewhere species of *Lathromeris* and *Oligosita* have been observed to develop as larval parasitoids of cecidomyiids (Diptera). Many species oviposit directly into more or less exposed host eggs and some may even attempt to oviposit into anything that has the same size and shape as an egg, e.g. dried globules of sap. A few trichogrammatids parasitise the eggs of aquatic hosts, such as Dytiscidae, Notonectidae or Odonata, while the egg is beneath the surface of the water. These aquatic species (e.g. *Prestwichia aquatica*) search for hosts by apparently swimming underwater. Many species of trichogrammatids are of interest because of their widespread use in the biological control of various insect pests, especially Lepidoptera.

Trichogrammatids frequently pose problems with their identification because of their small size. The easiest key to use to identify most genera of trichogrammatids occurring in Britain is that given in Peck *et al.* (1964; see general introduction to Chalcidoidea); a more difficult, but comprehensive key is provided in Doutt & Viggiani (1968). The latter key can be greatly simplified for use with the British fauna if it is modified to include only those few genera recorded for the British Isles. Also useful are Peck, Boucek & Hoffer (1964) (Central European genera); Voegele & Pintureau (1982) (species of *Trichogramma*); Nagarkatti & Nagaraja (1977) (biosystematic literature on *Trichogramma* and *Trichogrammatoidea*).

Bakkendorf, O. 1934. Biological investigations on some Danish hymenopterous egg-parasites, especially in homopterous and heteropterous eggs, with taxonomic remarks and descriptions of new species. *Entomologiske Meddelelser* 19: 1–135.
An excellent and interesting account of the biologies of several species of trichogrammatids; includes good habitus illustrations of the species.

Blood, B.N. 1923. Notes on Trichogrammatinae taken around Bristol. *Annual Report and Proceedings of the Bristol Naturalists Society (4)* 5: 253–8.
Diagnostic notes on several British species of trichogrammatid, including the description of five new species.

Blood, B.N. & Kryger, J.P. 1928. New genera and species of Trichogrammidae with remarks upon the genus *Asynacta* (Hym. Trichogr.). *Entomologiske Meddelelser* 16: 203–22.
Descriptions of five British species of Trichogrammatidae with excellent habitus figures of the taxa.

Doutt, R.L. & Viggiani, G. 1968. The classification of the Trichogrammatidae (Hymenoptera: Chalcidoidea). *Proceedings of the California Academy of Sciences* 35: 477–586.
A key to world genera of Trichogrammatidae with a check-list of species for each genus giving citation of original description. Many useful illustrations, which should help in recognition of the genera. A useful publication but only for those intending a serious study of the family.

Hincks, W.D. 1957. Notes on some species of Trichogrammatidae (Hym.: Chalcidoidea) omitted from the 'Check List of British Insects'. *Journal of the Society for British Entomology* 5: 215–18.
Provides some notes that may help recognition of British species of trichogrammatid.

Kryger, J.P. 1919. The European Trichogrammiinae. *Entomologiske Meddelelser (2)* 7: 257–346.

Provides keys to species; may help in the identification of many British species provided that it is used in conjunction with an up-to-date listing of species with synonymies, etc. such as Fitton *et al.* (1978); see General Introduction to Hymenoptera.

Kryger, J.P. 1920. Further investigations upon the European Trichogramminae. *Entomologiske Meddelelser (2)* 8: 183–91.

Provides a key that should enable the user to separate almost all British species.

Nagarkatti, S. & Nagaraja, H. 1977. Biosystematics of *Trichogramma* and *Trichogrammatoidea* species. *Annual Review of Entomology* 22: 157–76.

A review of the biosystematics of *Trichogramma* which may be helpful in suggesting the best way to separate the different British species of the genus.

Nowicki, S. 1935. Descriptions of new genera and species of the family Trichogrammidae (Hym. Chalcidoidea) from the Palearctic region, with notes – I. *Zeitschrift für Angewandte Entomologie* 21: 566–96.

No keys to species, but with several good illustrations and can be used to identify the British species of several genera, e.g. *Mirufens, Ufens* and *Monorthochaeta*.

Nowicki, S. 1936. Descriptions of new genera and species of the family Trichogrammidae (Hym. Chalcidoidea) from the Palearctic region, with notes – II. *Zeitschrift für Angewandte Entomologie* 2: 115–48.

No keys to species, but may be helpful in identifying species of *Oligosita, Chaetostrichella* (as *Brachista*), *Aphelinoidea*, and perhaps other genera that have not yet been recorded as British.

Voegelé, J. & Pintureau, B. 1982. Caractérisation morphologique des groupes et espèces du genre *Trichogramma* Westwood. *Colloques de l'INRA* 9: 45–75.

A contentious key to all the world's species of *Trichogramma* more or less entirely based on characters of the male genitalia. May be helpful in separating species, but any identification should be treated cautiously.

Family Mymaridae: 'fairy flies'

A moderately sized family with 93 recorded British species. All species are internal, solitary (rarely gregarious) parasitoids of the eggs of other insects. Most species seem to parasitise eggs in concealed situations such as those embedded in plant tissue, placed under scales bracts or in soil. Mymarids are not particularly host-specific, and species within one genus may parasitise eggs of insects belonging to several families. The most common hosts are eggs of auchenorrhynchous Homoptera, but eggs of other Hemiptera (Tingidae, Miridae and perhaps even Coccoidea) together with those of Coleoptera (especially Curculionidae and Dytiscidae) and Psocoptera are commonly parasitised. Some mymarids (e.g. *Caraphractus cinctus*) parasitise submerged eggs of aquatic insects (e.g. Dytiscidae) and are capable of swimming underwater, using their wings as paddles (Jackson, 1966).

Pupation of mymarids takes place within the host eggshell and overwintering is normally as a mature larva in the host egg.

The members of this family are a favourite with amateur microscopists because of their very delicate appearance. However, as with trichogrammatids the identification of mymarids is difficult because of their generally small size. Identification of the British genera is probably easiest with the key included in Peck *et al.* (1964; see general introduction to Chalcidoidea), but Annecke & Doutt (1961) and Schauff (1984) will also be of use. On the other hand, identification of the species presents several problems, particularly for those genera that are badly in need of revision, e.g. *Anaphes.*

Annecke, D.P. 1961. The genus *Mymar* Curtis (Hymenoptera, Mymaridae). *South African Journal of Agricultural Science* 4: 543–52.

> Provides a key to, and diagnoses of, the world species of *Mymar* allowing for easy separation of the two recorded British species.

Annecke, D.P. & Doutt, R.L. 1961. The genera of the Mymaridae. Hymenoptera: Chalcidoidea. *Entomology Memoir of the Department of Agriculture of the Union of South Africa* 5: 1–71.

> One of the most popular works for those seriously studying mymarids and includes a very useful, although outdated, key to the world genera of Mymaridae.

Bakkendorf, O. 1934. Biological investigations on some Danish hymenopterous egg-parasites, especially in homopterous and heteropterous eggs, with taxonomic remarks and descriptions of new species. *Entomologiske Meddelelser* 19: 1–135.

> An excellent, illustrated account of the biologies of several species which also includes good habitus illustrations of several genera of mymarids along with a key to some species of *Polynema.*

Cheke, R.A. & Turner, B.D. 1973. Two new species of *Alapus* Westwood (Hym. Mymaridae) parasitising eggs of Psocoptera. *The Entomologist* 106: 279–83.

> Description of a new British species of *Alaptus.*

Chiappini, E. 1989. Review of the European species of the genus *Anagrus* Haliday (Hymenoptera Chalcidoidea). *Bollettino di Zoologia Agraria e di Bachicoltura (II)* 21: 85–119.

> Includes a well-illustrated key to all European species.

Chiappini, E., Triapitsyn, S. & Donev, S.V. 1996. Key to the Holarctic species of *Anagrus* Haliday (Hymenoptera: Mymaridae) with a review of the Nearctic and Palaearctic species (other than European) and description of new taxa. *Journal of Natural History* 30: 551–95.

> A key to the Holarctic species of *Anagrus* which includes all British species.

Debauche, H.R. 1948. Étude sur les Mymarommidae et les Mymaridae de la Belgique (Hym., Chalcidoidea). *Mémoires du Musée Royal d'Histoire Naturelle de Belgique* 108: 248pp.

> A very useful study of the mymarids of Belgium and probably a classic in the study of European taxa. Includes well-illustrated keys to the genera and species, although several of the species are misidentified and much of the nomenclature is now out of date.

Enock, F. 1910. A record. The battledore-wing fly. *Mymar pulchellus. Knowledge* 33: 256.
   Photographs and description of *Mymar pulchellum*.

Enock, F. 1911. A notable record. Capture of a new species of *Mymar. Knowledge* 34: 271–3.
   Photographs and description of *Mymar regalis*.

Hincks, W.D. 1950. Notes on some British Mymaridae (Hym.). *Transactions of the Society for British Entomology* 10: 167–207.
   A key to British genera related to *Polynema* with a key to the British species of *Polynema*. Includes some illustrations.

Hincks, W.D. 1952. The British species of the genus *Ooctonus* Haliday, with a note on some recent work on the fairy flies (Hym., Mymaridae). *Transactions of the Society for British Entomology* 11(7): 153–63.
   A review of the British species of *Ooctonus* with a key to species.

Hincks, W.D. 1959. The British species of the genus *Alaptus* Haliday in Walker (Hym., Chalc., Mymaridae). *Transactions of the Society for British Entomology* 13(8): 137–48.
   Provides a key to five of the six or seven British species of *Alaptus*.

Hincks, W.D. 1960a. A new British species of the genus *Alaptus* Haliday (Hym., Mymaridae). *The Entomologist* 93: 170–2.

Hincks, W.D. 1960b. Some additions to the British Mymaridae (Hym., Chalcidoidea). *Entomologist's Monthly Magazine* 95: 210–16.
   Includes some note on separating species of *Anaphes* and the description of two new species of *Polynema*.

Huber, J.T. 1986. Systematics, biology, and hosts of the Mymaridae and Mymarommatidae (Insecta: Hymenoptera): 1758–1984. *Entomography* 4: 185–243.
   A very useful and informative review of the family as a whole, providing a summary of published works on their taxonomy, distribution and hosts.

Jackson, D.J. 1966. Observations on the biology of *Caraphractus cinctus* Walker (Hymenoptera: Mymaridae), a parasitoid of the eggs of Dytiscidae (Coleoptera). *Transactions of the Royal Entomological Society of London* 118(2): 23–49.
   An account of the biology of this interesting species. Included here because of its unusual biology.

Kryger, J.P. 1950. The European Mymaridae comprising the genera known up to c. 1930. *Entomologicke Meddelelser* 26: 1–97.
   A useful general introduction to mymarids with a key to European genera.

Matthews, M.J. 1986. The British species of *Gonatocerus* Nees (Hymenoptera: Mymaridae), egg parasitoids of Homoptera. *Systematic Entomology* 11: 213–29.
   Keys all recorded British species of *Gonatocerus*.

Schauff, M.E. 1984. The Holarctic genera of Mymaridae (Hymenoptera: Chalcidoidea). *Memoirs of the Entomological Society of Washington* 12: 1–67.
   A review of the classification of the family with a well-illustrated key to all Holarctic genera.

Soyka, W. 1949. Monographie der *Mymar*-Gruppe mit den Gattungen *Mymar* Curtis, *Synanaphes* Soyka, *Ferrièrella* Soyka, *Anaphoidea* Girault, *Hofnederia* Soyka, *Fulmekiella* Soyka und *Yungaburra* Girault (Hymenoptera, Chalcidoidea, Mymaridae). *Revista de Entomologia, Rio de Janeiro* **20**: 301–422.

A revision of the European genera near *Anaphes*, including keys to genera and species. Although Soyka's taxonomy is based on a typological concept resulting in oversplitting of genera and species, the keys may help in preliminary sorting of taxa.

Soyka, W. 1956. Monographie der Polynemagruppe. *Abhandlungen der Zoologisch-Botanischen Gesellschaft in Wien* **19**: 1–115.

A revision of the European genera near *Polynema*, including keys to genera and species. As with above reference, Soyka's taxonomy is based on a typological concept resulting in oversplitting of genera and species, but the keys may help in preliminary sorting of taxa.

Superfamily Mymarommatoidea: Family Mymarommatidae

A very small group with only 13 species recorded worldwide and a single representative in the British Isles. The biology of any species belonging to this group is unknown, but a specimen has been reared from a bracket fungus and in New Zealand the species are commonly associated with weevils in mosses. Most commonly found in Malaise trap catches, but can be very difficult to spot because of their very small size.

Mymarommatids are most likely to be confused with mymarids, having a generally similar appearance. The sole British species can be recognised by the good illustrations provided by Blood & Kryger (1922, 1936).

Blood, B.N. & Kryger, J.P. 1922. A new mymarid from Brockenhurst. *Entomologist's Monthly Magazine* **58**: 229–30.

Description of the male of *Palaeomymar anomalum* with a whole insect figure.

Blood, B.N. & Kryger, J.P. 1936. *Petiolaria anomala* Bl. & Kr. (Hym., Chalcid): description of the female. *Journal of the Society for British Entomology* **1**: 115–16, Plate III.

Description of the female of *Palaeomymar anomalum* with a good whole insect figure.

Class Arachnida: the spiders, harvestmen, pseudoscorpions and mites

The Arachnida is a class of arthropods possessing the same status as the class Insecta. Arachnids are characterised by having a basic body plan comprising a cephalothorax (combined head and thorax), an abdomen, mouthparts consisting of a pair of chelicerae, a pair of sensory or grasping pedipalps, and eight legs. Unlike many insects, arachnids never possess wings. Besides the subclass Acari (mites and ticks), the class Arachnida contains ten living orders of which three are represented in the British Isles: Araneae (spiders), Opiliones (harvestmen) and Pseudoscorpiones (pseudoscorpions). (There is also one established population of a single species of scorpion, *Euscorpius flavicaudis*, Order Scorpiones, at a dockland site in SE England.) In world terms, the number of known species of spiders is approximately 35 000; the harvestmen number about 3000, the pseudoscorpions about 1600, and the mites and ticks over 45 000.

Societies, associations and journals

The British Arachnological Society publishes both a *Bulletin* and a *Newsletter*, which are distributed together to members three times a year. Details are available from: S.H. Hexter, BAS Membership Treasurer, 71 Havant Road, London E17 3JE.

Working alongside the Biological Records Centre, Monks Wood, the British Arachnological Society organises the **Spider Recording Scheme**, the **Opiliones Recording Scheme** and the **Pseudoscorpion Recording Scheme**. The aims of the schemes are to produce up-to-date distribution atlases of British arachnids based on records from each 10 km square in the country. Details are obtainable from:

D.R. Nellist, Spider Recording Scheme, 198A Park Street Lane, Park Street, St Albans, Herts AL2 2AQ.

P.D. Hillyard, Opiliones Recording Scheme, c/o The Natural History Museum, Cromwell Road, London SW7 5BD.

G. Legg, Pseudoscorpion Recording Scheme, c/o The Booth Museum of Natural History, 194 Dyke Road, Brighton, East Sussex BN1 5AA.

The Field Studies Council and The Scottish Field Studies Association run a

variety of courses including some on spiders and other arachnids at their res-
idential field centres. Details from:

Field Studies Council, Preston Montfort, Shrewsbury, Shropshire SY4 1HW.

Scottish Field Studies Association, Pinewoods, 10 Stormont Place, Scone, Perth PH2 6SR.

Order Pseudoscorpiones: the pseudoscorpions

(26 species in 6 families)

PAUL D. HILLYARD

Pseudoscorpions, or false-scorpions, are not commonly seen because of their secretive habits and small size (1–4 mm). They resemble tiny scorpions with their relatively large, grasping pedipalps, but they lack the scorpion's tail with its venomous sting. However, pseudoscorpions do have poison glands opening on the fixed digit of each of the pedipalps. Unusually among the arachnids silk is produced from the mouthparts where the silk glands open on the fixed digit of the chelicerae. This silk is used to construct protective chambers and brood sacs. In pseudoscorpions the reproductive biology involves sperm transfer by means of a spermatophore. Later the eggs are carried by the female in a brood sac until they hatch.

Pseudoscorpions occur throughout the British Isles but are probably most easily found in deep leaf litter in deciduous woodland. Numbers as high as 900 have been recorded in beech litter with a surface area of one square metre (the commonest species: *Neobisium muscorum, Chthonius ischnocheles* and *Roncus lubricus)*. In extra rich habitats such as compost or manure heaps, these three species are joined by *Lamprochernes* spp. This genus is especially likely to be seen holding on to the legs of blowflies to aid their dispersal (phoresy). Under tree bark, species such as *Chernes cimicoides* and *Dendrochernes cyrneus* can also be found in high densities. In barns, *Cheiridium museorum* and in bird houses *Dinocheirus panzeri* may reach numbers even higher than those species found in leaf litter. By contrast, the recently recorded species *Larca lata*, from a hollow oak at Windsor, is probably the rarest and most endangered of the British pseudoscorpions (Judson & Legg, 1996).

The 26 species currently recorded from the British Isles are placed in six families: Chthoniidae (6 species); Neobisiidae (5 species); Larcidae (1 species); Cheiridiidae (1 species); Chernetidae (10 species); and Cheliferidae (3 species).The main key available (Legg & Jones, 1988) identifies the species (adults) but several closely related species may be difficult to separate, e.g. *Lamprochernes nodosus/L. chyzeri* and *Allochernes powelli/A. wideri*. Identifications are usually determined by a complex of characters, including the arrangement of setae and trichobothria. Unfortunately, genitalic characters are likely to be inconclusive. Care is needed when studying the patterns of setae and trichobothria as some may be missing. Certain chernetids, e.g. *Pselaphochernes dubius*, can be recognised by their setae, which are multi-tipped.

References

Beier, M. 1963. Ordnung Pseudoscorpionidea (Afterskorpione). *Bestimmungsbucher Bedenfauna Europas* 1: 1–313, Akademie-Verlag, Berlin.
This book (in German) is the essential key work covering the wider European fauna and thus should be consulted if any suspected new species is discovered in Britain.

Elliot, P., King, P.E., Morgan, C.I., Pugh, P.J.A., Smith, A. & Wheeler, S.L.A. 1990. Chelicerata, Uniramia, and Tardigrada. *In* Hayward, P.J. & Ryland, J.S. (Eds) *The marine fauna of the British Isles and north-west Europe.* Vol. 1, *Introduction and protozoans to arthropods*, pp. 553–627. Oxford University Press, Oxford.
This paper includes a key to the species of pseudoscorpions found in the marine littoral zone.

Harvey, M.S. 1988. The systematics and biology of pseudoscorpions. *Australian Entomological Society Miscellaneous Publications* 5: 75–85.
A useful, general work, more up-to-date than Weygoldt (1969).

Judson, M.L.I. 1990. On the presence of *Chthonius halberti* Kew and *Chthonius ressli* Beier in France, with remarks on the status of *Kewochthonius* Chamberlin and *Neochthonius* Chamberlin (Arachnida, Chelonethida, Chthoniidae). *Bulletin du Museum d'Histoire Naturelle*, Paris (4) **11** (A; 3): 593–603.
This paper changes the generic combination of *Kewochthonius halberti* back to *Chthonius halberti* Kew, 1916.

Judson, M.L.I. & Legg, G. 1996. Discovery of the pseudoscorpion *Larca lata* (Garypoidea, Larcidae) in Britain. *Bulletin of the British Arachnological Society* **10** (6): 205–10.
This paper adds the species *Larca lata* (Hansen, 1884) and the family Larcidae to the British fauna.

Legg, G. & Jones, R.E. 1988. Pseudoscorpions. *Synopses of the British Fauna* (New Series) **40**: 159pp. Linnean Society of London.
The current monograph on British pseudoscorpions presenting biology, systematics and species identification; the distribution maps of the Biological Records Centre are included.

Weygoldt, P. 1969. *The biology of pseudoscorpions.* Harvard University Press, Cambridge, Mass.
This book describes the little-known but fascinating biology and behaviour of pseudo-scorpions of the world.

Order Opiliones: the harvestmen

(24 species in 6 families)

PAUL D. HILLYARD

Harvestmen, also known as harvest-spiders or daddy-long-legs, are mostly long-legged creatures which resemble spiders but differ in a number of respects. Instead of the spiders' narrowly joined, two-part body, harvestmen have a body with the cephalothorax and abdomen fused into one piece. The chelicerae are composed of three segments and harvestmen lack both silk and poison glands; they are not venomous as spiders are and are virtually harmless, although they may defend themselves with secretions from a pair of odoriferous glands. Harvestmen have only one pair of simple eyes whereas spiders typically have eight. During reproduction, the sexes engage in direct copulation, which is most unusual among the Arachnida. The eggs are laid in damp soil or in other niches among woody vegetation.

Harvestmen are found throughout Britain but their species diversity is greatest in the south. In general, they are most in evidence during late summer (harvest-time), hence their popular English name. The long-legged species tend to stride across vegetation but not all are long-legged. Some of the short-legged species, e.g. *Trogulus tricarinatus* and *Anelasmocephalus cambridgei*, remain hidden in the soil or under stones. Some species are often abundant in particular habitats, e.g. *Mitopus morio* on moorlands; *Odiellus spinosus* in urban areas; *Leiobunum rotundum* in hedgerows; and *Paroligolophus agrestis* in undergrowth generally. A species introduced from northern Spain, *Centetostoma bacilliferum*, has been recorded recently from a number of sites close to Plymouth, where it now appears to be established (Smithers & Hogg, 1991).

The 24 species currently recorded from the British Isles are placed in six families: Nemastomatidae (3 species); Trogulidae (2 species); Sabaconidae (1 species); Sclerosomatidae (1 species); Phalangiidae (13 species); and Leiobunidae (4 species). The main key available (Hillyard & Sankey, 1989) identifies the species (adults) primarily by morphological features such as tubercles and denticulae. However, several species may need experience to separate, e.g. *Paroligolophus agrestis/P. meadii* and *Leiobunum rotundum/L. blackwalli*. In these and other species, where there is doubt, it is necessary to dissect out the internal genitalia in order to make comparisons with the corresponding figures in the key.

References

Hillyard, P.D. & Sankey, J.H.P. 1989. Harvestmen. *Synopses of the British Fauna* (New Series) **4** (2nd edn): 120pp. Linnean Society of London.

Definitive keys and notes for the identification of 23 British species (adults) together with information on biology and further reading; includes also the distribution maps of the Biological Records Centre.

Martens, J. 1978. Weberknechte, Opiliones. *Die Tierwelt Deutschlands* **64**: 464pp., Jena.

This book, in German, is the current monograph covering identification, systematics, distribution and biology of the 110 species of harvestman found in Europe north of the Mediterranean. It should be consulted if any suspected new species are discovered in Britain.

Smithers, P. & Hogg, M. 1991. *Centetostoma*, a harvestman new to Britain. *Newsletter of the British Arachnological Society* **60**: 8.

A more detailed and formal description of this harvestman, under the name of *Centetostoma bacilliferum* (Simon, 1879), is in preparation by Hillyard, Smithers & Hogg.

Order Araneae: the spiders

(*ca.* 650 species in 34 families)

PAUL D. HILLYARD

Spiders are the best known of all the arachnids but they need to be distinguished from the harvestmen. The spider's body is in two parts: cephalothorax and abdomen, joined by a narrow pedicel. Eight legs are attached to the cephalothorax, which also carries a pair of chelicerae composed of two segments including a movable fang (absent in harvestmen), up to eight simple eyes, and a pair of pedipalps which, in the adult male, carry accessory sex organs ('palps'). Before mating, the male fills his syringe-like palps with a drop of sperm from his genital opening. Like the female's genital opening (the epigyne) this opening is on the underside of the abdomen. The abdomen houses a number of silk glands ending in spinnerets through which the silk emerges. As is well known, spiders are able to bite; they inject venom through the chelicerae from the pair of venom glands. In Britain, the sizes of spiders range from the numerous but tiny 'money' spiders (1–4 mm body length) to the largest species including fishing spiders, *Dolomedes* spp. (up to 24 mm body length), and the largest house spiders, *Tegenaria* spp., which have modest body lengths but their legs can span as much as five and three quarter inches (145 mm).

Spiders, as a group, are ubiquitous and although many are nocturnal there are invariably some to be found in any habitat and at all seasons. A rough division can be used to distinguish the web-builders from the hunting kinds that do not build webs. However, all spiders produce silk, for, besides making webs, silk is used for various purposes including making nests, egg cocoons and safety lines. Probably the web-builders in general are the most familiar to us: e.g. the orb-weavers *Araneus diadematus,* in gardens everywhere, and *Zygiella x-notata,* often found around window frames; and the cobweb weavers, *Tegenaria* spp., in outbuildings virtually everywhere. As might be expected with such a large group, there are many rare species of spiders, e.g. the lace-web weaver, *Eresus sandaliatus,* from a single heath in Dorset, and the fishing spider, *Dolomedes plantarius,* from Redgrave and Lopham Fens in East Anglia, and Pevensey Levels in Sussex.

In Britain 34 of the world's 105 families of spiders are found. To identify a spider it is necessary to first establish the family. Many families are best characterised by the number and arrangement of the eyes. At species level, examination of the detailed structures of the sex organs (male palp or female epigyne) is essential. In the case of immature specimens it is not reliable to go further than genus level. Among the references listed below, the three-volume *British Spiders* by Locket & Millidge (1951, 1953, 1974) gives Britain the unique ability to identify almost all of the country's spider fauna (and indeed the entire fauna with the aid of the other, more recent references also listed below).

The 34 families in Britain are:

Suborder Mygalomorphae: Atypidae
Suborder Araneomorphae: Eresidae, Oecobiidae, Uloboridae, Amaurobiidae, Dictynidae, Oonopidae, Scytodidae, Dysderidae, Segestriidae, Pholcidae, Salticidae, Hahniidae, Oxyopidae, Zodariidae, Lycosidae, Pisauridae, Zoridae, Argyronetidae, Anyphaenidae, Mimetidae, Agelenidae, Philodromidae, Thomisidae, Heteropodidae, Gnaphosidae, Clubionidae, Liocranidae, Nesticidae, Theridiidae, Tetragnathidae, Theridiosomatidae, Araneidae, Linyphiidae.

References

Ashmole, N.P. & Merrett, P. 1981. *Lepthyphantes antroniensis* Schenkel,, a spider new to Britain (Araneae: Linyphiidae). *Bulletin of the British Arachnological Society* 5: 189–236.
   This paper describes a species that is not covered in any of the other references listed here.
Bristowe, W.S. 1958 (reprint 1971). *The world of spiders.* Collins New Naturalist, London.
   A highly entertaining account of British spiders which also has some value for identification. Currently out of print.
Felton, C. 1997. *Gnaphosa nigerrima* L. Koch (Araneae: Gnaphosidae) a spider new to Britain. *Bulletin of the British Arachnological Society* 10: 311–12.
   This paper describes a species that is not covered in any of the other references listed here.

Foelix, R.F. 1996. *Biology of spiders* (2nd edn). Oxford University Press, New York.

An essential textbook on biology and behaviour for students of spiders anywhere.

Hillyard, P.D. 1997. *Collins Gem photoguide to spiders*. Harper Collins Publishers, London.

A handy, pocket-sized introduction to spiders of the world, which includes illustrated descriptions of 60–70 common British species.

Locket, G.H. & Millidge, A.F. 1951–3 (reprint 1975). *British spiders*. Vols 1 & 2. Ray Society, London.

The first two volumes were updated by volume 3 (with Merrett, P., 1974). *British spiders* remains the definitive identification manual for almost the entire British list but a considerable number of new species have been added and nomenclatural changes have occurred since the publication of volume 3 (see Platnick, 1989).

Merrett, P. 1979. Changes in distribution of British spiders, and recent advances in knowledge of distribution. *Bulletin of the British Arachnological Society* 4: 366–76.

This paper lists ten species that had been added to the British list since volume 3 of *British Spiders* (Locket, Millidge & Merrett, 1974).

Merrett, P. 1985. A check list of British spiders. *Bulletin of the British Arachnological Society* 6: 381–403.

A further benchmark detailing additions and changes to the British list since the publication of volume 3 of *British Spiders* (Locket *et al.*, 1974).

Merrett, P. 1995. Eighteen hundred new county records of British spiders. *Bulletin of the British Arachnological Society* 10: 15–18.

Although records are being collected on a 10 km square basis for the spider recording scheme, this author continues to keep the county records up-to-date.

Merrett, P. & Stevens, R.A. 1995. A new genus and species of linyphiid spider from south-west England (Araneae: Linyphiidae). *Bulletin of the British Arachnological Society* 10: 118–20.

This paper describes *Nothophantes horridus* Merrett & Stevens; it is not covered in any of the other references listed here.

Platnick, N.I. 1989. *Advances in spider taxonomy 1981–1987*. Edited by P. Merrett. Manchester University Press (Manchester and New York) in association with The British Arachnological Society.

This comprehensive catalogue, based on Pierre Bonnet's *Bibibliographia Araneorum* (1945–1961) together with the two later volumes (see below), performs a valuable function in keeping up-to-date the spider taxonomy of the world.

Platnick, N.I. 1993. *Advances in spider taxonomy 1988–1991*. Edited by P. Merrett. New York Entomological Society in association with The American Museum of Natural History.

See Platnick (1989).

Platnick, N.I. 1998. *Advances in spider taxonomy 1992–1995*. Edited by P. Merrett. New York Entomological Society in association with The American Museum of Natural History.

See Platnick (1989).

Roberts, M.J. 1995. *Spiders of Britain & northern Europe.* Harper Collins Publishers, London.

An excellent field guide with over 450 species described and beautifully illustrated (note: there are 650 species on the British list) together with information on biology and further reading;

Snazell, R. 1980. *Erigone aletris* Crosby & Bishop, a spider new to Britain (Araneae: Linyphiidae). *Bulletin of the British Arachnological Society* 5: 97–100.

This paper describes a species that is not covered in any of the other references listed here.

Subclass Acari: the mites and ticks

(*ca.* 1720 species in 185 families and 5 orders)

ANNE S. BAKER

The subclass Acari is the most morphologically and ecologically diverse group within the Arachnida. Free-living species have colonised terrestrial, marine and freshwater habitats throughout the world; unlike other arachnids, many species are plant-parasitic or live in association with other animals. The acarine body is typically sac-like without subdivisions, and the mouthparts form a discrete structure, which articulates with the body slightly anteroventrally. The exoskeleton has hair-like extensions (setae), whose shape, number and position are used extensively to distinguish between taxa, sexes and life stages. The acarine life cycle comprises a maximum of five active stages after the egg, namely a six-legged larva, and eight-legged protonymph, deutonymph, tritonymph and adult (male and female).

The higher classification of the Acari is in a state of flux, but seven orders (given the rank of suborder by some authors) are normally recognised. The mite fauna of the British Isles, represented by the orders Astigmata, Mesostigmata, Oribatida and Prostigmata, is incompletely known and both newly recorded and undescribed species are frequently discovered. It is often necessary to refer to revisions of the world fauna of families or genera to identify species. The Astigmata includes free-living fungivores or detritivores as well as vertebrate parasites among its members. Some species that infest food products in domestic and commercial stores (family Acaridae and Glycyphagidae) or occur in house dust (family Pyroglyphidae) can cause respiratory and skin disorders in humans and animals that come into contact with them. The scabies or itch mite, *Sarcoptes scabiei* (family Sarcoptidae), is a well-known astigmatid mite which burrows into the cornified outer layer of the skin of a wide range of mammals. Mesostigmatids are mostly free-living predators of other small arthropods that live in soil and decomposing organic matter. Some species of the family Phytoseiidae are used commercially to control pest mites, particularly in glasshouses and orchards, while those of other families are parasites of vertebrates and invertebrates, such as the bee parasite *Varroa jacobsoni*. The

Prostigmata is the most diverse of the mite orders and includes predatory, algivorous and fungivorous free-living species, and parasites and commensals of vertebrates and invertebrates. Among its members are the spider-mites (family Tetranychidae) and gall mites (family Eriophyidae), which can cause serious damage to crop plants, and the harvest mite *Neotrombicula autumnalis*, which can greatly irritate walkers and picnickers through its lymph-feeding habit. Oribatid mites occur in large numbers in the soil, particularly in the upper layers in deciduous forests, although some species live on or periodically migrate to the aerial parts of plants or trees. They feed on micro-organisms and contribute to the breakdown of plant material and subsequent recycling of nutrients.

Ticks make up the order Ixodida and all are blood-feeding ectoparasites of vertebrates. They can be distinguished from mites by the finger-like central process of the mouthparts, which is armed with recurved teeth. The most common and geographically widespread species in the British Isles, *Ixodes ricinus* (the sheep tick), is the main vector of *Borrelia burgdorferi* s. lat., the causative agent of Lyme borreliosis, in Europe.

General references

Evans, G.O. 1992. *Principles of acarology.* CAB International, Wallingford.
  The bulk of this book comprises a comprehensive review of functional morphology and general biology; the concluding chapter on classification contains keys to orders and suborders or superfamilies.

Evans, G.O., Sheals, J.G., & Macfarlane, D. 1961. *The terrestrial Acari of the British Isles. An introduction to their morphology, biology, and classification.* Vol. 1, *Introduction and biology.* British Museum (Natural History), London.
  Although aspects of the classification and biology are now out of date, this is still a useful general account of the British fauna.

Hyatt, K.H. 1993. The acarine fauna of the Isles of Scilly. *Cornish Studies* Second Series 1: 120–61.
  Gives an annotated checklist of 169 species.

Krantz, G.W. 1978. *A manual of acarology* (2nd edn). Oregon State University, Corvallis, Oregon.
  Introductory sections on collecting methods, morphology and biology are given, but this is principally a general systematic account and gives keys to families of Astigmata (including hypopi), Prostigmata, Mesostigmata and Ixodida and superfamilies of Oribatida. A good companion book to Evans (1992).

Turk, F.A. 1953. A synonymic catalogue of British Acari. *Annals and Magazine of Natural History* (12)**6**: 1–26, 81–99.
  The classification used is out of date and many nomenclatural changes have been made since this checklist was compiled, but it is still a useful record of the British fauna.

Host–habitat associations

Donisthorpe, H.St.J.K. 1927. Chapter 14. Acarina (mites). *In*: *The guests of British ants: their habits and life-histories*, pp. 202–17. Routledge, London.
Although this is not a taxonomic account and the classification in the species list is out of date, it contains useful data on associations and distribution.

Fain, A. & Hyland, K.E. 1962. The mites parasitic in the lungs of birds. The variability of *Sternostoma tracheacolum* Lawrence, 1948, in domestic and wild birds. *Parasitology* 52: 401–24.
Includes a list of the species recorded and details of their host associations and distribution.

Fain, A., Guérin, B. & Hart, B.J. 1990. *Mites and allergic disease*. Allerbio, Varennes en Argonne, France.
A comprehensive book about mites in house dust around the world; it includes keys to species of the Pyroglyphidae and the common Acaridae, plus chapters on their biology and control measures.

Hughes, A.M. 1976. *The mites of stored food and houses*. Ministry of Agriculture, Fisheries and Food Technical Bulletin No. 9. HMSO, London.
Based on British habitats, this book includes keys to and/or descriptions of species predominantly of the astigmatid families Acaridae, Histiostomatidae (=Anoetidae), Pyroglyphidae, Carpoglyphidae, Chortoglyphidae and Glycyphagidae, but also common Mesostigmata, Oribatida and Prostigmata; also gives details of biology and worldwide distribution.

Hyatt, K.H. 1990. Mites associated with terrestrial beetles in the British Isles. *Entomologist's Monthly Magazine* 126(1512–1515): 133–47.
Gives a table of host associations, an annotated checklist and an extensive bibliography.

Redfern, M. & Askew, R.R. 1992. *Plant galls*. Naturalists' Handbook No. 17. Richmond Publishing, Slough.
Includes the common eriophyid species found in the British Isles.

Treat, A.E. 1975. *Mites of moths and butterflies*. Comstock Publishing Associates, Ithaca, New York.
A comprehensive guide to the study and identification of mites associated with Lepidoptera.

Yunker, C. 1973. Mites. *In* Flynn, R.J. (Ed.) *Parasites of endothermal laboratory animals*, pp. 425–92. Iowa State University Press, Ames.
Gives diagnoses of major species of veterinary importance, most of which have been recorded in Britain; includes useful biological data and an extensive bibliography.

Order Astigmata (=Acaridida)

Fain, A. 1965. A review of the family Epidermoptidae Trouessart parasitic on the skin of birds (Acarina: Sarcoptiformes). *Verhandelingen der Koninklijke Akademie van Wetenshappen* 27(84): Part I (text) 1–176; Part II (figures) 1–144.
Covers the world fauna; includes keys to species, biological data and host lists.

Fain, A. 1967. Les hypopes parasites des tissus cellulaires des oiseaux (Hypodectidae: Sarcoptiformes). *Bulletin de l'Institut Royal des Sciences Naturelles de Belgique* **43**: 1–139.
Covers the world fauna; includes keys to species and host associations.

Fain, A. 1971. Les Listrophorides en afrique au sud du Sahara (Acarina: Sarcoptiformes) II. Familles Listrophoridae et Chirodiscidae. *Acta Zoologica et Pathologica Antverpiensia* **54**: 1–231.
Covers species recorded from the British Isles; includes keys to species and host lists.

Fain, A. 1975. Nouveaux taxa dans les Psoroptinae. Hypothèse sur l'origine de ce groupe (Acarina, Sarcoptiformes, Psoroptidae). *Acta Zoologica et Pathologica Antverpiensia* **61**: 57–84.
Gives keys to the world genera of the family Psoroptidae and to species of *Caparinia*.

Fain, A. 1979. Les Listrophorides d'amerique neotropicale (Acarina: Astigmates). II Famille Atopomelidae. *Bulletin de l'Institut Royal des Sciences Naturelles de Belgique* (Entomologie) **51**: 1–158.
Covers species recorded from the British Isles; includes keys to species and host lists.

Fain, A. 1981. Notes on the Hyadesiidae Halbert, 1915 and Algophagidae Fain, 1974, nov. tax. (Acari, Astigmata), with a redescription of *Hyadesia curassaviensis* Viets, 1936 and *H. sellai* Viets, 1937. *Acarologia* **22**: 47–61.
A key to world species is given.

Fain, A. 1981. Notes on the genus *Laminosioptes* Megnin, 1880 (Acari: Astigmata) with description of three new species. *Systematic Parasitology* **2**: 123–32.
Includes a key to world species (females only).

Fain, A. 1982. Cinq espèces du genre *Schwiebea* Oudemans, 1916 (Acari, Astigmata) dont trois nouvelles decouvertes dans des sources du sous-sol de la ville de Vienne (Autriche) au cours des travaux du Metro. *Acarologia* **23**: 359–71.
Includes a key to world species.

Fain, A. & Bafort, J. 1964. Les acariens de la famille Cytoditidae (Sarcoptiformes). Description de sept espèces nouvelles. *Acarologia* **6**: 504–28.
Gives a key to world species (females only) and host records.

Fain, A. & Elsen, P. 1967. Les acariens de la famille Knemidokoptidae producteurs de gale chez les oiseaux. *Acta Zoologica Pathologica Antverpiensia* **45**: 1–142.
Covers the world fauna; gives keys to species and host records.

Fain, A., Guérin, B. & Hart, B.J. 1990.
See Host/habitat associations. Gives keys to world species of the family Pyroglyphidae.

Fain, A., Munting, A.J. & Lukoschus, F. 1970. Les Myocoptidae parasites des rongeurs en Hollande et en Belgique (Acarina: Sarcoptiformes). *Acta Zoologica Pathologica Antverpiensia* **50**: 67–172.
Covers species recorded from the British Isles; includes keys to species and host lists.

Gaud, J. & Atyeo, W.T. 1996. Feather mites of the world (Acarina: Astigmata): the supraspecific taxa. *Annales. Musée Royal de l'Afrique Centrale*, Sciences Zoologiques **277**: Part I (Text) 1–193; Part II (Illustrations) 1–436.
Includes keys to genera and an extensive bibliography to taxonomic and biological works.

Gaud, J., Atyeo, W.T. & Barré, N. 1985. Les Acariens du genre *Megninia* (Analgidae) parasites de *Gallus gallus*. *Acarologia* 26: 171–82.
  Keys the adults of species found on chickens around the world.

Griffiths, D.A. 1970. A further systematic study of the genus *Acarus* L., 1758 (Acaridae, Acarina) with a key to species. *Bulletin of the British Museum (Natural History) (Zoology)* 19: 89–120.
  Covers the world fauna.

Hughes, A.M. 1976.
  See Host/habitat associations.

Hughes, R.D. & Jackson, C.G. 1958. A review of the Anoetidae (Acari). *Virginia Journal of Science* 9: 5–198.
  This is a comprehensive account of the Anoetidae (=Histiostomatidae) including study methods, biology and a key to species of the world; it covers taxa from a wider range of habitats than Hughes (1976).

Klompen, J.S.H. 1992. Phylogenetic relationships in the mite family Sarcoptidae (Acari: Astigmata). *Miscellaneous Publications, Museum of Zoology, University of Michigan* No. 180: 1–154.
  Covers the world fauna; gives keys to all life stages of each species.

Michael, A.D. 1901 & 1903. *British Tyroglyphidae*. Vols 1 & 2. Ray Society, London.
  The classification is out of date, but the beautiful illustrations allow species recognition in many instances and show details of the immature stages and internal anatomy. Includes species now classified in the families Acaridae, Histiostomatidae, Glycyphagidae, Chortoglyphidae, Algophagidae and Carpoglyphidae.

OConnor, B.M. 1982. Astigmata. *In* Parker, S.P. (Ed.) *Synopsis and classification of living organisms* 2, pp. 146–69. McGraw-Hill, New York.
  Uses a modern classification and is a useful source of general information and literature.

Philips, J.R. 1990. Acarina: Astigmata (Acaridida). *In* Dindal, D.L. (Ed.) *Soil biology guide*, pp. 757–778. John Wiley, New York.
  Although based on the North American fauna, many taxa that occur in the British Isles are covered and an extensive bibliography is included. Gives keys to adults and hypopi of families and genera of Acaridae, Anoetidae, Chaetodactylidae, Glycyphagidae and Saproglyphidae, and to adults of Pyroglyphidae.

Scheucher, R. 1957. Systematik und Ökologie der deutschen Anoetiden. *In* Stammer, H.J. (Ed.) *Beiträge zur Systematik und Ökologie mitteleuropäischer Acarina* 1(1), pp. 233–384. Geest & Portig, Leipzig.
  Gives a more detailed coverage of the European fauna than Hughes & Jackson (1958) or Hughes (1976); includes a key to species.

Sweatman, G.K. 1957. Life history, non-specificity and revision of the genus *Chorioptes*, a parasitic mite of herbivores. *Canadian Journal of Zoology* 35: 641–89.
  The world species of *Chorioptes* are reviewed.

Sweatman, G.K. 1958. On the life history and validity of the species in *Psoroptes*, a genus of mange mites. *Canadian Journal of Zoology* 36: 905–929.
  The world species of *Psoroptes* are reviewed.

Turk, E. & Turk, F. 1957. Systematik und Ökologie der Tyroglyphiden Mitteleuropas. *In* Stammer, H.J. (Ed.) *Beiträge zur Systematik und Ökologie mitteleuropäischer Acarina* 1(1), pp. 1–231. Geest & Portig, Leipzig.

Gives a comprehensive taxonomic account of the European fauna of free-living Astigmata; includes a key to species now classified in the families Acaridae, Winterschmidtiidae (=Saproglyphidae), Carpoglyphidae, Glycyphagidae and Chortoglyphidae.

Order Mesostigmata (=Gamasida)

Beglyarov, G.A. 1981. Key for identification of the predacious mites Phytoseiidae (Parasitiformes, Phytoseiidae) in the fauna of the SSSR. 1. *Informatsionnyi Byulleten, Leningrad: MOBB* (2): 1–97. 2. *Informatsionnyi Byulleten, Leningrad: MOBB* (3): 1–47. [In Russian.]

Complements Karg (1993) and Chant (1959); includes keys to species.

Bhattacharyya, S.K. 1963. A revision of the British mites of the genus *Pergamasus* Berlese *s. lat.* (Acari: Mesostigmata). *Bulletin of the British Museum (Natural History)* (Zoology) **11**: 133–242.

Contains keys to species.

Bolger, T. 1990. Acari of the families Macrochelidae and Veigaiidae (Mesostigmata) recorded from Ireland. *Bulletin of the Irish Biogeographical Society* **13**(1): 29–43.

Gives distributional records of Irish species.

Chant, D.A. 1959. Phytoseiid mites (Acarina: Phytoseiidae). Part I. Bionomics of seven species in southeastern England. Part II. A taxonomic review of the family Phytoseiidae with descriptions of 38 new species. *Canadian Entomologist* (suppl.) **2**: 1–166.

The classification is somewhat out of date and the British fauna is only partly covered; however, the keys are still useful, particularly for identifying phytoseiids found in orchards.

Chant, D.A. & McMurtry, J.A. 1994. A review of the subfamilies Phytoseiinae and Typhlodrominae. *International Journal of Acarology* **20**: 223–310.

A comprehensive treatment of the world fauna; includes keys to species groups.

de Moraes, G.J., McMurtry, J.A. & Denmark, H.A. 1986. *A catalog of the mite family Phytoseiidae. References to taxonomy, synonymy, distribution and habitat.* EMBRAPA, Brasilia, Brazil.

A valuable source of information and literature on the world fauna.

Evans, G.O. 1955. A revision of the family Epicriidae (Acarina: Mesostigmata). *Bulletin of the British Museum (Natural History)* (Zoology) **3**: 169–200.

Keys the species of *Epicrius* and *Berlesiana.*

Evans, G.O. 1955. British mites of the genus *Veigaia* Oudemans (Mesostigmata: Veigaiaidae). *Proceedings of the Zoological Society of London* **125**: 569–86.

Includes a key to species (females only).

Evans, G.O. 1956. On the classification of the family Macrochelidae with particular reference to the subfamily Parholaspinae (Acarina: Mesostigmata). *Proceedings of the Zoological Society of London* 127: 345–77.
  Now given familial status (Parholaspididae); includes a key to species.

Evans, G.O. 1958. A revision of the British Aceosejinae (Acarina: Mesostigmata). *Proceedings of the Zoological Society of London* 131: 177–229.
  Classified in the family Ascidae; includes a key to species.

Evans, G.O. 1959. The genera *Cyrthydrolaelaps* Berlese and *Gamasolaelaps* Berlese (Acarina: Mesostigmata). *Acarologia* 1: 201–15.
  Classified in the family Veigaiidae; includes a key to species.

Evans, G.O. 1969. A new mite of the genus *Thinoseius* Halbert, (Gamasina: Eviphidiidae) from the Chatham Islands. *Acarologia* 11: 505–14.
  Includes a key to the world species.

Evans, G.O. & Hyatt, K.H. 1956. British mites of the genus *Pachylaelaps* Berlese (Gamasina: Pachylaelaptidae). *Entomologist's Monthly Magazine* 92: 118–29.
  Familial spelling now Pachylaelapidae; includes a key to species (females and some males).

Evans, G.O. & Hyatt, K.H. 1960. A revision of the Platyseiinae (Mesostigmata: Aceosejidae) based on material in the collection of the British Museum (Natural History). *Bulletin of the British Museum (Natural History)* (Zoology) 6: 25–101.
  Now classified in the family Ascidae; includes a key to species.

Evans, G.O. & Till, W.M. 1965. Studies on the British Dermanyssidae (Acari: Mesostigmata). Part I. External morphology. *Bulletin of the British Museum (Natural History)* (Zoology) 13: 247–94.
  Describes the external morphology of all active life stages of species now classified in the families Dermanyssidae, Laelapidae and Macronyssidae.

Evans, G.O. & Till, W.M. 1966. Studies on the British Dermanyssidae (Acari: Mesostigmata). Part II. Classification. *Bulletin of the British Museum (Natural History)* (Zoology) 14: 107–370.
  Covers the same families as above; gives keys to species.

Evans, G.O. & Till, W.M. 1979. Mesostigmatic mites of Britain and Ireland (Chelicerata: Acari-Parasitiformes). An introduction to their morphology and classification. *Transactions of the Zoological Society of London* 35: 139–270.
  Includes keys to families; *Varroa jacobsoni*, recorded in the British Isles after publication, is dealt with in Karg (1993) and Krantz (1978).

Fain, A. & Hyland, K.E. 1964.
  See Host/habitat associations.

Furman, D.P. & Dailey, M.D. 1980. The genus *Halarachne* (Acari: Halarachnidae), with the description of a new species from the Hawaiian monk seal. *Journal of Medical Entomology* 17: 352–9.
  Includes a key to world species.

Hirschmann, W. & Wisniewski, J. 1993. Die Uropodiden der Erde. *Acarologie. Schriftenreihe für Vergleichende Milbenkunde* 40: 1–466.
  Covers the world fauna; is the most comprehensive work on uropods available.

Hyatt, K.H. 1956. British mites of the genera *Halolaelaps* Berlese and Trouessart and *Saprolaelaps* Leitner (Gamasina: Rhodacaridae). *Entomologist's Gazette* 7: 7–26.
Includes a key to species.

Hyatt, K.H. 1956. British mites of the genus *Pachyseius* Berlese, 1910 (Gamasina: Neoparasitidae). *Annals and Magazine of Natural History* (12) 9: 1–6.
Includes a key to species.

Hyatt, K.H. 1980. Mites of the subfamily Parasitinae (Mesostigmata: Parasitidae) in the British Isles. *Bulletin of the British Museum (Natural History)* (Zoology) 38: 237–378.
Includes a key to species.

Hyatt, K.H. 1987. Mites of the genus *Holoparasitus* Oudemans, 1936 (Mesostigmata: Parasitidae) in the British Isles. *Bulletin of the British Museum (Natural History)* (Zoology) 52: 139–64.
Includes keys to species (adults and deutonymphs).

Hyatt, K.H. & Emberson, R.M. 1988. A review of the Macrochelidae (Acari: Mesostigmata) of the British Isles. *Bulletin of the British Museum (Natural History)* (Zoology) 54: 63–125.
Includes keys to species (females only for *Macrocheles*).

Karg, W. 1993. Acari (Acarina), Milben Parasitiformes (Anactinochaeta), Cohors Gamasina Leach, Raubmilben. *Tierwelt Deutschlands* 59: 1–523.
Includes keys to *ca.* 1000 species of free-living and ectoparasitic Gamasina recorded in Europe.

Kostiainen, T.S. & Hoy, M.A. 1996. *The Phytoseiidae as biological control agents of pest mites and insects: a bibliography (1960–1994)*. University of Florida, Monograph 17: 355pp.
Complements de Moraes *et al.* (1986).

Rudnick, A. 1960. A revision of the mites of the family Spinturnicidae (Acarina). *University of California Publications in Entomology* 17: 157–284.
Gives a key to world species.

Sheals, J.G. 1958 A revision of the British species of *Rhodacarus* and *Rhodacarellus*. *Annals and Magazine of Natural History*, (13) 1: 298–304.
Includes a key to species.

Skorupski, M. & Luxton, M. 1996. Mites of the family Zerconidae Canestrini, 1891 (Acari: Parasitiformes) from the British Isles, with descriptions of two new species. *Journal of Natural History* 30(12): 1815–32.
Includes a key to species.

Order Oribatida (= Cryptostigmata)

Balogh, J. 1972. *The oribatid genera of the world*. Akadémiai Kiàdó, Budapest.
Gives keys to genera.

Balogh, J. & Balogh, P. 1990. Identification key to the genera of the Galumnidae Jacot, 1925 (Acari: Oribatei). *Acta Zoologica Hungarica* 36: 1–23.
Covers the world fauna.

Balogh, J. & Mahunka, S. 1983. Primitive oribatids of the Palaearctic region. *Soil mites of the world* 1: 1–372.
Keys species of Palaeosomata, Arthronota, Parhyposomata, Mixonomata and Holosomata.

Colloff, M.J. 1993. A taxonomic revision of the oribatid mite genus *Camisia* (Acari, Oribatida) *Journal of Natural History* 27: 1325–408.
Contains a key to the world fauna of *Camisia*.

Evans, G.O. 1952. British mites of the genus *Brachychthonius* Berl., 1910. *Annals and Magazine of Natural History* (12)5: 227–39.
Includes a key to species.

Luxton, M. 1987. Mites of the genus *Malaconothrus* (Acari: Cryptostigmata) from the British Isles. *Journal of Natural History* 21: 199–206.
Includes a key to species.

Luxton, M. 1987. The oribatid mites (Acari: Cryptostigmata) of J.E. Hull. *Journal of Natural History* 21: 1273–91.
Interprets 133 of Hull's identifications of British oribatids; includes a key to species of *Mycobates*.

Luxton, M. 1987. The British oribatid mites (Acari: Cryptostigmata) of Warburton and Pearce. *Journal of Natural History* 21: 1359–65.
Interprets 41 identifications of British oribatids made by Warburton and Pearce in 1904 and 1905.

Luxton, M. 1987. Oribatid mites (Acari: Cryptostigmata) from the Isle of Man. *Naturalist* 112: 67–77.
Includes a species list, key to British species of *Quadroppia* and redescriptions of *Cultroribula juncta* (Michael) and *Quadroppia quadricarinata* (Michael).

Luxton, M. 1989. Oribatid mites (Acari: Cryptostigmata) of Orkney. *Naturalist* 114: 85–91.
Includes an annotated species list.

Luxton, M. 1989. Michael's British damaeids (Acari: Cryptostigmata). *Journal of Natural History* 23: 1367–72.
Includes a key to species of British Damaeidae.

Luxton, M. 1990. Oribatid mites (Acari: Cryptostigmata) from the Isles of Scilly. *Naturalist* 115: 7–11.
Gives an annotated species list.

Luxton, M. 1990. Oribatid mites from Holy Island. *Transactions of the Natural History Society of Northumbria* 55: 144–146.
Includes an annotated species list.

Luxton, M. 1990. Oribatid mites (Acari: Cryptostigmata) from Jersey. *Annual Bulletin Société Jersiaise* 25: 360–6.
Includes an annotated species list.

Luxton, M. 1996. Oribatid mites of the British Isles. A checklist and notes on biogeography (Acari, Oribatida). *Journal of Natural History* 30: 803–22.
Lists the 135 genera and 303 species recorded; includes an extensive bibliography.

Michael, A.D. 1884, 1888. *British Oribatidae.* Vols 1 & 2. Ray Society, London.
The classification is out of date, but the beautiful illustrations of adults and immatures allow species recognition in many instances.

Niedbala, W. 1992 *Phthiracaroidea (Acari, Oribatida). Systematic studies.* Elsevier, Amsterdam & Polish Scientific Publishers, Warsaw.
Comprehensive treatment of the world fauna; includes keys to species and remarks on zoogeography and ecology.

Norton, R.A. 1990. Acarina: Oribatida. *In* Dindal, D.L. (Ed.) *Soil biology guide*, pp. 779–803. John Wiley, New York.
Although based on the North American fauna, many taxa that occur in the British Isles are covered and an extensive bibliography is included; a key to families is provided.

Seniczak, M. 1975. Revision of the family Oppiidae Grandjean 1953 (Acarina; Oribatei). *Acarologia* 27: 331–45.
Worldwide revision.

Order Prostigmata (=Actinedida)

Amrine, J.W. 1996. *Keys to the world genera of the Eriophyoidea (Acari: Prostigmata).* Indira Publishing House, Michigan.
Also provides a review of terminology, a taxonomic synopsis of the genera and a comprehensive bibliography.

André, H.M. 1979. A generic revision of the family Tydeidae (Acari: Actinedida). 1. Introduction, paradigms and general classification. *Annales de la Société Royale Zoologique de Belgique* 108(3–4): 189–208.

André, H.M. 1980. A generic revision of the family Tydeidae (Acari: Actinedida). 4. Generic descriptions, keys and conclusions. *Bulletin et Annales de la Société Royale Entomologique de Belgique* 116: 103–30; 139–68.
Includes keys to genera and lists species classified in each genus.

André, H.M. 1981. A generic revision of the family Tydeidae (Acari: Actinedida). 2. Organotaxy of the idiosoma and gnathosoma. *Acarologia* 22: 31–46. 3. Organotaxy of the legs. *Acarologia* 22: 165–78.
Details the characters used in the above classification and revision.

Atyeo, W.T. & Tuxen, S.L. 1962. The Icelandic Bdellidae (Acarina). *Journal of the Kansas Entomological Society* 35: 281–98.
Species recorded in the British Isles are included; provides a key to species.

Bolland, H.R. 1986. Review of the systematics of the family Camerobiidae (Acari, Raphignathoidea). I. The genera *Camerobia, Decaphyllobius, Tillandsobius* and *Tycherobius. Tijdschrift voor Entomologie* 129: 191–215.
Includes keys to genera and species.

Cook, D.R. 1974. Water mite genera and subgenera. *Memoirs American Entomological Institute* 21: 1–860.
Keys and diagnoses families, subfamilies and genera.

Cross, E.A. 1965. The generic relationships of the family Pyemotidae (Acarina: Trombidiformes). *Kansas University Scientific Bulletin* 45: 29–275.
Includes taxa now classified in the family Pygmephoridae and Acarophenacidae; keys the world genera.

Davis, R., Flechtmann, C.H.W., Boczek, J.H. & Barké, H.E. 1982. *Catalogue of erioph-yid mites (Acari: Eriophyoidea).* Warsaw Agricultural University Press, Warsaw.
A useful information source on host plant associations, geographical distribution and taxonomy of the eriophyid species of the world.

Dusbábek, F. 1969. Generic revision of the myobiid mites (Acarina: Myobiidae) parasitic on bats. *Folia Parasitologica* 16: 1–17.
Includes a key to world genera of bat-associated myobiids.

Fain, A. 1965. Les Ereynetidae de la collection Berlese à Florence; designation d'une espèce type pour le genre *Ereynetes* Berlese. *Redia* 49: 87–111.
Includes a list of species of *Ereynetes.*

Fain, A. 1975. Observations sur les Myobiidae parasites des rongeurs. Evolution par-allèle hôtes-parasites (Acariens: Trombidiformes). *Acarologia* 16: 441–75.
Includes keys to species of *Radfordia* and *Myobia* and host lists.

Fain, A. & Lukoschus, F.S. 1976. Observations sur les Myobiidae d'insectivores avec description de taxa nouveaux (Acarina: Prostigmates). *Acta Zoologica et Pathologica Antverpiensia* 66: 121–88.
Keys world genera associated with insectivores; lists species of Myobiidae and their host records.

Giesen, K.M.T. 1990. A review of the parasitic mite family Psorergatidae (Cheyletoidea: Prostigmata: Acari) with hypotheses on the phylogenetic rela-tionships of species and species groups. *Zoologische Verhandelingen* 259: 1–69.
Includes a key to world species.

Gledhill, T. & Viets, K.O. 1976. A synonymic and bibliographic check-list of the fresh-water mites (Hydrachnellae and Limnohalacaridae, Acari) recorded from Great Britain and Ireland. *Occasional Publications of the Freshwater Biological Association* 1: 59pp.
Cites papers in which British records of each species appear; Green & Macquitty (1987) should be used for up-to-date information on Limnohalacaridae.

Goff, M.L., Loomis, R.B. & Wrenn, W.J. 1986. A chigger bibliography 1758–1984 (Acari: Trombiculidae): systematics, biology and ecology. *Bulletin of the Society of Vector Ecologists* 11(1): 1–177.
A comprehensive source of information about larvae of the family Trombiculidae.

Gonzalez-Rodriguez, R.H. 1965. A taxonomic study of the genera *Mediolata, Zetzellia* and *Agistemus* (Acarina: Stigmaeidae). *University of California Publications in Entomology* 41: 1–64.
Includes keys to genera of Stigmaeidae and to females of the three genera in the title.

Green, J. & Macquitty, M. 1987 Halacarid mites. *Synopsis of the British Fauna. The Linnean Society of London* 36: 1–178.
Includes introductory chapters on collecting methods and biology, but mainly comprises keys to and descriptions of the 66 species found.

Helle, W. & Sabelis, M.W. (Eds) 1985. *World crop pests. Spider mites, their biology, natural enemies and control.* Elsevier, Amsterdam.
This is the most comprehensive book on the Tetranychidae available; includes detailed chapters on anatomy, systematics (including keys to genera), study techniques, physiol-ogy and host plant associations.

Hirst, S. 1919. *Studies on Acari I. The genus* Demodex *Owen.* British Museum (Natural History), London.
Includes useful morphometric data for *Demodex* from British host species.

Hopkins, C.C. 1961. A key to the water mites (Hydracarina) of the Flatford area. *Field Studies* 1: 1–20.

Kethley, J.B. 1970. A revision of the family Syringophilidae (Prostigmata: Acarina). *Contributions of the American Entomological Institute* 5(6): 1–76.
Includes a key to world species.

Kethley, J.B. 1982. Prostigmata. *In* Parker, S.P. (Ed.) *Synopsis and classification of living organisms* 2, pp. 117–45. McGraw-Hill, New York.
Uses a modern classification and is a useful source of general information and further literature.

Kethley, J. 1990. Acarina: Prostigmata (Actinedida). *In* Dindal, D.L. (Ed.) *Soil biology guide*, pp. 667–756. John Wiley, New York.
Although based on the North American fauna (including Parasitengona), many taxa that occur in the British Isles are covered; a key to families and certain subfamilies and an extensive bibliography are provided.

Krczal, H. 1959. Systematik und Ökologie der Pyemotiden. *In* Stammer, H.J. (Ed.) *Beiträge zur Systematik und Ökologie mitteleuropäischer Acarina* 1(2), pp. 385–625. Geest & Portig, Leipzig.
Gives a more detailed coverage of the European fauna than Cross (1965); includes a key to species (some are now classified in the families Acarophenacidae and Pygmephoridae).

Lindquist, E.E. 1986. The world genera of the Tarsonemidae (Acari: Heterostigmata): a morphological, phylogenetic, and systematic revision, with a reclassification of family-group taxa in the Heterostigmata. *Memoirs of the Entomological Society of Canada* No. 136: 1–517.
Includes keys to genera and subgenera.

Lindquist, E.E., Sabelis, M.W. & Bruin, J. 1996. *World crop pests. Eriophyoid mites, their biology, natural enemies and control* 6: 790pp. Elsevier, Amsterdam.
This is the most comprehensive book on the Eriophyoidea available; includes detailed chapters on anatomy, systematics (including keys to genera and selected species of economic importance), study techniques, physiology and host plant associations.

Luxton, M. 1987. Mites of the family Cryptognathidae Oudemans, 1902 (Prostigmata) in the British Isles. *Entomologist's Monthly Magazine* 123: 113–15.
Includes a key to genera and reviews the British fauna.

Mahunka, S. 1965. Identification keys for the species of the family Scutacaridae (Acari: Tarsonemini). *Acta Zoologica Academiae Scientiarium Hungaricae* 11: 353–401.
Based on the world fauna; provides keys to genera and species.

Mahunka, S. 1969. Beiträge zur Tarsonemini-Fauna Ungarns, VI. (Acari, Trombidiformes). *Opuscula Zoologica Instituti Zoolsystematici Universitatis Budapestinensis* 9: 363–72.
The key to *Siteroptes* is more extensive than that of Krczal (1959); includes the species recorded from the British Isles.

Mahunka, S. 1970. Consideration on the systematics of the Tarsonemina and the description of new European taxa (Acari: Trombidiformes) *Acta Zoologica Academiae Scientiarium Hungaricae* 16: 137–74.
Keys the families and genera of Pyemotoidea and Pygmephoroidea, including those that occur in the British Isles.

Newell, I.M. 1957. Studies on the Johnstonianidae (Acari, Parasitengona). *Pacific Science* 11: 396–466.
Includes keys to species of larvae.

Prasad, V. & Cook, D.R. 1972. The taxonomy of water mite larvae. *Memoirs of the American Entomological Institute* 18: 1–326.
Includes keys to families and genera.

Pritchard, A.E. & Baker, E.W. 1955. A revision of the spider mite family Tetranychidae. *Memoirs of the Pacific Coast Entomological Society* 2: 1–472.
Based on the world fauna; provides keys to species which include British representatives.

Rack, G. 1975. Bibliographica gefundene Arten der Gattung *Pygmephorus* (Acarina, Pygmephoridae). *Mitteilungen aus dem Hamburgischen Zoologischen Museum et Institute* 72: 157–76.
Includes a key to species of *Pygmephorus* s.str.

Regenfuss, H. 1968. Untersuchungen zur Morphologie, Systematik und Ökologie der Podapolipidae (Acarina, Tarsonemini). *Zeitschrift für Wissenschaftliche Zoologie* 177: 183–282.
Includes key to species of *Eutarsopolipus* and *Dorsipes*.

Schaarschmidt, L. 1959. Systematik und Ökologie der Tarsonemiden. *In* Stammer, H.J. (Ed.) *Beiträge zur Systematik und Ökologie mitteleuropäischer Acarina* 1(2), pp. 713–823. Geest & Portig, Leipzig.
Gives a detailed coverage of the European fauna, including a key to species.

Smiley, R.L. 1992. *The predatory mite family Cunaxidae (Acari) of the world with a new classification.* Indira Publishing House, West Bloomfield, Michigan, USA.
The most comprehensive and up-to-date systematic treatment of the Cunaxidae available.

Smith Meyer, M.K.P. 1979. The Tenuipalpidae (Acari) of Africa with keys to the world fauna. *Entomology Memoirs of the Department of Agricultural and Technical Services of the Republic of South Africa* No. 50: 1–135.
Contains useful keys.

Smith Meyer, M.K.P. & Ueckerman, E.A. 1987. A taxonomic study of some Anystidae. *Entomology Memoir of the Department of Agriculture & Water Supply of the Republic of South Africa* No. 68: 1–37.
Contains a key to genera and selected species.

Soar, C.D. & Williamson, W. 1925–1929. *The British Hydracarina.* 3 vols. Ray Society, London.
These volumes contain the most detailed taxonomic treatment of the British freshwater mite fauna available; they should be used in conjunction with Gledhill & Viets (1976).

Southcott, R.V. 1961. Studies on the systematics and biology of the Erythraeoidea (Acarina), with a critical revision of the genera and subfamilies. *Australian Journal of Zoology* 9: 367–610.

Includes keys to genera of larval and postlarval Erythraeidae and Smarididae.

Southcott, R.V. 1986. Studies on the taxonomy and biology of the subfamily Trombidiinae (Acarina: Trombidiidae), with a critical revision of the genera. *Australian Journal of Zoology* (suppl.) **123**: 1–126.

Keys to larval and postlarval stages of genera.

Southcott, R.V. 1987. The classification of the mite families Trombellidae and Johnstonianidae and related groups, with the description of a new larva (Acarina: Trombellidae *Nothrotrombidium*) from North America. *Transactions of the Royal Society of South Australia* 111(1–2): 25–42.

Provides keys to genera (larvae).

Southcott, R.V. 1992. Revision of the larvae of *Leptus* Latreille (Acarina: Erythraeidae) of Europe and North America, with descriptions of post-larval instars. *Zoological Journal of the Linnean Society* **105**: 1–153.

Gives keys to species and observations on host specificity and temporal distribution.

Southcott, R.V. 1993. Revision of the taxonomy of the larvae of the subfamily Eutrombidiinae (Acarina: Microtrombidiidae). *Invertebrate Taxonomy* 7: 885–959.

Gives keys to world species.

Southcott, R.V. 1994. Revision of the larvae of the Microtrombidiinae (Acarina: Microtrombidiidae), with notes on life histories. *Zoologica* 48(144): 1–155.

Includes a key to genera and species descriptions.

Strandtmann, R.W. 1971. The eupodid mites of Alaska (Acarina: Prostigmata). *Pacific Insects* 13: 75–118.

The higher classification is somewhat out of date; includes keys to genera currently classified in the families Eupodidae, Penthaleidae, Penthalodidae and Rhagidiidae, and to some species of *Eupodes, Cocceupodes* and *Penthalodes*.

Summers, F.M. 1960. *Eupalopsis* and eupalopsellid mites (Acarina: Stigmaeidae, Eupalopsellidae). *Florida Entomologist* 43: 119–38.

Clarifies the status of the family Eupalopsellidae; includes species recorded from the British Isles.

Summers, F.M. 1962. The genus *Stigmaeus* (Acarina: Stigmaeidae). *Hilgardia* 33: 491–537.

Provides a key to species (females) of *Stigmaeus*; includes species recorded from the British Isles.

Summers, F.M. & Price, D.W. 1961. New and redescribed species of *Ledermuellaria* from North America (Acarina: Stigmaeidae). *Hilgardia* 31: 369–81.

Provides a key to species (females) of *Ledermuellaria*; includes species recorded from the British Isles.

Volgin, V.I. 1987. *Acarina of the family Cheyletidae of the world*. Amerind Publishing Co. Pvt. Ltd, New Delhi.

Provides keys to species; includes those classified in the family Cheyletiellidae by other authors.

Zacharda, M. 1980. Soil mites of the family Rhagidiidae (Actinedida: Eupodoidea). Morphology, systematics, ecology. *Acta Universitatis Carolinae.* (Biol.) **1978**: 489–785.

Includes a key to world species.

Zhang, Z.-Q. & Xin, J.-L. 1992. Review of larval *Allothrombium* (Acari: Trombidioidea), with description of a new species ectoparasitic on aphids in China. *Journal of Natural History* **26**: 383–93.

Includes a key to species.

Order Ixodida (= Metastigmata): Ticks

Arthur, D.R. 1963. *British ticks.* Butterworth, London.

Describes adult and immature stages of 22 species (*Hyalomma marginatum* not included) and gives host, distributional and biological data about each.

Hillyard, P. D. 1996. Ticks of north-west Europe. *Synopsis of the British Fauna (Field Studies Council)* **52**: 178pp.

As well as keying and describing the adults of British ticks, this book also contains details of the distribution, biology and disease relationships of each species, sections on collection and control methods and an extensive bibliography based on the worldwide literature.

Martyn, K. 1988. *Provisional atlas of the ticks (Ixodoidea) of the British Isles.* Biological Records Centre, Natural Environment Research Council, Monks Wood, UK.

Gives a distribution map plus host and biological notes for the 23 species that occur in the British Isles

Snow, K.R. 1978. *Identification of larval ticks found on small mammals in Britain.* The Mammal Society, Reading.

Keys and/or describes the larvae of *Ixodes festai, I. hexagonus, I. ricinus, I. trianguliceps, I. arvicolae* and *D. reticulatus.*

Index

This index includes all the higher taxonomic categories mentioned in the book, from orders down to families, but page numbers are given only for the main occurrences of those names. It therefore also acts as a complete alphabetic list of the higher taxa of British insects and arachnids (except for the numerous families of mites).

Acalyptratae 173, 188
Acanthosomatidae 55
Acari 320, 330
Acartophthalmidae 173, 191
Acerentomidae 23
Acrididae 39
Acroceridae 172, 180, 181
Aculeata 197, 206
Adelgidae 56, 62, 64
Adelidae 146
Adephaga 82, 91
Aderidae 83, 126, 127
Aeolothripidae 52
Aepophilidae 55
Aeshnidae 31
Agelenidae 327
Agromyzidae 173, 188, 193
Alexiidae 83
Aleyrodidae 56, 67, 68
Aleyrodoidea 56, 66
Alucitidae 146
Alucitoidea 146
Alydidae 55, 58
Amaurobiidae 327
Amblycera 48
Anisolabiidae 41
Anisopodidae 172, 175, 177
Anisopodoidea 172
Anisoptera 31
Anobiidae 82, 119
Anoplura 48
Anthicidae 83, 90, 126
Anthocoridae 55, 57, 58
Anthomyiidae 173, 174, 186, 187
Anthomyzidae 173, 188
Anthribidae 83, 88, 133, 134

Anyphaenidae 327
Aphelinidae 198, 293, 308
Aphelocheiridae 55
Aphididae 56, 62
Aphidoidea 56, 61
Aphrophoridae 56
Apidae 198, 217
Apioninae 83, 134
Apocrita 197, 198, 206, 227
Apoidea 198, 214
Arachnida 320
Aradidae 55
Araneae 320, 326
Araneidae 327
Araneomorphae 327
Archaeognatha 21, 25, 26
Arctiidae 146, 162
Argidae 197, 201
Argyronetidae 327
Arthropleona 22
Aschiza 173, 184
Asilidae 172, 180, 181, 182
Asiloidea 172, 181
Asilomorpha 172, 180, 182
Asteiidae 173, 189
Asterolecaniidae 56, 70
Atelestidae 172, 183, 185
Athericidae 172, 181
Attelabidae 83, 134
Atypidae 327
Auchenorrhyncha 54, 55, 59
Aulacidae 198, 228
Aulacigastridae 173, 192

Baetidae 28
Beraeidae 142

Berytidae 55
Bethylidae 198, 223, 224
Bibionidae 172, 175, 176
Bibionoidea 172
Bibionomorpha 172, 176
Biphyllidae 83, 122
Blaberidae 43
Blasticotomidae 197, 200
Blastobasidae 146
Blattellidae 43
Blattidae 43
Blattodea 42
Bolitophilidae 172, 177
Bombycoidea 146
Bombyliidae 172, 180, 181
Boreidae 140
Bostrichidae 82, 119
Bostrichiformia 82
Bostrichoidea 82, 119
Bothrideridae 83
Brachycentridae 142
Brachycera 171, 172, 180
Brachypteridae 82
Braconidae 198, 228, 246
Braulidae 173, 188, 192
Brentidae 83, 134
Bruchidae 83, 128, 130
Bucculatricidae 146
Buprestidae 82, 114
Buprestoidea 82, 114
Byrrhidae 82, 112
Byrrhoidea 82, 89, 112
Byturidae 83, 122

Caeciliidae 45
Caenidae 28
Calliphoridae 173, 186
Calophyidae 56, 69
Calopterygidae 31
Calyptratae 173, 186
Camillidae 173, 190
Campichoetidae 173
Campodeidae 24
Canacidae 173
Cantharidae 82, 115
Cantharoidea 82, 115
Capniidae 34

Carabidae 82, 85, 92
Carnidae 173, 188
Carnioidea 173
Cecidomyiidae 172, 176
Cephidae 197, 204
Cerambycidae 83, 128
Ceraphronidae 198, 288
Ceraphronoidea 198, 287
Ceratocombidae 55
Ceratophyllidae 195
Ceratopogonidae 172, 173, 178
Cercopidae 56
Cercopoidea 56, 59
Cerylonidae 83, 90, 122
Chalcididae 198, 295
Chalcidoidea 198, 292
Chamaemyiidae 173, 188
Chaoboridae 172, 175, 178
Cheiridiidae 322
Cheliferidae 322
Chernetidae 322
Chironomidae 172, 175, 178
Chironomoidea 172
Chloroperlidae 34
Chloropidae 173, 188
Choreutidae 146
Chrysididae 198, 223, 225
Chrysidoidea 198, 223
Chrysomelidae 83, 91, 128, 130, 132
Chrysomeloidea 83, 128
Chrysopidae 77
Chthoniidae 322
Chyromyidae 173
Cicadellidae 56, 59
Cicadidae 56
Cicadoidea 56, 59
Cicindelidae 82, 89, 93, 94
Ciidae 83, 126
Cimbicidae 197, 201
Cimicidae 55
Cimicomorpha 55, 58
Cixiidae 56
Clambidae 82, 111
Cleridae 82, 120
Cleroidea 82, 120
Clubionidae 327
Clusiidae 173, 191, 193

Coccidae 56, 70, 73
Coccinellidae 83, 122
Coccoidea 56, 61, 70
Coelopidae 173, 191
Coenagrionidae 31
Coleophoridae 146, 147, 153
Coleoptera 80
Collembola 21, 23
Colydiidae 83, 90, 126
Coniopterygidae 77
Conopidae 173, 184, 185
Conopoidea 173
Cordulegastridae 31
Corduliidae 31
Coreidae 55, 58
Corixidae 55, 57
Corticariidae 83
Corylophidae 83, 122
Cosmopterigidae 146, 156
Cossidae 146, 153, 162
Cossoidea 146
Cryptococcidae 56, 70, 72
Cryptophagidae 83, 85, 122
Ctenuchidae 146
Cucujidae 82, 122
Cucujiformia 82, 122
Cucujoidea 82, 122
Culicidae 172, 173, 175, 178
Culicoidea 172
Culicomorpha 172, 178
Curculionidae 83, 134
Curculionoidea 83, 133
Cydnidae 55
Cylindrotomidae 172
Cynipidae 198, 266, 267
Cynipoidea 198, 266
Cyphoderidae 22

Dascillidae 82
Dascilloidea 82
Delphacidae 56, 61
Dermaptera 40
Dermestidae 82, 90, 118
Dermestoidea 82, 118
Derodontidae 82, 118
Diadociidae 172
Diapriidae 198, 271

Diaspididae 56, 70, 73
Diastatidae 173, 191
Dictynidae 327
Diopsoidea 173
Diplura 21, 24, 25
Diprionidae 197, 201
Dipsocoridae 55
Dipsocoromorpha 55, 58
Diptera 171
Ditomyiidae 172, 177
Dixidae 172, 175, 179
Dolichopodidae 172, 183
Douglasiidae 146
Drepanidae 146, 163
Drepanoidea 146
Drilidae 82, 115
Drosophilidae 173, 188, 189, 191
Dryinidae 198, 223
Dryomyzidae 173, 193
Dryopidae 82, 90, 113
Dryopoidea 82, 112
Dysderidae 327
Dytiscidae 82, 90, 95

Echinophthiriidae 48
Ecnomidae 142
Ectopsocidae 45
Elachistidae 146, 147, 157
Elasmidae 198, 310
Elateridae 82, 115
Elateriformia 82
Elateroidea 82, 84, 85, 115
Elenchidae 139
Elipsocidae 45
Elmidae 82, 90, 113
Embolemidae 198, 223, 224
Empididae 172, 182, 183
Empidoidea 172, 182, 183
Encyrtidae 198, 292, 304
Enderleinellidae 48
Endomychidae 83, 90, 122, 125
Endromidae 146
Entomobryidae 22
Entomobryoidea 22
Eosentomidae 23
Epermeniidae 146, 158
Epermenioidea 146, 158

Ephemerellidae 28
Ephemeridae 28
Ephemeroptera 27
Ephydridae 173, 188
Ephydroidea 173
Epipsocidae 45
Eresidae 327
Eriococcidae 56, 70, 72
Eriocraniidae 145
Eriocranoidea 145
Erotylidae 83, 90, 122
Ethmiidae 146, 155, 156
Eucharitidae 198, 299
Eucinetidae 82, 111
Eucinetoidea 82, 111
Eucnemidae 82, 115, 116
Eulophidae 198, 311
Eupelmidae 198, 302
Eurytomidae 198, 295
Evaniidae 198, 227
Evanioidea 198, 227

Fanniidae 173, 186, 187
Figitidae 198, 268
Forficulidae 41
Formicidae 197, 209
Fulgoroidea 56, 59

Gasterophilidae 173, 174, 186
Gasteruptiidae 198, 228
Gelechiidae 146, 148, 153, 154, 156
Gelechioidea 146, 147, 155
Geometridae 146, 163
Geometroidea 146, 147, 163
Georissidae 82, 90, 97
Geotrupidae 82, 110
Gerridae 55, 57
Gerromorpha 55, 58
Glossata 145
Glossosomatidae 142
Glyphipterigidae 146, 153
Gnaphosidae 327
Goeridae 142
Gomphidae 31
Goniodidae 48
Gracillariidae 146
Gryllidae 39

Gryllotalpidae 39
Gyrinidae 82, 90, 92, 96
Gyropidae 48

Haematopinidae 48
Hahniidae 327
Halictophagidae 139
Haliplidae 82, 91, 92, 95
Hebridae 55
Helcomyzidae 173, 191
Heleomyzidae 173, 193
Heliodinidae 146, 153, 162
Heliozelidae 146, 153
Helomyzidae 173
Heloridae 198, 270, 276
Hemerobiidae 77
Hemiptera 54
Hepialidae 145
Hepialoidea 145, 146
Heptageniidae 28
Herminiidae 146
Hesperiidae 146, 168
Hesperioidea 146
Heteroceridae 82, 90, 113
Heteropodidae 327
Heteroptera 55, 56
Hippoboscidae 173, 186, 187
Hippoboscoidea 173, 186
Histeridae 82, 90, 96, 98
Histeroidea 82, 96, 98
Histrichopsyllidae 195
Homoptera 54
Homotomidae 56, 69
Hoplopleuridae 48
Hybotidae 172, 182, 183
Hydraenidae 82, 90, 98, 99
Hydrochidae 82, 90, 97
Hydrometridae 55
Hydrophilidae 82, 90, 96, 97
Hydrophiloidea 82, 89, 96
Hydropsychidae 142
Hydroptilidae 142
Hygrobiidae 82, 89, 91
Hymenoptera 196
Hypocopridae 83
Hypogastruridae 22

Ibaliidae 198, 266, 268
Ichneumonidae 198, 229
Ichneumonoidea 198, 228
Incurvariidae 146
Incurvarioidea 146
Ischnocera 48
Ischnopsyllidae 195
Isotomidae 22
Issidae 56

Kateretidae 82, 123, 124
Kermesidae 56, 70
Keroplatidae 172, 177

Labiduridae 41
Labiidae 41
Lachesillidae 45
Laemobothriidae 48
Laemophloeidae 82, 122
Lagriidae 83, 126, 127
Lampyridae 82, 115, 117
Larcidae 322, 323
Lasiocampidae 146, 151, 161
Latridiidae 83, 122, 124
Lauxaniidae 173, 190
Lauxanioidea 173
Leiobunidae 324
Leiodidae 82, 99, 101
Lepidopsocidae 45
Lepidoptera 145
Lepidostomatidae 142
Lepismatidae 25
Leptinidae 82, 101
Leptoceridae 142
Leptophlebiidae 28
Leptopodomorpha 55, 58
Leptopsyllidae 195
Lestidae 31
Leuctridae 34
Libellulidae 31
Limacodidae 146
Limnephilidae 142
Limnichidae 82, 90, 113
Limoniidae 172
Linognathidae 48
Linyphiidae 327
Liocranidae 327

Liposcelidae 45
Lonchaeidae 173, 188, 190
Lonchopteridae 173, 184, 185
Lonchopteroidea 173
Lucanidae 82, 110, 111
Lycaenidae 146
Lycidae 82, 115, 116
Lycosidae 327
Lyctidae 82, 119, 120
Lygaeidae 55
Lymantriidae 146
Lymexylidae 82, 121, 122
Lymexyloidea 82, 120, 121
Lyonetiidae 146

Machilidae 26
Margarodidae 56, 70
Mecoptera 140
Megalopodidae 83, 128, 130, 132
Megaloptera 78
Megamerinidae 173, 188, 189
Megaspilidae 198, 290
Melandryidae 83, 125, 126, 128
Meloidae 83, 126
Melyridae 82, 120, 121
Membracidae 56, 59
Membracoidea 56, 59
Menoponidae 48
Merophysiidae 83, 125
Mesopsocidae 45
Mesoveliidae 55
Micropezidae 173, 188, 189, 190
Microphoridae 172, 182, 183
Microphysidae 55
Micropterigidae 145, 153
Micropterigoidea 145
Microsporidae 82, 96
Milichiidae 173, 188, 192
Mimetidae 327
Miridae 55, 57, 59
Molannidae 142
Momphidae 146, 155, 156
Monotomidae 82, 122
Mordellidae 83, 125, 126
Muscidae 173, 186, 188
Muscoidea 173, 186
Muscomorpha 172, 180, 183

Mutillidae 197, 207
Mycetobiidae 172, 177
Mycetophagidae 83, 90, 126, 127
Mycetophilidae 172, 176, 177
Mycteridae 83, 126
Mygalomorphae 327
Mymaridae 198, 316
Mymarommatidae 198, 319
Mymarommatoidea 198, 319
Myrmeleontidae 77
Myxophaga 82, 96

Nabidae 55, 58
Nanophyinae 83, 134
Naucoridae 55
Neanuridae 22
Neelidae 22
Neelipleona 22
Nemastomatidae 324
Nematocera 172, 174
Nemestrinoidea 172
Nemonychidae 83, 134
Nemouridae 34
Neobisiidae 322
Nepidae 55, 57
Nepomorpha 55, 58
Nepticulidae 145, 153, 154
Nepticuloidea 145, 154
Nerioidea 173
Nesticidae 327
Neuroptera 76
Nitidulidae 82, 90, 122
Noctuidae 146, 162, 164
Noctuoidea 146, 164
Noteridae 82, 90, 91, 92, 95
Notodontidae 146, 148, 162
Notonectidae 55, 57
Nycteribiidae 173, 186, 187
Nymphalidae 146, 168

Odiniidae 173, 190
Odonata 30
Odontoceridae 142
Oecobiidae 327
Oecophoridae 146, 154, 155
Oedemeridae 83, 126
Oestridae 173, 174, 186

Oestroidea 173, 186
Oncopoduridae 22
Onychiuridae 22
Oonopidae 327
Opetiidae 173, 185
Opiliones 320, 324
Opomyzidae 173, 188, 191
Opomyzoidea 173
Opostegidae 145, 154
Ormyridae 198, 299
Ortheziidae 56, 70
Orthoptera 38
Orussidae 197, 199, 204
Osmylidae 77
Oxyopidae 327

Pachytroctidae 45
Pallopteridae 173, 188, 190
Pamphilidae 197, 200
Panorpidae 140
Papilionidae 146
Papilionoidea 146
Parasitica 197, 198, 227
Pediciidae 172
Pediculidae 48
Peltidae 82, 121
Pentatomidae 55
Pentatomomorpha 55
Perilampidae 198, 300
Peripsocidae 45
Periscelidae 173
Perlidae 34
Perlodidae 34
Phaeomyzidae 173
Phalacridae 83, 122
Phalangiidae 324
Phasmatidae 37
Phasmida 36
Philodromidae 327
Philopotamidae 142
Philopteridae 48
Philotarsidae 45
Phlaeothripidae 52
Phloiophilidae 82, 120, 121
Pholcidae 327
Phoridae 173, 184, 185
Phryganeidae 142

Phthiraptera 47
Phylloxeridae 56, 62
Pieridae 146
Piesmatidae 55
Piophilidae 173, 188, 189, 192
Pipunculidae 173, 184, 185
Pisauridae 327
Platycnemididae 31
Platygastridae 198, 280
Platygastroidea 198, 279
Platypezidae 173, 184, 185
Platypezoidea 173
Platypodidae 83, 134
Platystomatidae 173, 189
Plecoptera 33
Pleidae 55
Poduridae 22
Poduroidea 22
Polycentropodidae 142
Polyphaga 83, 96
Polyplacidae 48
Pompilidae 197, 211
Potamanthidae 28
Proctotrupidae 198, 270, 277
Proctotrupoidea 198, 270
Prodoxidae 146
Protentomidae 23
Protura 23
Pselaphidae 82, 99, 110
Psephenidae 82, 90, 113
Pseudococcidae 56, 70, 71
Pseudopomyzidae 173
Pseudoscorpiones 320, 322
Psilidae 173, 188, 190
Psocidae 45
Psocomorpha 45
Psocoptera 44
Psoquillidae 45
Psychidae 146, 154
Psychodidae 172, 175, 177
Psychodoidea 172
Psychodomorpha 172, 177
Psychomyiidae 142
Psyllidae 56, 69
Psyllipsocidae 45
Psylloidea 56, 61, 68
Pteromalidae 198, 293, 301

Pterophoridae 146, 159
Pterophoroidea 146, 147
Pthiridae 48
Ptiliidae 82, 83, 99, 100
Ptinidae 82, 119
Ptychopteridae 172, 173, 175, 178
Ptychopteroidea 172
Ptychopteromorpha 172, 178
Pulicidae 195
Pyralidae 146, 158, 159
Pyraloidea 146, 147, 158
Pyrochroidae 83, 126
Pyrrhocoridae 55, 58
Pythidae 83, 126

Raphidiidae 79
Raphidioptera 79
Raymondionymidae 83, 134
Reduviidae 55, 56
Rhagionidae 172, 180, 181, 182
Rhaphidophoridae 39
Rhinophoridae 173, 186, 187
Rhipiphoridae 83, 126
Rhizophagidae 82, 122
Rhopalidae 55, 58
Rhyacophilidae 142
Ricinidae 48
Roeslerstammiidae 146

Sabaconidae 324
Saldidae 55, 58
Salpingidae 83, 126
Salticidae 327
Sapygidae 197, 208
Sarcophagidae 173, 186, 187
Saturniidae 146
Scaphidiidae 82, 103
Scarabaeidae 82, 110
Scarabaeiformia 82
Scarabaeoidea 82, 110
Scathophagidae 173, 186, 187
Scatopsidae 172, 175, 177
Scatopsoidea 172
Scelionidae 198, 279, 284
Scenopinidae 172, 180, 181
Schizophora 173, 184, 186, 188
Schreckensteiniidae 146

Schreckensteinioidea 146
Sciaridae 172, 176
Sciaroidea 172
Sciomyzidae 173, 188, 192
Sciomyzoidea 173
Scirtidae 82, 90, 111
Scirtoidea 82, 111
Sclerosomatidae 324
Scoliidae 197, 208
Scolytidae 83, 134
Scorpiones 320
Scraptiidae 83, 125, 126
Scutelleridae 55
Scydmaenidae 82, 99, 102
Scythrididae 146, 155
Scytodidae 327
Segestriidae 327
Sepsidae 173, 188, 191, 192
Sericostomatidae 142
Sesiidae 146, 163
Sesioidea 146, 163
Sialidae 78
Signiphoridae 198, 308
Silphidae 82, 99, 102
Silvanidae 83, 122
Simuliidae 172, 173, 178, 179
Siphlonuridae 28
Siphonaptera 194
Siricidae 197, 203
Sisyridae 76, 77
Sminthuridae 22
Sminthurididae 22
Sphaeriidae 82, 96
Sphaeritidae 82, 98
Sphaeroceridae 173, 192
Sphaeroceroidea 173
Sphaeropsocidae 45
Sphecidae 198, 214
Sphindidae 82, 122, 125
Sphingidae 146, 162
Staphylinidae 82, 91, 99, 103
Staphyliniformia 82
Staphylinoidea 82, 99
Stenocephalidae 55, 58
Stenomicridae 173, 190, 192
Stenopsocidae 45
Sternorrhyncha 54, 55, 56, 61

Stratiomyidae 172, 173, 180, 181, 182
Stratiomyoidea 172
Strepsiptera 139
Strongylophthalmyiidae 173
Stylopidae 139
Symphypleona 22
Symphyta 197, 199
Syrphidae 173, 184, 185
Syrphoidea 173

Tabanidae 172, 173, 180, 181, 182
Tabanoidea 172, 181
Tabanomorpha 172, 180, 181
Tachinidae 173, 186, 187
Taeniopterygidae 34
Tanypezidae 173, 189, 190
Tenebrionidae 83, 126
Tenebrionoidea 83, 125
Tenthredinidae 197, 200, 201
Tephritidae 173, 188, 193
Tephritoidea 173
Tethinidae 173, 191
Tetracampidae 198, 310
Tetragnathidae 327
Tetratomidae 83, 126
Tetrigidae 39
Tettigometridae 56, 59
Tettigoniidae 39
Thaumaleidae 172, 179
Therevidae 172, 180, 181, 182
Theridiidae 327
Theridiosomatidae 327
Thomisidae 327
Thripidae 52
Throscidae 82, 115
Thysanoptera 51
Thysanura 21, 25
Tineidae 146
Tineoidea 146, 154
Tingidae 55, 57, 58
Tiphiidae 197, 206
Tipulidae 172, 175
Tipuloidea 172
Tipulomorpha 172, 175
Tischeriidae 146
Tischerioidea 146
Tomoceridae 22

Tortricidae 146, 157
Tortricoidea 146, 147, 157
Torymidae 198, 297
Trichoceridae 172, 175, 177
Trichoceroidea 172
Trichodectidae 48
Trichogrammatidae 198, 314
Trichopsocidae 45
Trichoptera 141
Trigonalidae 198, 227
Trigonaloidea 198, 227
Trimenoponidae 48
Triozidae 56, 69
Troctomorpha 45
Trogidae 82, 110
Trogiidae 45
Trogiomorpha 45
Trogossitidae 82, 120
Trogulidae 324

Ulidiidae 173
Uloboridae 327

Veliidae 55
Vermipsyllidae 195
Vespidae 197, 212
Vespoidea 197, 206

Xiphydriidae 197, 203
Xyelidae 197, 200
Xylomyidae 172, 180, 181
Xylophagidae 172, 173, 180, 181
Xylophagoidea 172
Xylophagomorpha 172, 181

Yponomeutidae 146, 153
Yponomeutoidea 146, 155
Ypsolophidae 146

Zeugloptera 145
Zodariidae 327
Zoridae 327
Zygaenidae 146
Zyga⌢⌢oidea 146
Zygoptera 31